Lecture Notes in Computer Science 11394

Commenced Publication in 1973
Founding and Former Series Editors:
Gerhard Goos, Juris Hartmanis, and Jan van Leeuwen

More information about this series at http://www.springer.com/series/7407

Sudebkumar Prasant Pal
Ambat Vijayakumar (Eds.)

Algorithms and Discrete Applied Mathematics

5th International Conference, CALDAM 2019
Kharagpur, India, February 14–16, 2019
Proceedings

 Springer

Editors
Sudebkumar Prasant Pal
Indian Institute of Technology
Kharagpur, India

Ambat Vijayakumar
Cochin University of Science
and Technology
Cochin, India

ISSN 0302-9743   ISSN 1611-3349 (electronic)
Lecture Notes in Computer Science
ISBN 978-3-030-11508-1   ISBN 978-3-030-11509-8 (eBook)
https://doi.org/10.1007/978-3-030-11509-8

Library of Congress Control Number: 2018967448

LNCS Sublibrary: SL1 – Theoretical Computer Science and General Issues

This Springer imprint is published by the registered company Springer Nature Switzerland AG
The registered company address is: Gewerbestrasse 11, 6330 Cham, Switzerland

Dedicated to Professor Subir Kumar Ghosh on the Occasion of His 65th Birthday

Prof. Subir Kumar Ghosh is an internationally renowned expert on geometric graphs, visibility algorithms, and other areas of computational geometry. He was a Professor of Computer Science at the Tata Institute of Fundamental Research (TIFR), Mumbai, untill July 2015. Since then, he has been Professor in the Department of Computer Science at the Ramakrishna Mission Vivekananda Educational and Research Institute (RKMVERI), Belur, West Bengal (formerly known as RKM Vivekananda University). He is a Fellow of the Indian Academy of Sciences.

After his doctorate from TIFR in computational geometry during the mid-1980s, Prof. Ghosh contributed around 65 papers in the fields of computational geometry, discrete applied mathematics, geometric graph theory, algorithms (sequential, parallel, on-line and approximation), and robot motion planning. Some of his discoveries are considered landmarks in the respective areas. In fact, Prof. Ghosh has solved some outstanding open problems with his doctoral students and collaborators. Moreover, his famous conjectures on visibility graphs and art gallery problems have generated many research activities.

His widely acclaimed book *Visibility Algorithms in the Plane*, published by Cambridge University Press in 2007, is used as a textbook in universities of Europe, North America, and India. Prof. Ghosh has visited several reputed universities and research institutes around the world and has delivered invited lectures in national and international conferences, workshops, and in research institutes and universities.

An excellent teacher, Prof. Ghosh is also passionate about improving algorithms and discrete mathematics education in India. In a span of eight years (2008–2015), Prof. Ghosh conducted 22 introductory workshops on graph and geometric algorithms for teachers and students of engineering colleges and universities in different states and union territories of India, at the undergraduate, postgraduate, and doctoral levels. The workshops have been successful in motivating students toward algorithmic research. He has also organized seven international schools in India for doctoral students of algorithms and discrete mathematics. In 2015, he initiated the new international conference on algorithms and discrete applied mathematics (CALDAM).

Preface

This volume contains the papers presented at CALDAM 2019: the 5th International Conference on Algorithms and Discrete Applied Mathematics held during February 14–16, 2019, in Kharagpur. CALDAM 2019 was organized by the Department of Computer Science and Engineering, Indian Institute of Technology, Kharagpur, and the Association for Computer Science and Discrete Mathematics (ACSDM). The conference had papers in the areas of algorithms, graph theory, combinatorics, computational geometry, discrete geometry, and computational complexity. The 86 submissions had authors from 13 different countries. Each submission received at least one detailed review and nearly all were reviewed by three Program Committee members. The committee decided to accept 22 papers. The program also included three invited talks by Janos Pach, Matthew Katz, and David Mount.

The first CALDAM was held in February 2015 at the Indian Institute of Technology, Kanpur, and had 26 papers selected from 58 submissions from 10 countries. The second edition was held in February 2016 at the University of Kerala, Thiruvananthapuram (Trivandrum), and had 30 papers selected from 91 submissions from 13 countries. The third edition was held in February 2017 at Birla Institute of Technology and Science, Pilani (BITS Pilani), K. K. Birla Goa Campus, Goa, and had selected 32 papers from 103 submissions from 18 countries. The fourth edition was held in February 2018 at the Department of Computer Science Engineering, Indian Institute of Technology, Guwahati (IIT Guwahati), and had 23 papers from 68 submissions from 12 countries.

CALDAM 2019 included a special session on February 14, 2019, at the end of the first day of the conference, in honor of Prof. Subir Ghosh on the occasion of his 65th birthday, and for completing 40 years of active research and teaching activities to date.

We would like to thank all the authors for contributing high-quality research papers to the conference. We express our sincere thanks to the Program Committee members and the external reviewers for reviewing the papers within a very short period of time. We thank Springer for publishing the proceedings in the *Lecture Notes in Computer Science* series. We thank the invited speakers Janos Pach, Matthew Katz, and David Mount for accepting our invitation. We thank the Organizing Committee chaired by Rogers Mathew from Indian Institute of Technology, Kharagpur, for the smooth functioning of the conference. We thank the chair of the Steering Committee, Subir Ghosh, for his active help, support, and guidance throughout. We thank our sponsors Google Inc., the National Board of Higher Mathematics (NBHM), the Department of Atomic Energy, and Microsoft Research India, for their financial support. We also thank Springer for its support for the Best Paper Presentation Awards. We thank the EasyChair conference management system, which was very effective in handling the entire reviewing process.

February 2019

Sudebkumar Prasant Pal
Ambat Vijayakumar

Organization

Program Committee

Amitabha Bagchi	Indian Institute of Technology, Delhi, India
Niranjan Balachandran	Indian Institute of Technology Bombay, India
Partha Bhowmick	Indian Institute of Technology, Kharagpur, India
Bostjan Bresar	University of Maribor, Slovenia
Manoj Changat	University of Kerala, India
Sandip Das	Indian Statistical Institute, Kolkata, India
Ajit Diwan	Indian Institute of Technology, Bombay, India
Zachary Friggstad	University of Alberta, Canada
Daya Gaur	University of Lethbridge, Canada
Partha P. Goswami	Institute of Radio Physics and Electronics, University of Calcutta, Kolkata, India
Sathish Govindarajan	Indian Institute of Science Bangalore, India
Petr Hlineny	Masaryk University, Brno, Czech Republic
Subrahmanyam Kalyanasundaram	Indian Institute of Technology Hyderabad, India
Gyula Katona	Alfred Renyi Institute of Mathematics, Budapest, Hungary
Ramesh Krishnamurti	Simon Fraser University, Canada
Van Bang Le	Universität Rostock, Institut fur Informatik, Germany
Andrzej Lingas	Lund University, Sweden
Anil Maheshwari	Carleton University, Canada
Kazuhisa Makino	Kyoto University, Japan
Bodo Manthey	University of Twente, The Netherlands
Shashank K. Mehta	Indian Institute of Technology Kanpur, India
Bojan Mohar	Simon Fraser University, Canada
Apurva Mudgal	Indian Institute of Technology Ropar, India
N. S. Narayanaswamy	Indian Institute of Technology Madras, India
Sudebkumar Prasant Pal (Chair)	Indian Institute of Technology, Kharagpur, India
Bhawani Panda	Indian Institute of Technology Delhi, India
Michiel Smid	Carleton University, Canada
Joachim Spoerhase	University of Wurzburg, Germany
C. R. Subramanian	The Institute of Mathematical Sciences, Chennai, India
Ambat Vijayakumar (Chair)	Cochin University of Science and Technology (CUSAT), Kerala, India

Organizing Committee

Palash Dey	IIT Kharagpur, India
Swami Dhyanagamyananda	Ramakrishna Mission Vivekananda Educational and Research Institute, India
Pritee Khanna	PDPM Indian Institute of Information Technology, Design and Manufacturing Jabalpur, India
Sasanka Roy	ISI Kolkata, India
Rogers Mathew (Chair)	IIT Kharagpur, India
Arti Pandey	IIT Ropar, India
Tathagata Ray	BITS-Pilani, Hyderabad Campus, India
Rishi Ranjan Singh	IIT Bhilai, India
Dinabandhu Pradhan	IIT (ISM) Dhanbad, India
Swagato Sanyal	IIT Kharagpur, India

Steering Committee

Subir Kumar Ghosh (Chair)	Ramakrishna Mission Vivekananda Educational and Research Institute, India
Janos Pach	Ecole Polytechnique Federale De Lausanne (EPFL), Lausanne, Switzerland
Nicola Santoro	School of Computer Science, Carleton University, Canada
Swami Sarvattomananda	Ramakrishna Mission Vivekananda Educational and Research Institute, India
Peter Widmayer	Institute of Theoretical Computer Science, ETH Zürich, Switzerland
Chee Yap	Courant Institute of Mathematical Sciences and New York University, USA

Additional Reviewers

Aravind, N. R.	Francis, Maria	Levcopoulos, Christos
Ağaoğlu, Deniz	Ganesan, Ashwin	Masařík, Tomáš
Balakrishnan, Kannan	Goyal, Shuchita	Masopust, Tomas
Baswana, Surender	Hoeksma, Ruben	Mathew, Rogers
Biswas, Ranita	Iranmanesh, Ehsan	Muthu, Rahul
Chakraborty, Dibyayan	Jansson, Jesper	M. V. Panduranga Rao
Chaplick, Steven	Kare, Anjeneya Swami	N. R., Aravind
Dey, Hiranya	Kern, Walter	Nandakumar, Satyadev
Di Stefano, Gabriele	Khodamoradi, Kamyar	Nandi, Soumen
Dourado, Mitre	Klein, Rolf	Padinhatteeri, Sajith
Debski, Michał	Kowaluk, Miroslaw	Pal, Shyamosree
Firman, Oksana	Lakshmanan S., Aparna	Paul, Subhabrata

Polishchuk, Valentin
Pratihar, Sanjoy
Rajendraprasad, Deepak
Raman, Venkatesh
Roy, Bodhayan

Sarkar, Apurba
Sen, Sagnik
Simon, Sunil
Sreenivasaiah, Karteek
Sritharan, R.

Tewari, Raghunath
Uniyal, Sumedha
Venkitesh, S.
Çağirici, Onur

Short Papers

Guarding a Polygon from Its Boundary

Matthew J. Katz

Department of Computer Science, Ben-Gurion University of the Negev,
Beer-Sheva 8410501, Israel
matya@cs.bgu.ac.il

Abstract. In this talk, we survey some recent and not-so-recent results dealing with guarding a polygon from its boundary. In particular, we discuss problems in which the goal is to place as few guards as possible on the boundary of a polygon, either at vertices or at arbitrary points, in order to guard the polygon's vertex set or its entire boundary. The v2v version is especially interesting, since it is equivalent to finding a minimum dominating set in the visibility graph of the polygon. Among other results, we present a local-search-based PTAS for guarding the vertices of a weakly-visible polygon from its vertices. Previously, only a constant-factor approximation algorithm (by Bhattacharya, Ghosh and Roy, 2017) was known for this problem.

Strings and Order

János Pach

École Polytechnique Fédérale de Lausanne
and Rényi Institute of Hungarian Academy of Sciences
pach@cims.nyu.edu

Given a family of sets, $\mathcal{C}$, the *intersection graph* of $\mathcal{C}$ is the graph, whose vertices correspond to the elements of $\mathcal{C}$, and two vertices are joined by an edge if and only if the corresponding sets have a nonempty intersection. The complement of the intersection graph of $\mathcal{C}$ is called the *disjointness graph* of $\mathcal{C}$. A *string* (or *curve*) is the image of a continuous function $\phi : [0, 1] \to \mathbb{R}^2$.

A *string graph* is an intersection graph of strings. The study of string graphs was initiated by Benzer [1], who investigated certain genetic structures in cells, and shortly after by Sinden [15], for the design of integrated circuits (chips). The recognition of string graphs is an NP-complete problem [13]. There exist string graphs on n vertices such that no matter how we represent them by strings, two of them will intersect exponentially many times [5]. Every planar graph is the intersection graphs of segments (straight-line strings) [2]. The number of string graphs on n vertices is $2^{\left(\frac{3}{4} + o(1)\right)\binom{n}{2}}$ [11]. Almost all string graphs on n vertices are intersection graphs of convex sets in the plane. The vertex sets of almost all string graphs can be partitioned into 5 cliques so that there are two of them not connected by any edge [8]. The vertex set of every string graph G with m edges can be partitioned into three parts $V(G) = A \cup S \cup B$ such that $|A|, |B| \leq 2|V(G)|/3$, $|S| = O(\sqrt{m})$, and no vertex of A is connected to any vertex of B by an edge [6].

Partially ordered sets seem to play an important role in the study of string graphs. An *incomparability graph* is a graph whose vertices correspond to the elements of a partially ordered set, with two vertices being joined by an edge if and only if they are incomparable. It is known that every incomparability graph is a string graphs [7, 12, 14]. A partial converse of this theorem is also true: Given a collection of n strings such that their intersection graph has at least cn^2 edges, we can shorten each string in such a way that their intersection graph becomes an incomparability graph and still has at least $c'n^2$ edges [4].

A string is said to be *x-monotone* if every vertical line intersects it in at most one point. A string is called a *grounded* if one of its endpoints lies on the y-axis and the rest of the curve lies in the nonnegative half-plane $\{x \geq 0\}$. A graph G is called *magical* if there are two total orderings, $<_1$ and $<_2$, on its vertex set such that any three distinct

Research partially supported by Swiss National Science Foundation grants no. 200020-162884 and 200021-175977.

vertices $a, b, c \in V(G)$ with $a <_1 b <_1 c$ satisfy the following condition: if $ab, bc \in E(G)$ and $ac \notin E(G)$, then $b <_2 a$ and $b <_2 c$.

Theorem 1. [10] *A graph is the disjointness graph of a finite collection of grounded x-monotone curves if and only if it is magical.*

A graph G is said to be *double-magical* if there exist three total orderings $<_1, <_2, <_3$ on $V(G)$ such that G is the union of two graphs, G^1 and G^2, with $V(G^1) = V(G^2) = V(G)$, where G^1 is magical with respect to the orders $<_1, <_2$ and G^2 is magical with respect to the orders $<_1, <_3$.

Theorem 2. [10] *A graph is the disjointness graph of a finite collection of x-monotone curves, each of which intersects the y-axis, if and only if it is double-magical.*

These characterizations can be used to deduce some coloring properties of disjointness graphs of curves (that is, complements of string graphs).

Theorem 3. [9, 10] *There exist disjointness graphs G_k of collections x-monotone curves containing no clique of size k, such that their chromatic numbers satisfy $\chi(G_k) = \Omega(k^4)$. The order of magnitude of this bound, as $k \to \infty$, cannot be improved.*

It is a major unsolved problem to decide whether there exists a constant $\varepsilon > 0$ such that every string graph (and, therefore, every disjointness graph of strings) has a clique of size at least n^ε or an independent set of size at least $\varepsilon > 0$. For intersection graphs of x-monotone curves, this follows from Theorem 3. This problem is closely related to the celebrated Erdős-Hajnal conjecture [3].

Theorems 1–3 are joint work with István Tomon.

References

1. Benzer, S.: On the topology of the genetic fine structure. Proc. Nat. Acad. Sci. USA **45**(11), 1607–1620 (1959)
2. Chalopin, J., Goncalves, D., Ochem, P.: Planar graphs have 1-string representations. Discrete Comput. Geom. **43**(3), 626–647 (2010)
3. Erdős, P., Hajnal, A.: Ramsey-type theorems. Discrete Appl. Math. **25**, 37–52 (1989)
4. Fox, J., Pach, J.: String graphs and incomparability graphs. Adv. Math. **230**, 1381–1401 (2012)
5. Kratochvíl, J., Matoušek, J.: String graphs requiring exponential representations. J. Combin. Theory Ser. B **53**(1), 1–4 (1991)
6. Lee, J.R.: Separators in region intersection graphs. arXiv:1608.01612
7. Lovász, L.: Perfect graphs. In: Selected Topics in Graph Theory, vol. 2, pp. 55–87. Academic Press, London (1983)
8. Pach, J., Reed, B.A., Yuditsky, Y.: Almost all string graphs are intersection graphs of plane convex sets. In: Symposium on Computational Geometry, pp. 68:1–14 (2018)
9. Pach, J., Törőcsik, J.: Some geometric applications of Dilworth's theorem. Discrete Comput. Geom. **12**(1), 1–7 (1994)
10. Pach, J., Tomon, I.: On the chromatic number of disjointness graphs of curves. arXiv:1811. 09158
11. Pach, J., Tóth, G.: How many ways can one draw a graph? Combinatorica **26**, 559–576 (2006)

12. Pach, J., Tóth, G.: Comments on fox news. Geombinatorics **15**, 150–154 (2006)
13. Schaefer, M., Sedgwick, E., Štefankovič, D.: Recognizing string graphs in NP. J. Comput. Syst. Sci. **67**(2), 365–380 (2003)
14. Sidney, J.B., Sidney, S.J., Urrutia, J.: Circle orders, n-gon orders and the crossing number. Order **5**(1), 1–10 (1988)
15. Sinden, F.W.: Topology of thin film RC-circuits. Bell Syst. Tech. J. **45**, 1639–1662 (1966)

Contents

New Directions in Approximate Nearest-Neighbor Searching

David M. Mount[✉]

Department of Computer Science and Institute for Advanced Computer Studies,
University of Maryland, College Park, MD 20742, USA
mount@umd.edu

Abstract. Approximate nearest-neighbor searching is an important retrieval problem with numerous applications in science and engineering. This problem has been the subject of many research papers spanning decades of work. Recently, a number of dramatic improvements and extensions have been discovered. In this paper, we will survey some of recent techniques that underlie these developments. In particular, we discuss local convexification, Macbeath regions, Delone sets, and how to apply these concepts to develop new data structures for approximate polytope membership queries and approximate vertical ray-shooting queries.

Keywords: Approximate nearest-neighbor searching ·
Convexification · Macbeath regions ·
Geometric algorithms and data structures

1 Introduction

Nearest-neighbor searching is a fundamental retrieval problem with numerous applications in fields such as machine learning, data mining, data compression, and pattern recognition. A set of n points, called *sites*, is preprocessed into a data structure such that, given any query point q, it is possible to report the site that is closest to q. The most common formulation involves points in $\mathbb{R}^d$ under the Euclidean distance. Unfortunately, the best solution achieving $O(\log n)$ query time uses roughly $O(n^{d/2})$ storage space [22], which is too high for many applications. This has motivated the study of approximations. Given an approximation parameter $\varepsilon > 0$, the objective is to return any site whose distance from q is within a factor of $1 + \varepsilon$ of the distance to the true nearest neighbor. This is called ε-*approximate nearest-neighbor searching* (ε-ANN). This problem has been the subject of many research papers, and it remains a topic of active study.

In this paper, we will survey some of recent techniques on efficient approximate nearest-neighbor searching. The results presented here are joint with

Research supported by NSF grant CCF–1618866.

S. P. Pal and A. Vijayakumar (Eds.): CALDAM 2019, LNCS 11394, pp. 1–15, 2019.
https://doi.org/10.1007/978-3-030-11509-8_1

Ahmed Abdelkader, Sunil Arya, and Guilherme da Fonseca. Throughout, we assume that the sites reside in real d-dimensional space, $\mathbb{R}^d$, and the dimension d is a constant.[1] We treat both n and ε as asymptotic quantities. Thus, asymptotic notation may conceal constant factors that grow exponentially with the dimension.

Let us begin with some of the history on this problem. Over two decades ago, we proposed a data structure for answering ε-ANN queries through the use of a balanced quadtree-like partition tree, which achieves $O((1/\varepsilon)^d \log n)$ query time with $O(n)$ space [12]. Various improvements were proposed by others, including Bespamyatnikh [17], Duncan *et al.* [24] and Chan [20], the best of which offers a query time of $O(\log n + 1/\varepsilon^{d-1})$ with $O(n)$ space. In addition to the fact that the space is optimal, a nice feature of these solutions is that the preprocessing is independent of ε, implying that once the data structure has been built, queries can be answered for any error bound ε. Despite these strengths, the result still suffers from the fact that the constant factor in the query time grows roughly as $1/\varepsilon^{d-1}$, which is suboptimal.

From the perspective of query time, the most efficient approach for ε-ANN searching in fixed dimensions is based on the concept of the *approximate Voronoi diagram* (AVD), which was introduced by Har-Peled [26]. It achieves a query time of roughly $O(\log(n/\varepsilon))$ with roughly $O(n/\varepsilon^d)$ space. In joint work with Arya and Malamatos, we showed how to generalize the AVD to achieve space-time tradeoffs [8–10]. Despite subsequent work on the problem (see, e.g., [7,10]), the storage requirements needed to achieve logarithmic query time remained essentially unchanged for over 15 years. In joint work with Arya and da Fonseca we finally overcame this obstacle by presented a data structure that achieves a tradeoff between space and query time such that the product of these two quantities is roughly $O(n/\varepsilon^{d/2})$ [4,6]. At one extreme, we achieve $O(\log(n/\varepsilon))$ query time with $O(n/\varepsilon^{d/2})$ space and, at the other extreme, we achieve roughly $O(\log n + 1/\varepsilon^{d/2})$ query time with $O(n)$ space. In comparison with Har-Peled's original AVD, this matches the query time while reducing the exponent in the ε-dependencies by roughly half.

These improvements arose from a combination of two components. The first is the use of the AVD data structure to subdivide space into simple regions, called *cells*, that have nice separation properties with respect to the sites. The second is the application of new space-efficient techniques for answering *vertical ray-shooting queries* approximately in convex bodies. These two components are mediated through a well-known technique in computational geometry, called the *lifting transformation*, which reduces the problem of nearest-neighbor searching in d-dimensional space to ray-shooting in a convex body in $(d+1)$-dimensional space [15].

While much progress has been made on the problem of Euclidean approximate nearest-neighbor searching, the results do not extend readily to other

---

[1] There is a significant literature on approximate nearest-neighbor searching in spaces of high dimension, but the techniques are very different, and it will be beyond the scope of this paper to discuss them.

metrics. Unlike Har-Peled's original AVD, which could be readily adapted to a variety of metrics, these recent data structures rely on the lifting transformation, which applies only to the Euclidean metric. This raises the question of whether these new, highly efficient data structures can be applied to non-Euclidean distance functions.

Through the use of *minimization diagrams*, Har-Peled and Kumar introduced a powerful technique for nearest-neighbor searching which can be applied to non-Euclidean distances [27]. Their approach can even be applied to situations where each site is associated with its own distance function. For each site p_i, let $f_i : \mathbb{R}^d \to \mathbb{R}^+$ be the associated distance function. Let $F_{\min}$ denote the pointwise minimum of these functions, that is, the *lower-envelope function*. Clearly, approximating the value of $F_{\min}$ at a query point q is equivalent to approximating the nearest neighbor of q. Har-Peled and Kumar proved that ε-ANN searching over a wide variety of distance functions (including additive and multiplicatively weighted sites) could be cast in this form.

In recent joint work with Abdelkader, Arya, and da Fonseca, we have presented a more efficient approach for ε-ANN searching involving non-Euclidean distance functions, including convex scaling distance functions and Bregman divergences [1]. Subject to certain admissibility conditions on these distance functions, our approach allows us to answer ε-ANN queries in logarithmic time using roughly $O(n/\varepsilon^{d/2})$ space, which nearly matches the best known result for the Euclidean metric.

These recent developments have employed a variety of new techniques, which are described in diverse publications. In this paper, we will present in one place a summary of the key ideas underlying these data structures. The material presented here is a coalescence of results from a number of sources, including [1,2,4,6,7,10].

2 Preliminaries

In this section we present a number of basic definitions and results, which will be used throughout the paper. We consider the real d-dimensional space, $\mathbb{R}^d$, where d is a fixed constant. Let O denote the origin of $\mathbb{R}^d$. Given a vector $v \in \mathbb{R}^d$, let $\|v\|$ denote its Euclidean length, and let $\langle \cdot, \cdot \rangle$ denote the standard inner product. Given two points $p, q \in \mathbb{R}^d$, the Euclidean distance between them is $\|p - q\|$.

Given a function $f : \mathbb{R}^d \to \mathbb{R}$, its *graph* is the set of $(d + 1)$-dimensional points $(x, f(x))$, its *epigraph* is the set of points on or above the graph, and its *hypograph* is the set of points on or below the graph (where the $(d + 1)$-st axis is directed upwards).

The gradient and Hessian of a function generalize the concepts of the first and second derivative to a multidimensional setting. The *gradient* of f, denoted ∇f, is defined as the vector field $\left(\frac{\partial f}{\partial x_1}, \ldots, \frac{\partial f}{\partial x_d}\right)^\mathsf{T}$. The gradient vector points in a direction in which the function grows most rapidly, and it is orthogonal to the level surface. For any point x and any unit vector v, the rate of change of f along v is given by the dot product $\nabla f(x) \cdot v$. The *Hessian* of f at x, denoted

$\nabla^2 f(x)$, is a $d \times d$ matrix of second-order partial derivatives at x. For twice continuously differentiable functions, $\nabla^2 f(x)$ is symmetric, implying that it has d (not necessarily distinct) real eigenvalues.

Given a d-vector v, let $\|v\|$ denote its length under the *Euclidean norm*, and the *Euclidean distance* between points p and q is $\|q - p\|$. Given a $d \times d$ matrix A, its *spectral norm* is $\|A\| = \sup\{\|Ax\| / \|x\| : x \in \mathbb{R}^d$ and $x \neq 0\}$. Since the Hessian is a symmetric matrix, it follows that $\|\nabla^2 f(x)\|$ is the largest absolute value attained by the eigenvalues of $\nabla^2 f(x)$. A twice continuously differentiable function on a convex domain is convex if and only if its Hessian matrix is positive semidefinite in the interior of the domain. It follows that all the eigenvalues of the Hessian of a convex function are nonnegative.

Let K be a convex body in $\mathbb{R}^d$, represented as the intersection of m closed halfspaces. The boundary of K will be denoted by ∂K. For $0 < \kappa \le 1$, we say that K is in κ-*canonical form* if $B(\kappa/2) \subseteq K \subseteq B(1/2)$. Clearly, such a body has a diameter between κ and 1.

It is well known that in $O(m)$ time it is possible to compute a non-singular affine transformation T such that $T(K)$ is in $(1/d)$-canonical form [5,26]. Further, if a convex body P is within Hausdorff distance ε of $T(K)$, then $T^{-1}(P)$ is within Hausdorff distance at most $d\varepsilon$ of K. (Indeed, this transformation is useful, since the resulting approximation is directionally sensitive, being more accurate along directions where K is skinnier.) Therefore, for the sake of approximation with respect to Hausdorff distance, we may assume that K has been mapped to canonical form, and ε is scaled by a factor of $1/d$. Because we assume that d is a constant, this transformation will only affect the constant factors in our analysis.

Assuming that K is in κ-canonical form for some constant κ, let us define K_δ to be the convex body that results by translating each of K's bounding hyperplanes outwards by distance δ. It is easy to show that for such a convex body K, the Hausdorff distance between K and K_δ is $\Theta(\delta)$, where the hidden constant factor depends on κ.

3   Separating and Lifting

Let us begin with two fundamental constructions in approximate nearest-neighbor searching: subdividing space to achieve separation and the lifting transformation. Let us assume that we are given a set of n sites in $\mathbb{R}^d$, and let ε be the desired approximation parameter.

To establish the separation, we apply a simple quadtree-based technique to subdivide space into regions, called *cells*. This subdivision results from a height-balanced variant of a quadtree, called a *balanced box decomposition tree* (or BBD tree) [11]. The separation properties are essentially the same as those of the Euclidean AVD data structure of [10]. For any cell w, the sites can be partitioned into three subsets, any of which may be empty. First, a single site may lie within w (see Fig. 1(a)). Second, a subset of sites, called the *outer cluster*, are well-separated from the cell. Specifically, they lie outside a ball O_w centered at the center of w and whose radius is a suitable constant times the diameter of w (see Fig. 1(b)). Finally, there may be a dense cluster of points, called the *inner cluster*, that lies within a ball I_w that is well-separated from the cell (see Fig. 1(c)). Specifically, the distance between ball I_w and the cell w is $\Omega(1/\varepsilon)$ times the diameter of I_w. After locating the cell containing the query point, the approximate nearest neighbor is computed independently for each of these subsets, and the overall closest is returned. The resulting data has size $O(n \log \frac{1}{\varepsilon})$ and the cell containing a query point can be located in logarithmic time.

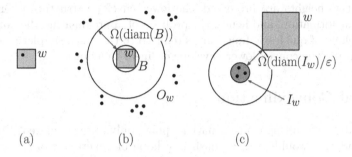

(a)   (b)   (c)

Fig. 1. AVD separation properties.

Next, let us discuss the lifting transformation. We assume the Euclidean metric. Recall that answering an ε-ANN query for a point q is equivalent to approximating the vertical distance from q to the lower envelope of the distance functions. For each site p_i, let $f_i(q) = \|q - p_i\|^2$ denote the squared Euclidean distance from q to p_i. For all sufficiently small ε, computing an ε-approximate nearest neighbor in the Euclidean distance is roughly equivalent to computing an $(\varepsilon/2)$-approximate nearest neighbor in the squared Euclidean distance. Our objective is to compute an $(\varepsilon/2)$-approximation to the value of $\mathcal{F}_{\min}(q)$, where $\mathcal{F}_{\min}$ is the lower envelope of the functions f_i.

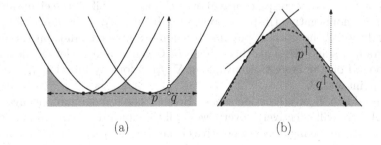

(a)   (b)

Fig. 2. The lifting transformation (shown in a vertically inverted form).

Unfortunately, the lower envelope can be unwieldy (see Fig. 2(a)). In the case of the Euclidean metric, there is a well-known trick, called the *lifting transfor-*

mation which can be used to reduce this envelope to convex form. The minimization diagram is a set in $\mathbb{R}^{d+1}$. Each point in this space can be represented as $p = (x_1, \ldots, x_d, z)$, where z stores the distance value. Consider the transformation that maps such a point to $p^\uparrow = (x_1, \ldots, x_d, z - \sum_i x_i^2)$. This transformation maps all the points on the $z = 0$ coordinate hyperplane to a downward-curving paraboloid. It is easily verified that the each of the distance functions f_i is transformed to a linear function that is tangent to this paraboloid. Thus, the hypograph of the lower envelope (which is nonconvex) is mapped to the lower envelope of a set of n linear functions (which is convex) (see Fig. 2(b)). Since relative heights are preserved, the lower envelope structure is preserved by this transformation, and hence, the problem of approximating the nonconvex lower envelope of quadratic functions is reduced to the much simpler problem of approximating the convex lower envelope of linear functions.

4   Local Convexification

Unfortunately, the lifting transformation applies only to the (squared) Euclidean metric, and this would seem to dash any hope of generalizations beyond the Euclidean case. In joint work with Abdelkader, Arya, and da Fonseca, we discovered a novel way of circumventing this limitation [1]. They key observation is that the convex approximation techniques developed in [6] do not only apply to convex polytopes, they can be applied to any convex body, even one with curved boundaries. This provides an additional degree of flexibility. Rather than applying a transformation to linearize the various distance functions, one can go a bit overboard and "convexify" them.

To make this more formal, let $\mathcal{F} = \{f_1, \ldots, f_m\}$ be a collection of functions, and let $\mathcal{F}_{\min}$ denote its lower envelope. We make two assumptions. First, we restrict the functions to a bounded convex domain, which for our purposes may be taken to be a closed Euclidean ball B in $\mathbb{R}^d$. Second, let us assume that the functions are smooth, implying in particular that each function f_i has a well defined gradient ∇f_i and Hessian $\nabla^2 f_i$ for every point of B. It is well known that a function f_i is convex (resp., concave) over B if and only if all the eigenvalues of $\nabla^2 f_i(x)$ are nonnegative (resp., nonpositive). Intuitively, if the functions f_i are sufficiently well-behaved it is possible to compute upper bounds on the norms of the gradients and Hessians throughout B. Given $\mathcal{F}$ and B, let Λ^+ denote an upper bound on the largest eigenvalue of $\nabla^2 f_i(x)$ for any function $f_i \in \mathcal{F}$ and for any point $x \in B$.

If we add to f_i any function whose Hessian has a maximum eigenvalue at most $-\Lambda^+$, we will effectively "overpower" all the upward curving terms, resulting in a function having only nonpositive eigenvalues, that is, a concave function. The lower envelope of concave functions is concave, and so techniques for convex approximation (such as Lemma 5) can be applied to the hypograph of the resulting lower-envelope function.

More precisely, let $p \in \mathbb{R}^d$ and $r \in \mathbb{R}$ denote the center point and radius of our domain B, respectively. Define a function ϕ (which depends on B and Λ^+)

to be

$$\phi(x) \;=\; \frac{\Lambda^+}{2}\left(r^2 - \sum_{j=1}^{d}(x_j - p_j)^2 \right) \;=\; \frac{\Lambda^+}{2}(r^2 - \|x - p\|^2).$$

(See Fig. 3(b).) It is easy to verify that ϕ evaluates to zero along B's boundary and is positive within B's interior. Also, for any $x \in \mathbb{R}^d$, the Hessian of $\|x - p\|^2$ (as a function of x) is a $d \times d$ diagonal matrix $2I$, and therefore $\nabla^2\phi(x) = -\Lambda^+ I$. Now, define

$$\widehat{f_i}(x) = f_i(x) + \phi(x), \quad \text{for } 1 \le i \le m, \text{ and}$$
$$\widehat{F}_{\min}(x) = \min_{1 \le i \le m} \widehat{f_i}(x) = \mathcal{F}_{\min}(x) + \phi(x).$$

(See Fig. 3(c).)

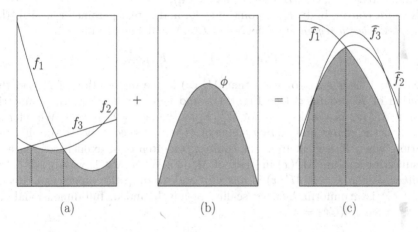

(a)   (b)   (c)

Fig. 3. Converting lower envelopes to convex form through local convexification.

Because all the functions are subject to the same offset at each point x, $\widehat{F}_{\min}$ preserves the relevant combinatorial structure of $\mathcal{F}_{\min}$, and in particular f_i yields the minimum value to $\mathcal{F}_{\min}(x)$ at some point x if and only if $\widehat{f_i}$ yields the minimum value to $\widehat{F}_{\min}(x)$. Absolute vertical errors are preserved as well. Observe that $\widehat{F}_{\min}(x)$ matches the value of $\mathcal{F}_{\min}$ along B's boundary and is larger within its interior. Also, since $\nabla^2\phi(x) = -\Lambda^+ I$, it follows from elementary linear algebra that each eigenvalue of $\nabla^2\widehat{f_i}(x)$ is smaller than the corresponding eigenvalue of $\nabla^2 f_i(x)$ by Λ^+. Thus, all the eigenvalues of $\widehat{f_i}(x)$ are nonpositive, and so $\widehat{f_i}$ is concave over B. In turn, this implies that $\widehat{F}_{\min}$ is concave, as desired. We will show that, when properly applied, relative errors are nearly preserved, and hence approximating the convexified lower envelope yields an approximation to the original lower envelope.

The general technique of *convexification* has been developed in the context of nonlinear approximation [3,16]. We refer to our application of this technique as *local convexification*. In [1], we show that through the application of local convexification, it is possible to obtain efficient ε-ANN data structures for a number of non-Euclidean distance functions, such as convex scaling distances and Bregman divergences (subject to suitable admissibility assumptions).

5   Macbeath Regions

Next, let us consider the question of how to approximate convex bodies in $\mathbb{R}^d$. The fundamental geometric construct underlying the recent improvements in convex approximation data structure is a concept dating from the 1950's, developed by A. M. Macbeath. It provides an elegant way to sample points economically and define local approximations of a convex body. They have found uses in diverse areas (see, e.g., Bárány's survey [14]).

Given a convex body K, a point $x \in K$, and a real parameter $\lambda \geq 0$, the λ-*scaled Macbeath region* at x, denoted $M_K^\lambda(x)$, is defined to be

$$x + \lambda((K - x) \cap (x - K)).$$

When $\lambda = 1$, it is easy to verify that $M_K^1(x)$ is the intersection of K and the reflection of K around x (see Fig. 4(a)), and hence it is centrally symmetric about x. $M_K^\lambda(x)$ is a scaled copy of $M_K^1(x)$ by the factor λ about x. We refer to x and λ as the *center* and *scaling factor* of $M_K^\lambda(x)$, respectively. To simplify the notation, when K is clear from the context, we often omit explicit reference in the subscript and use $M^\lambda(x)$ in place of $M_K^\lambda(x)$. When $\lambda < 1$, we say $M^\lambda(x)$ is *shrunken*. When $\lambda = 1$, $M^1(x)$ is *unscaled* and we drop the superscript. Recall that if C^λ is a uniform λ-factor scaling of any bounded, full-dimensional set $C \subset \mathbb{R}^d$, then $\mathrm{vol}(C^\lambda) = \lambda^d \cdot \mathrm{vol}(C)$.

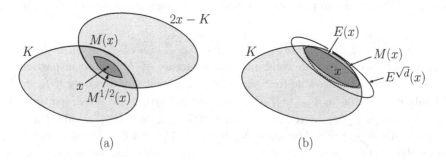

(a)   (b)

Fig. 4. (a) Macbeath regions and (b) Macbeath ellipsoids.

An important property of Macbeath regions, which we call *expansion-containment*, is that if two shrunken Macbeath regions overlap, then an appropriate expansion of one contains the other (see Fig. 5(a)). The following is a generalization of results of Ewald, Rogers and Larman [25] and Brönnimann, Chazelle, and Pach [19]. Our generalization allows the shrinking factor λ to be adjusted, and shows how to adjust the expansion factor β of the first body to cover an α-scaling of the second body, e.g., the center point only (see Fig. 5(b)).

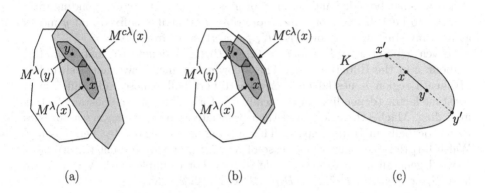

(a)   (b)   (c)

Fig. 5. (a)–(b) Expansion-containment per Lemma 1. (c) The Hilbert metric.

Lemma 1 (Expansion-Containment). *Let $K \subset \mathbb{R}^d$ be a convex body and let $0 < \lambda < 1$. If $x, y \in K$ such that $M^\lambda(x) \cap M^\lambda(y) \neq \emptyset$, then for any $\alpha \geq 0$ and $\beta = \frac{2+\alpha(1+\lambda)}{1-\lambda}$, $M^{\alpha\lambda}(y) \subseteq M^{\beta\lambda}(x)$ (see Fig. 5(a)).*

5.1 Delone Sets and the Hilbert Metric

An important concept in the context of metric spaces involves coverings and packings by metric balls [23]. Given a metric f over $\mathcal{X}$, a point $x \in \mathcal{X}$, and real $r > 0$, define the ball $B_f(x, r) = \{y \in \mathcal{X} : f(x, y) \leq r\}$. For $\varepsilon, \varepsilon_p, \varepsilon_c > 0$, a set $X \subseteq \mathcal{X}$ is an:

ε-**packing:** If the balls of radius $\varepsilon/2$ centered at every point of X do not intersect.

ε-**covering:** If every point of $\mathcal{X}$ is within distance ε of some point of X.

$(\varepsilon_p, \varepsilon_c)$-**Delone Set:** If X is an ε_p-packing and an ε_c-covering.

Delone sets have been used in the design of data structures for answering geometric proximity queries in metric spaces through the use of hierarchies of nets, such as navigating nets [30], net trees [28], and cover trees [18].

In order to view a collection of Macbeath regions as a Delone set, it will be useful to introduce an underlying metric. The Hilbert metric [29] was introduced over a century ago by David Hilbert as a generalization of the Cayley-Klein model of hyperbolic geometry. A *Hilbert geometry* (K, f_K) consists of a convex domain K in $\mathbb{R}^d$ with the Hilbert distance f_K. For any pair of distinct points $x, y \in K$, the line passing through them meets ∂K at two points x' and y'. We label these points so that they appear in the order $\langle x', x, y, y' \rangle$ along this line (see Fig. 5(c)). The Hilbert distance f_K is defined as

$$f_K(x, y) = \frac{1}{2} \ln \left(\frac{\|x' - y\| \, \|x - y'\|}{\|x' - x\| \, \|y - y'\|} \right).$$

When K is not bounded and either x' or y' is at infinity, the corresponding ratio is taken to be 1. To get some intuition, observe that if x is fixed and y moves along a ray starting at x towards ∂K, $f_K(x, y)$ varies from 0 to ∞.

Given a point $x \in K$ and $r > 0$, let $B_H(x, r)$ denote the ball of radius r about x in the Hilbert metric. The following lemma shows that a shrunken Macbeath region is nested between two Hilbert balls whose radii differ by a constant factor (depending on the scaling factor). Thus, up to constant factors in scaling, Macbeath regions and their associated ellipsoids can act as proxies to metric balls in Hilbert space. This nesting was observed by Vernicos and Walsh [31] (for the conventional case of $\lambda = 1/5$), and we present the straightforward generalization to other scale factors. For example, with $\lambda = 1/5$, we have $B_H(x, 0.09) \subseteq M^{1/5}(x) \subseteq B_H(x, 0.21)$ for all $x \in K$.

Lemma 2. *Given a convex body $K \subset \mathbb{R}^d$, for all $x \in K$ and any $0 \le \lambda < 1$,*

$$B_H\left(x, \frac{1}{2} \ln (1 + \lambda)\right) \subseteq M^\lambda(x) \subseteq B_H\left(x, \frac{1}{2} \ln \frac{1 + \lambda}{1 - \lambda}\right).$$

5.2   Macbeath Ellipsoids

For the sake of efficient computation, it will be useful to approximate Macbeath regions by shapes of constant combinatorial complexity. We have opted to use ellipsoids.

Given a Macbeath region, define its associated *Macbeath ellipsoid* $E_K^\lambda(x)$ to be the maximum-volume ellipsoid contained within $M_K^\lambda(x)$ (see Fig. 4(c)). Clearly, this ellipsoid is centered at x and $E_K^\lambda(x)$ is an λ-factor scaling of $E_K^1(x)$ about x. It is well known that the maximum-volume ellipsoid contained within a convex body is unique, and Chazelle and Matoušek showed that it can be computed for a convex polytope in time linear in the number of its bounding halfspaces [21]. By John's Theorem (applied in the context of centrally symmetric bodies) it follows that $E_K^\lambda(x) \subseteq M_K^\lambda(x) \subseteq E_K^{\lambda\sqrt{d}}(x)$ [13].

Given a point $x \in K$ and $\delta > 0$, define $M_\delta(x)$ to be the (unscaled) Macbeath region with respect to K_δ (as defined in Sect. 2), that is, $M_\delta(x) = M_{K_\delta}(x)$. Let $E_\delta(x)$ denote the maximum volume ellipsoid contained within $M_\delta(x)$. As $M_\delta(x)$ is symmetric about x, $E_\delta(x)$ is centered at x. For any $\lambda > 0$, define $M_\delta^\lambda(x)$ and $E_\delta^\lambda(x)$ to be the uniform scalings of $M_\delta(x)$ and $E_\delta(x)$, respectively, about x by a factor of λ. By John's Theorem, we have

$$E_\delta^\lambda(x) \subseteq M_\delta^\lambda(x) \subseteq E_\delta^{\lambda\sqrt{d}}(x). \tag{1}$$

Two particular scale factors will be of interest to us. Define $M_\delta'(x) = M_\delta^{1/2}(x)$ and $M_\delta''(x) = M_\delta^{\lambda_0}(x)$, where $\lambda_0 = 1/(4\sqrt{d} + 1)$. Similarly, define $E_\delta'(x) =$

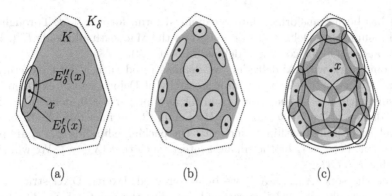

Fig. 6. A Delone set for a convex body. (Not drawn to scale.)

$E_\delta^{1/2}(x)$ and $E_\delta''(x) = E_\delta^{\lambda_0}(x)$ (see Fig. 6(a)). Given a fixed δ, let X_δ be any maximal set of points, all lying within K, such that the ellipsoids $E_\delta''(x)$ are pairwise disjoint for all $x \in X_\delta$.

These ellipsoids form a packing of K_δ (see Fig. 6(b)). The following lemma, which was proved in [2], shows that their suitable expansions cover K while being contained within K_δ (see Fig. 6(c)).

Lemma 3. *Given a convex body K in $\mathbb{R}^d$ and a set X_δ as defined above for $\delta > 0$,*

$$K \subseteq \bigcup_{x \in X_\delta} E_\delta'(x) \subseteq K_\delta.$$

In conclusion, if we treat the scaling factor λ in $E^\lambda(x)$ as a proxy for the radius of a metric ball, we have shown that X_δ is a $(2\lambda_0, 1/2)$-Delone set for K. By Lemma 2 this is also true in the Hilbert metric over K_δ up to a constant factor adjustment in the radii. (Note that the scale of the Hilbert balls does not vary with δ. What varies is the choice of the expanded body K_δ defining the metric.)

By John's Theorem, Macbeath regions and Macbeath ellipsoids differ by a constant scaling factor, both with respect to enclosure and containment. We remark that all the results of the previous two sections hold equally for Macbeath ellipsoids. We omit the straightforward, but tedious, details.

5.3 Approximate Polytope Membership (APM)

Given a convex body K in canonical form and a query point q, an *ε-approximate polytope membership query* (ε-APM) returns a positive result if q lies within K, a negative result if q's distance from K exceeds $\varepsilon \cdot \mathrm{diam}(K)$, and may return either result otherwise. The Macbeath-based Delone sets X_δ described above yield a simple data structure for answering ε-APM queries for a convex body K. We assume that K is represented as the intersection of m halfspaces (but our results generalize to arbitrary convex bodies). We may assume that in $O(m)$

time it has been transformed into κ-canonical form, for $\kappa = 1/d$. Throughout, we will assume that Delone sets are based on the Macbeath ellipsoids $E''_\delta(x)$ for packing and $E'_\delta(x)$ for coverage (defined in Sect. 5.2).

In joint work with Abdelkader, we demonstrated such a data structure in [2]. The data structure is based on a hierarchy of Delone sets of exponentially increasing accuracy. Define $\delta_0 = \varepsilon$, and for any integer $i \geq 0$, define $\delta_i = 2^i \delta_0$. Let X_i denote a Delone set for K_{δ_i}. By Lemma 3, we may take X_i to be any maximal set of points within K such that the packing ellipsoids $E''_\delta(x)$ are pairwise disjoint. Let $\ell = \ell_\varepsilon$ be the smallest integer such that $|X_\ell| = 1$. We will show below that $\ell = O(\log \frac{1}{\varepsilon})$.

Given the sets $\langle X_0, \ldots, X_\ell \rangle$, we build a rooted, layered DAG structure as follows. The nodes of level i correspond 1-1 with the points of X_i. The leaves reside at level 0 and the root at level ℓ. Each node $x \in X_i$ is associated with two things. The first is its *cell*, denoted cell(x), which is the covering ellipsoid $E'_\delta(x)$ (the larger hollow ellipsoids shown in Fig. 7). The second, if $i > 0$, is a set of *children*, denoted ch(x), which consists of the points $y \in X_{i-1}$ such that cell$(x) \cap$ cell$(y) \neq \emptyset$.

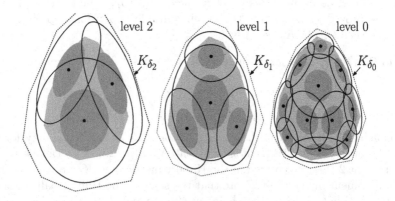

Fig. 7. Hierarchy of ellipsoids for answering APM queries.

To answer a query q, we start at the root and iteratively visit any one node $x \in X_i$ at each level of the DAG, such that $q \in$ cell(x). We know that if q lies within K, such an x must exist by the covering properties of Delone sets, and further at least one of x's children contains q. If q does not lie within any of the children of the current node, the query algorithm terminates and reports (without error) that $q \notin K$. Otherwise the search eventually reaches a node $x \in X_0$ at the leaf level whose cell contains q. Since cell$(x) \subseteq K_{\delta_0} = K_\varepsilon$, this cell serves as a witness to q's approximate membership within K. In [2], the following result establishes the algorithm's performance bounds.

Lemma 4 (Answering ε-APM Queries). *Given a convex body K in canonical form and $\varepsilon > 0$, there exists a data structure can answer ε-approximate polytope membership queries in time $O(\log \frac{1}{\varepsilon})$ and storage $O((\frac{1}{\varepsilon})^{d/2})$.*

Bringing us back to the original problem of minimization diagrams, it is possible to extend the above result from answering approxiamte membership queries with respect to a convex body to answering approximate ray-shooting queries with respect to the lower envelope of a collection of concave functions. Let $\mathcal{F} = \{f_1, \ldots, f_m\}$ be a family of convex functions, and recall that $\mathcal{F}_{\min}$ is its lower envelope. Consider the hypograph in $\mathbb{R}^{d+1}$ of $\mathcal{F}_{\min}$, and let us think of the $(d+1)$st axis as indicating the vertical direction. Answering ε-ANN queries at a point q is equivalent to approximating the result of a vertical ray shot upwards from the point $(q, 0) \in \mathbb{R}^{d+1}$ until it hits the lower envelope, where the allowed approximation error is $\varepsilon \cdot \mathcal{F}_{\min}(q)$. This is called an *$\varepsilon$-approximate vertical ray-shooting query* (ε-AVR). The following lemma, which is proved in [1], shows that if these functions are suitably normalized, such queries can be answered in essentially the same time and space complexities as ε-APM queries.

Lemma 5 (Answering ε-AVR Queries). *Consider a unit ball $B \subseteq \mathbb{R}^d$ and a family of concave functions $\mathcal{F} = \{f_1, \ldots, f_m\}$ defined over B such that for all $1 \leq i \leq m$ and $x \in B$, $f_i(x) \in [0, 1]$ and $\|\nabla f_i(x)\| \leq 1$. Then, for any $0 < \varepsilon \leq 1$, there is a data structure that can answer ε-AVR queries in time $O(\log \frac{1}{\varepsilon})$ and storage $O((\frac{1}{\varepsilon})^{d/2})$.*

5.4   Concluding Remarks

The study of data structures for approximate nearest-neighbor searching has proven to be remarkably rich and diverse. In this paper, we have demonstrated how concepts from nonlinear optimization (through convexification), analysis of convex bodies (through Macbeath regions), and sampling (through Delone sets and the Hilbert geometry) have informed the design of new, highly efficient data structures for this problem.

There are still many open problems that remain to be solved, including extensions to different metrics and different distance functions, more efficient algorithms for answering k-ANN queries (see [27]), and efficient construction of these data structures in dynamic contexts.

The author would like to express his profound gratitude to his collaborators, Ahmed Abdelkader, Sunil Arya, and Guilherme Dias da Fonseca.

References

1. Abdelkader, A., Arya, S., da Fonseca, G.D., Mount, D.M.: Approximate nearest neighbor searching with non-Euclidean and weighted distances. In: Proceedings of 30th Annual ACM-SIAM Symposium on Discrete Algorithms (2019)
2. Abdelkader, A., Mount, D.M.: Economical delone sets for approximating convex bodies. In: Proceedings of 16th Scandinavian Workshop Algorithm Theory, pp. 4:1–4:12 (2018)
3. Androulakis, I.P., Maranas, C.D., Floudas, C.A.: αBB: a global optimization method for general constrained nonconvex problems. J. Glob. Optim. **7**, 337–363 (1995)

4. Arya, S., da Fonseca, G.D., Mount, D.M.: Near-optimal ε-kernel construction and related problems. In: Proceedings of 33rd International Symposium on Computational Geometry, pp. 10:1–10:15 (2017). https://arxiv.org/abs/1604.01175

5. Arya, S., da Fonseca, G.D., Mount, D.M.: On the combinatorial complexity of approximating polytopes. Discrete Comput. Geom. **58**, 849–870 (2017)

6. Arya, S., da Fonseca, G.D., Mount, D.M.: Optimal approximate polytope membership. In: Proceedings of 28th Annual ACM-SIAM Symposium on Discrete Algorithms, pp. 270–288 (2017)

7. Arya, S., da Fonseca, G.D., Mount, D.M.: Approximate polytope membership queries. SIAM J. Comput. **47**, 1–51 (2018)

8. Arya, S., Malamatos, T.: Linear-size approximate Voronoi diagrams. In: Proceedings of 13th Annual ACM-SIAM Symposium on Discrete Algorithms, pp. 147–155 (2002)

9. Arya, S., Malamatos, T., Mount, D.M.: Space-efficient approximate Voronoi diagrams. In: Proceedings of the 34th Annual ACM Symposium on the Theory of Computing, pp. 721–730 (2002)

10. Arya, S., Malamatos, T., Mount, D.M.: Space-time tradeoffs for approximate nearest neighbor searching. J. Assoc. Comput. Mach. **57**, 1–54 (2009)

11. Arya, S., Mount, D.M.: Approximate range searching. Comput. Geom. Theory Appl. **17**, 135–163 (2000)

12. Arya, S., Mount, D.M., Netanyahu, N.S., Silverman, R., Wu, A.Y.: An optimal algorithm for approximate nearest neighbor searching fixed dimensions. J. Assoc. Comput. Mach. **45**(6), 891–923 (1998)

13. Ball, K.: An elementary introduction to modern convex geometry. In: Levy, S. (ed.) Flavors of Geometry, pp. 1–58. Cambridge University Press, Cambridge (1997). (MSRI Publications, vol. 31)

14. Bárány, I.: The technique of M-regions and cap-coverings: a survey. Rend. Circ. Mat. Palermo **65**, 21–38 (2000)

15. de Berg, M., Cheong, O., van Kreveld, M., Overmars, M.: Computational Geometry: Algorithms and Applications, 3rd edn. Springer, Heidelberg (2010). https://doi.org/10.1007/978-3-540-77974-2

16. Bertsekas, D.P.: Convexification procedures and decomposition methods for nonconvex optimization problems. J. Optim. Theor. Appl. **29**, 169–197 (1979)

17. Bespamyatnikh, S.N.: Dynamic algorithms for approximate neighbor searching. In: Proceedings of Eighth Canadian Conference on Computational Geometry, pp. 252–257 (1996)

18. Beygelzimer, A., Kakade, S., Langford, J.: Cover trees for nearest neighbor. In: Proceedings of 23rd International Conference on Machine Learning, pp. 97–104 (2006)

19. Brönnimann, H., Chazelle, B., Pach, J.: How hard is halfspace range searching. Discrete Comput. Geom. **10**, 143–155 (1993)

20. Chan, T.M.: Closest-point problems simplified on the RAM. In: Proceedings of 13th Annual ACM-SIAM Symposium on Discrete Algorithms, pp. 472–473 (2002)

21. Chazelle, B., Matoušek, J.: On linear-time deterministic algorithms for optimization problems in fixed dimension. J. Algorithms **21**, 579–597 (1996)

22. Clarkson, K.L.: A randomized algorithm for closest-point queries. SIAM J. Comput. **17**(4), 830–847 (1988)

23. Clarkson, K.L.: Building triangulations using ε-nets. In: Proceedings of 38th Annual ACM Symposium on Theory of Computing, pp. 326–335 (2006)

24. Duncan, C.A., Goodrich, M.T., Kobourov, S.: Balanced aspect ratio trees: combining the advantages of k-d trees and octrees. J. Algorithms **38**, 303–333 (2001)

25. Ewald, G., Larman, D.G., Rogers, C.A.: The directions of the line segments and of the r-dimensional balls on the boundary of a convex body in Euclidean space. Mathematika **17**, 1–20 (1970)
26. Har-Peled, S.: A replacement for Voronoi diagrams of near linear size. In: Proceedings of 42nd Annual IEEE Symposium on Foundations of Computer, pp. 94–103 (2001)
27. Har-Peled, S., Kumar, N.: Approximating minimization diagrams and generalized proximity search. SIAM J. Comput. **44**, 944–974 (2015)
28. Har-Peled, S., Mendel, M.: Fast construction of nets in low dimensional metrics, and their applications. SIAM J. Comput. **35**, 1148–1184 (2006)
29. Hilbert, D.: Über die gerade linie als kürzeste verbindung zweier punkte. Math. Ann. **46**, 91–96 (1895)
30. Krauthgamer, R., Lee, J.R.: Navigating nets: simple algorithms for proximity search. In: Proceedings of 15th Annual ACM-SIAM Symposium on Discrete Algorithms, pp. 798–807 (2004)
31. Vernicos, C., Walsh, C.: Flag-approximability of convex bodies and volume growth of Hilbert geometries (2016). https://hal.archives-ouvertes.fr/hal-01423693, hAL Archive (hal-01423693i)

The Induced Star Partition of Graphs

M. A. Shalu[1], S. Vijayakumar[1(✉)], and T. P. Sandhya[2]

[1] Indian Institute of Information Technology, Design and Manufacturing (IIITDM),
Kancheepuram, Chennai 600127, India
{shalu,vijay}@iiitdm.ac.in
[2] Department of Computing, The Hong Kong Polytechnic University,
Hung Hom, Kowloon, Hong Kong
sandhya2127@gmail.com

Abstract. Given a graph G, we call a partition $(V_1, V_2, \ldots, V_k)$ of its vertex set an induced star partition of G if each induced subgraph $G[V_i]$ is isomorphic to a $K_{1,r}$, $r \geq 0$. In this paper, we consider the problem of partitioning a graph into a minimum number of induced stars and its decision versions. This problem may be viewed as an amalgamation of the well-known dominating set problem and coloring problem. Although this problem coincides with the dominating set problem on K_3-free graphs, it resembles, in its hardness, the coloring problem on general graphs. We establish the following results: (1) Deciding whether a graph can be partitioned into k induced stars is NP-complete for each fixed $k \geq 3$ and has a polynomial time algorithm for each $k \leq 2$. (2) It is NP-hard to approximate the minimum induced star partition size within $n^{1-\epsilon}$ for all $\epsilon > 0$. (3) The decision version of the induced star partition problem is NP-complete for (a) subcubic bipartite planar graphs, (b) line graphs (a subclass of $K_{1,r}$-free graphs ($r \geq 3$)), (c) $K_{1,5}$-free split graphs and (d) co-tripartite graphs. (4) The minimum induced star partition problem has (a) an $\frac{r}{2}$-approximation algorithm for $K_{1,r}$-free graphs ($r \geq 2$) and (b) a 2-approximation algorithms for split graphs.

Keywords: Polynomial time · NP-completeness · Approximation algorithms

1 Introduction

Induced stars are a natural and elegant substructure of a graph, like independent sets, cliques, induced paths and induced cycles. For example, if a graph models a network, each induced star in the graph corresponds to a subnetwork that is a star network. In this paper, we consider the problem of partitioning a graph into a minimum number of induced stars and its decision versions.

The well-known dominating set problem is essentially about partitioning a graph into a minimum number of stars, not necessarily induced. In this sense, the induced star partition problem can be viewed as a restricted variant of the dominating set problem. Indeed, on triangle-free graphs both the problems become

© Springer Nature Switzerland AG 2019
S. P. Pal and A. Vijayakumar (Eds.): CALDAM 2019, LNCS 11394, pp. 16–28, 2019.
https://doi.org/10.1007/978-3-030-11509-8_2

the same. Consequently, from the literature on the dominating set problem, we learn many things about the induced star partition problem. For instance, it follows that the decision version is NP-complete for chordal bipartite graphs [15] and $(C_4, \ldots, C_{2t})$-free bipartite graphs $(t \geq 2)$ [7]. It also follows that the optimization version has no $c \log n$ approximation algorithm for some $c > 0$ and has efficient $O(\log n)$-approximation algorithms for K_3-free graphs [10,17]. It further follows that the problem has polynomial time algorithms for bipartite permutation graphs [2,8] and trees [3]. Since each induced star is an independent set with a *dominating* vertex, this problem can be viewed as a variant of the graph coloring problem as well. Our results suggest that the problem resembles the coloring problem on general graphs in its hardness.

Induced stars have been considered in [11] with the objective of generalizing the well-known maximum matching problem. It considers the problem of covering a maximum number of vertices of the input graph by vertex disjoint induced stars of the form $K_{1,i}, 1 \leq i \leq r$, where $r \geq 1$ is fixed. This problem is shown to have a polynomial time algorithm. In this paper, we use this algorithm for designing an $\frac{r}{2}$-approximation algorithm for the problem of finding a minimum induced star partition of $K_{1,r}$-free graphs $(r \geq 3)$.

The graph theory literature is replete with partition problems of many kinds. The problem of partitioning a graph into same size stars (not necessarily induced) is investigated for many natural subclasses of perfect graphs in [1]. The problem of partitioning a graph into induced paths of length t $(t \geq 3)$ is shown to be NP-complete even for bipartite graphs with maximum degree three in [14]. The problem of partitioning a graph into triangles is shown to be NP-complete even for tripartite graphs in [5]. The problem of finding a partition of the vertex set of a graph into a minimum number of sets where each set in the partition induces a matching (a 1-regular graph) is considered in [19]. For more results on different types of partitions, packings and coverings, one can see [12].

In this paper, we establish the following results: (1) Deciding whether a graph can be partitioned into k induced stars is NP-complete for each fixed $k \geq 3$ and has a polynomial time algorithm for each $k \leq 2$ (Sect. 3). (2) It is NP-hard to approximate the minimum induced star partition size within $n^{1-\epsilon}$ for all $\epsilon > 0$ (Sect. 4). (3) The decision version of the induced star partition problem is NP-complete for (a) subcubic bipartite planar graphs, (b) line graphs (a subclass of $K_{1,r}$-free graphs $(r \geq 3)$), (c) $K_{1,5}$-free split graphs and (d) co-tripartite graphs (Sect. 5). (4) The minimum induced star partition problem has (a) an $\frac{r}{2}$-approximation algorithm for $K_{1,r}$-free graphs $(r \geq 2)$; thus an exact algorithm for cluster graphs (or $K_{1,2}$-free graphs) and a 1.5-approximation algorithm for line graphs and co-bipartite graphs as they are $K_{1,3}$-free and (b) a 2-approximation algorithm for split graphs (Sect. 6).

2 Preliminaries

All graphs considered in this paper are simple, undirected and finite. For basic graph theory terminology, we refer to [18]. Let K_n, C_n, and P_n denote the complete graph, the cycle and the path on n vertices, respectively. Let $K_{m,n}$ denote the complete bipartite graph with independent bipartitions of sizes m and n. A *clique (independent set)* is a subset of vertices which are pairwise adjacent (respectively, non-adjacent) in G. The size of a maximum clique (maximum independent set) in G is denoted by $\omega(G)$ ($\alpha(G)$). A *k-(vertex) coloring* of a graph G is a partition $(V_1, V_2, \ldots, V_k)$ of $V(G)$ such that V_i is an independent set in G for $1 \leq i \leq k$. For a vertex v of a graph G, $N(v)$ denotes the set of all vertices adjacent to v in G and $N[v] = \{v\} \cup N(v)$. A set $D \subseteq V(G)$ is called a *dominating set* of G if $\cup_{v \in D} N[v] = V(G)$. Often we denote an edge $\{a, b\}$ in a graph as ab or ba. For a set $X \subseteq V(G)$, $G[X]$ denotes the graph induced by X in G. For a graph H, G is said to be *H-free* if no induced subgraph of G is isomorphic to H. The complement of a graph G is denoted by G^c.

A *star* is a graph isomorphic to $K_{1,r}$, $r \geq 0$, where $K_{1,0} = K_1$. If x is a vertex in a graph G and X is an independent set contained in $N(x)$, then the induced subgraph $G[\{x\} \cup X]$ is a star and it is denoted by (x, X) and is called an induced star with center x.

For a graph G, an induced star partition is a partition $(V_1, V_2, \ldots, V_k)$ of $V(G)$ such that for each $1 \leq i \leq k$, $G[V_i] \cong K_{1,r}$ for some $r \geq 0$. In this case, we also call $G[V_1], G[V_2], \ldots, G[V_k]$ an induced star partition of G. We may note that every graph admits an induced star partition (since for a graph G with $V(G) = \{v_1, v_2, \ldots, v_n\}$, $V_i = \{v_i\}$, $1 \leq i \leq n$, form an induced star partition of G). The minimum k for which a graph G admits an induced star partition with k sets is called the *induced star partition number of G* and is denoted $isp(G)$. We now formulate the following optimization and decision versions of the problem:

MIN INDUCED STAR PARTITION
Instance : A graph G.
Goal   : The minimum k for which G admits an induced k-star partition.

INDUCED STAR PARTITION
Instance : A graph G and a positive integer k.
Question : Does G admit an induced star partition of size k?

INDUCED k-STAR PARTITION
Instance : A graph G.
Question : Does G admit an induced star partition of size k?

We now formalize a related concept and prove a lemma about it.

Definition 1. *Let $x_1, x_2, \ldots, x_k$ be distinct vertices in G. Then the induced stars $(x_1, X_1), (x_2, X_2), \ldots, (x_k, X_k)$ are said to form an* induced star cover *of G if $\{x_1, \ldots, x_k\} \cup X_1 \cup \ldots \cup X_k = V(G)$.*

We emphasize that in an induced star cover, by definition, the centers of the stars are all distinct.

Lemma 1. *A graph G has an induced star partition of size k if and only if it has an induced star cover of size k.*

Proof. Any induced star partition is also an induced star cover. So, if G has an induced star partition of size k, it has an induced star cover of size k. Conversely, suppose that G has an induced star cover of size k, say $(x_1, X_1), \ldots, (x_k, X_k)$. Then $(x_1, X_1), (x_2, X_2'), \ldots, (x_k, X_k')$ form an induced star partition of size k, where $X_j' = X_j \setminus (\{x_1, \ldots, x_{j-1}\} \cup X_1 \cup \ldots \cup X_{j-1})$ for $2 \leq j \leq k$. □

We note that if G is a K_3-free graph, then $(x, N(x)) = G[N[x]]$ is an induced star in G for each vertex x. In fact, $\{x_1, \ldots, x_k\}$ is a dominating set in G if and only if $(x_1, N(x_1)), \ldots, (x_k, N(x_k))$ is an induced star cover of G. So, in the light of Lemma 1, it follows that INDUCED STAR PARTITION is essentially the same as DOMINATING SET on K_3-free graphs. Thus we have the following proposition. (The DOMINATING SET problem is, given a graph G and a positive integer k, to decide whether G has a dominating set of size k.)

Proposition 1. *For K_3-free graphs, the* INDUCED STAR PARTITION *problem is equivalent to the* DOMINATING SET *problem.*

3 Induced k-Star Partition

In this section, we establish that INDUCED k-STAR PARTITION is NP-complete for any $k \geq 3$ and has polynomial time algorithms for each $k \leq 2$. We first prove that INDUCED 3-STAR PARTITION is NP-complete (Theorem 1) by providing a reduction from the NP-complete 3-COLORING problem [9]. The 3-COLORING problem is, given a graph, to decide whether its vertex set can be partitioned into three independent sets.

Theorem 1. INDUCED 3-STAR PARTITION *is NP-complete for K_4-free graphs.*

Proof. Given a partition of the vertex set of a K_4-free graph into three sets, it can be verified in polynomial time whether each set in the partition induces a star. So, with such a partition as the certificate for a YES instance, the problem is seen to be in NP.

To show the NP-hardness, we reduce the NP-complete 3-COLORING of K_3-free graphs [9] to INDUCED 3-STAR PARTITION of K_4-free graphs. Let $G(V, E)$ be an instance of a 3-COLORING of triangle-free graphs. It is transformed to an INDUCED 3-STAR PARTITION instance $G'(V', E')$ as follows. Let $V' = V \cup \{x_1, x_2, x_3, y_1, y_2, y_3\}$ and $E' = E \cup \{x_i z : z \in V \text{ and } 1 \leq i \leq 3\} \cup \{x_1 y_1, x_2 y_2, x_3 y_3\}$. Since G is triangle-free, G' is K_4-free. We claim that G has a 3-coloring if and only if G' has an induced 3-star partition.

Assume that (V_1, V_2, V_3) is a 3-coloring of G. Then $(x_i, \{y_i\} \cup V_i)$ is an induced star in G' for $1 \leq i \leq 3$. Conversely, assume that $V' = (V_1', V_2', V_3')$ is an induced 3-star partition of G'. We note that $V_i' \cap \{y_1, y_2, y_3\} = 1$ for $1 \leq i \leq 3$. W.l.o.g., let $y_i \in V_i'$ for $1 \leq i \leq 3$. Since y_i is not adjacent to any vertex in $V \subseteq V(G')$ and y_i is adjacent to x_i alone, x_i must be the center of the star induced by V_i' in

G' for every $1 \leq i \leq 3$. Let $V_i = V_i' \cap V$ for each $1 \leq i \leq 3$. Then (V_1, V_2, V_3) is a 3-coloring of G as each V_i is an independent set and their union equals $V(G)$. Since 3-COLORING of triangle-free graphs is NP-complete [13], INDUCED 3-STAR PARTITION of K_4-free graphs is thus NP-complete. □

Let $k \geq 4$ be fixed. Then a polynomial time reduction from a K_4-free INDUCED 3-STAR PARTITION instance G to a K_4-free INDUCED k-STAR PARTITION instance can be constructed by simply adding $k - 3$ isolated vertices to G. So, we have the following more general theorem.

Theorem 2. *For each $k \geq 3$, INDUCED k-STAR PARTITION is NP-complete for K_4-free graphs.*

For a constant k, the problem of deciding whether an input graph has a dominating set of size k can be solved in polynomial time $O(n^k)$. So, from Proposition 1 and Theorem 2, we have the following dichotomy result.

Theorem 3. *Let $k \geq 3$ be fixed. Then INDUCED k-STAR PARTITION has a polynomial time algorithm for K_3-free graphs and it is NP-complete for K_4-free graphs.*

We now consider INDUCED k-STAR PARTITION for $k \leq 2$. The following lemma is obvious.

Lemma 2. *Let G be a graph with at least two vertices x and y and let $H = G \setminus \{x, y\}$. Then G can be partitioned into two induced stars with centers x and y if and only if (i) the induced subgraph H is a bipartite graph and (ii) there is a bipartition (X, Y) of $V(H)$ into independent sets such that (x, X) and (y, Y) are induced stars in G.*

We note that in the above lemma Condition (ii) implies Condition (i). This redundancy is still preferred for the clarity of exposition.

Lemma 3. *Let G be a graph with at least two vertices x and y and let $H = G \setminus \{x, y\}$. Then Conditions (i) and (ii) of Lemma 2 can be verified in polynomial time.*

Proof. Condition (i) is just about verifying whether the subgraph H is bipartite and it can clearly be done in polynomial time [18]. If Condition (i) does not hold, Condition (ii) also does not hold true. So, assume that Condition (i) holds; that is, that H is bipartite. We now have two cases.

Case 1: H is connected. In this case, $V(H)$ has a unique bipartition (X, Y) into independent sets, up to order and this partition can be computed in polynomial time. Therefore, Condition (ii) is true if either (x, X) and (y, Y) or (x, Y) and (y, X) are induced stars in G.

Case 2: H is disconnected. Let $H_1, H_2, \ldots, H_k$ be the connected components of H. Now, each H_i has a unique bipartition (X_i, Y_i) into independent sets, up to order. For every i, if (x, X_i) and (y, Y_i) or (x, Y_i) and (y, X_i) are induced stars in G, then Condition (ii) holds true; otherwise Condition (ii) fails. For instance, without loss of generality, if we assume that (x, X_i) and (y, Y_i) are induced stars in G for $1 \leq i \leq k$, then $X = \cup_{i=1}^{k} X_i$ and $Y = \cup_{i=1}^{k} Y_i$ provide a partition of $V(H)$ into two independent sets such that (x, X) and (y, Y) are induced stars in G. Since finding the connected components of a graph can also be done in polynomial time [18], it follows that Condition (ii) too can be verified in polynomial time. □

Theorem 4. *There is a polynomial time algorithm to decide whether a graph has a partition into at most two induced stars.*

Proof. Let G be a graph that it is not a star. Repeatedly consider pairs of vertices until a pair of vertices x and y that meets Conditions (i) & (ii) of Lemma 2 is found or until all pairs of vertices are exhausted. By Lemma 3, Conditions (i) & (ii) of Lemma 2 can be verified in polynomial time for any pair of vertices. As there are only polynomially many pairs of vertices, it follows that the overall running time is also a polynomial in terms of the input size. □

4 An Inapproximability Result

The MIN COLORING problem is, given a graph G, to find the minimum k for which G admits a k-coloring. It is known that it is NP-hard to approximate MIN COLORING within $n^{1-\epsilon}$ for all $\epsilon > 0$ [20]. In this section, we provide an (almost) optimum preserving reduction from MIN COLORING to MIN INDUCED STAR PARTITION. In fact, given a graph G, we construct a graph G' in polynomial time such that $isp(G') = \chi(G) + 1$. This implies the theorem below. Thus it improves the $c \log n$ ($c > 0$) inapproximability bound for K_3-free graphs that follows from the literature [10,17].

Theorem 5. *It is NP-hard to approximate* MIN INDUCED STAR PARTITION *to within* $n^{1-\epsilon}$ *for all* $\epsilon > 0$.

Proof. Let G be a graph with at least three vertices and let $V(G) = \{1, 2, \ldots, n\}$. We consider the graph $G + 0$ which is the same as G with a new vertex 0 which is joined to each vertex of G. We now construct a graph G' from $G + 0$ as follows. Corresponding to each vertex i of $G + 0$, G' will have a distinct set of n vertices $V_i = \{v_{i1}, v_{i2}, \ldots, v_{in}\}$, $0 \leq i \leq n$. Thus $V(G') = \{v_{ij} \mid 0 \leq i \leq n; 1 \leq j \leq n\}$. The edge set of G' is as follows: $E(G') = \{v_{ik} v_{jl} \mid ij \in E(G + 0), 1 \leq k, l \leq n\}$. It may be noted that each V_i, $0 \leq i \leq n$, is an independent set in G' and that, for $0 \leq i < j \leq n$, $V_i \cup V_j$ induces a complete bipartite graph if $ij \in E(G + 0)$ and is an independent set otherwise. We know, from Lemma 1, that a graph has an induced k-star partition if and only if it has an induced k-star cover. We will prove that G has a k-coloring if and only if G' has an induced $(k+1)$-star cover. This, in particular, implies that $isp(G') = \chi(G) + 1$.

Suppose that $(U_1, U_2, \ldots, U_k)$ is a k-coloring of G. Then $X_j = \cup_{p \in U_j} V_p$ is an independent set in G' for $1 \leq j \leq k$. It is obvious that these X_j's are disjoint and their union equals $\cup_{i=1}^n V_i$. We also note that each vertex in V_0 is adjacent to every vertex in $\cup_{i=1}^n V_i$. Thus $(v_{01}, X_1), (v_{02}, X_2), \ldots, (v_{0k}, X_k)$ along with (v_{11}, V_0) gives an induced $(k + 1)$-star cover of G'.

Conversely, suppose that G' has an induced $(k + 1)$-star cover, say $(x_1, X_1), (x_2, X_2), \ldots, (x_{k+1}, X_{k+1})$, where the centers $x_1, x_2, \ldots, x_{k+1}$ are all distinct by definition. W.l.o.g., we assume the following: (1) If X_j intersects exactly $V_{j_1}, V_{j_2}, \ldots, V_{j_{r_j}}$, then $X_j = V_{j_1} \cup V_{j_2} \cup \ldots \cup V_{j_{r_j}}$. (2) X_j's are disjoint. [If v_{pk} and $v_{ql} \in X_j$, then $V_p \cup V_q$ is an independent set and x_j is adjacent to all vertices in $V_p \cup V_q$; so, we can modify X_j so that $V_p \cup V_q \subseteq X_j$; hence we can assume (1). Since X_j's can always be made disjoint, we can assume (2).]

We must exhibit a k-coloring of G. If $k \geq n$, G on n vertices obviously has a k-coloring. If $k = n-1$, then G has a k-coloring if and only if G is not the complete graph. So, let us assume that $k \leq n - 2$ or $k + 1 \leq n - 1$. So, we have that the number of center vertices in our star cover is at most $n-1$. This means that each V_i has a non-center vertex and therefore intersects some X_j. From assumptions (1) and (2) above, we then have that X_j's form a partition of $V(G')$, where each X_j is a union of some V_i's. We note that each vertex in V_0 is adjacent to every vertex outside V_0. So, if $V_0 \subseteq X_j$ for some j, then $X_j = V_0$ as X_j is an independent set. Assume w.l.o.g. that $X_{k+1} = V_0$. It then follows that there is a partition $(U_1, U_2, \ldots, U_k)$ of $\{1, 2, \ldots, n\}$ such that $X_j = \cup_{p \in U_j} V_p$, $1 \leq j \leq k$. This means, from the construction of G', that each U_j is an independent set in G. Thus G has a k-coloring, namely $(U_1, U_2, \ldots, U_k)$. $\qquad\square$

5   NP-Completeness Results

It follows from the literature on DOMINATING SET that INDUCED STAR PARTITION is NP-complete for chordal bipartite graphs [15] and $(C_4, \ldots, C_{2t})$-free bipartite graphs $(t \geq 2)$ [7]. In this section, we prove that INDUCED STAR PARTITION is also NP-complete for (a) subcubic bipartite planar graphs, (b) line graphs, (c) $K_{1,5}$-free split graphs and (d) co-tripartite graphs. We may note that a graph in which each vertex has degree at most three is called a *subcubic graph*.

We begin by proving a lemma about dominating sets. This lemma is used in the proof of Theorem 6.

Lemma 4. *Let G be any graph and let G' be the graph obtained from G by replacing each edge in G by a P_5. Then G has a dominating set of size k if and only if G' has a dominating set of size $k + m$, where $m = |E(G)|$.*

Proof. Let G be any graph and let G' be the graph obtained from G by replacing each edge in G by a P_5; i.e., G' is obtained from G by subdividing each edge thrice. We will prove that G has a dominating set of size k if and only if G' has a dominating set of size $k + m$, where $m = |E(G)|$. Let D be a dominating set of G of size k. We now add more vertices to D to get a dominating set of G' of size $k + m$ as follows. Consider an edge ab in G and suppose that (a, x, y, z, b) is the path in G' that replaces the edge ab. We modify D as follows: If either $a, b \in D$ or $a, b \notin D$, then we add y to D. If $a \in D$ and $b \notin D$, then we add z to D. If

$b \in D$ and $a \notin D$, then we add x to D. Repeating this for each edge of G results in a set D of size $k + m$. It is easy to see that D is a dominating set for G'.

Conversely, let D be a dominating set of G' of size $k + m$. We now modify D to obtain a dominating set of G of size at most k. Consider the path (a, x, y, z, b) in G' that corresponds to an edge ab in G. Of x, y, z, if exactly one is in D, we simply remove it from D. If $x, y \in D$ but $z \notin D$, we remove x, y from D and add a to D if it is not in D. If $y, z \in D$ but $x \notin D$, then we remove y, z from D and add b to D if it is not in D. If $x, y, z \in D$, we remove them from D and add a and b to D if they are not there. Thus in all the cases, we reduce the size of D by at least one. Repeating this for each P_5 in G' corresponding to an edge in G, we get a set $D \subseteq V(G)$ of size at most k. It is easy to see that D is a dominating set for G. □

Theorem 6. INDUCED STAR PARTITION *is NP-complete for subcubic bipartite planar graphs.*

Proof. We note that DOMINATING SET for subcubic planar graphs is NP-complete [9]. By a simple polynomial time reduction from this problem, we prove that DOMINATING SET for subcubic bipartite planar graphs is NP-Complete. Given a subcubic planar graph G, we replace each edge by a P_5 (that is, we subdivide each edge thrice) to get a subcubic bipartite planar graph G'. From Lemma 4, it then follows that G has a dominating set of size k if and only if G' has a dominating set of size $k + m$, where $m = |E(G)|$. Thus it follows that DOMINATING SET for subcubic bipartite planar graphs is NP-complete. Since subcubic bipartite planar graphs are K_3-free, from Proposition 1, the theorem follows. □

We note that line graphs are especially important in the context of induced star partition since they are $K_{1,3}$-free. We prove the following theorem.

Theorem 7. INDUCED STAR PARTITION *is NP-complete for line graphs.*

Proof. Let G be any graph and let $L(G)$ be its line graph. It is NP-complete to decide whether $E(G)$ can be partitioned into P_4's (not necessarily induced) [6]. We note that each P_4 in G corresponds to a $K_{1,2}$ in $L(G)$, and conversely. In fact, $E(G)$ can be partitioned into P_4's if and only if $L(G)$ can be partitioned into $[|E(G)|/3]$ induced stars, each isomorphic to $K_{1,2}$ (as $L(G)$ is $K_{1,3}$-free). So, it follows that INDUCED STAR PARTITION is NP-complete for line graphs. □

Being $K_{1,3}$-free, line graphs form a subclass of $K_{1,r}$-free for each $r \geq 3$. So, we have the following important theorem as a corollary.

Theorem 8. INDUCED STAR PARTITION *is NP-complete for $K_{1,r}$-free graphs, where $r \geq 3$.*

Theorem 9. INDUCED STAR PARTITION *is NP-complete for $K_{1,5}$-free split graphs.*

Proof. Given a partition of the vertex set of a split graph, it can be verified in polynomial time whether each set in the partition induces a star. Thus, with such a partition as the certificate for a YES instance, the problem can be easily seen to be in NP.

We now prove that INDUCED STAR PARTITION is NP-hard for $K_{1,5}$-free split graphs by providing a polynomial time reduction from the NP-complete EXACT3COVER [9]. We may note that the EXACT3COVER problem is, given a set $X = \{x_1, x_2, \ldots, x_{3n}\}$ of $3n$ elements and a collection $\{S_1, S_2, \ldots, S_m\}$ of 3-element subsets of X, to decide whether there are n sets in the collection of sets whose union is X.

Let $(X = \{x_1, x_2, \ldots, x_{3n}\}, \{S_1, S_2, \ldots, S_m\})$ be an EXACT3COVER instance. It is transformed into an INDUCED STAR PARTITION instance (G, k') as follows. The graph G we construct has $V(G) = X \cup \{s_1, \ldots, s_m, s_{m+1}, \ldots, s_{2m}\} \cup \{t_1, t_2, \ldots, t_{m-n}\}$ and $E(G) = \{s_i s_j \mid 1 \leq i < j \leq 2m\} \cup \{s_i x \mid 1 \leq i \leq m, x \in S_i \subseteq X\} \cup \{s_{m+n+i} t_i \mid 1 \leq i \leq m - n\}$ (Fig. 1). The graph G can be partitioned into the clique $\{s_1, s_2, \ldots, s_{2m}\}$ and the independent set $X \cup \{t_1, t_2, \ldots, t_{m-n}\}$; therefore it is a split graph. The parameter k' is set to m. We now argue that $(X, \{S_1, S_2, \ldots, S_m\})$ is a YES instance of EXACT3COVER if and only if (G, m) is a YES instance of INDUCED STAR PARTITION.

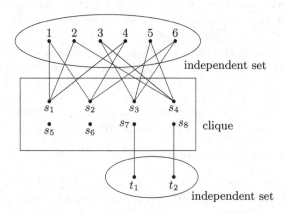

Fig. 1. The split graph constructed from the EXACT3COVER instance $(X, \{S_1, S_2, S_3, S_4\})$, where $X = \{1, 2, 3, 4, 5, 6\}$, $S_1 = \{1, 2, 4\}$, $S_2 = \{1, 4, 6\}$, $S_3 = \{3, 5, 6\}$, $S_4 = \{2, 3, 5\}$.

Suppose that $(X, \{S_1, S_2, \ldots, S_m\})$ has an exact cover. This implies that there are n sets, say $S_1, S_2, \ldots, S_n$, whose union equals X. The corresponding sets $\{s_i, s_{m+i}\} \cup S_i$ for $1 \leq i \leq n$ and the sets $\{s_{n+j}, s_{m+n+j}, t_j\}$ for $1 \leq j \leq m - n$ provide a partition of G into m induced stars. Conversely, suppose that G has a partition into m induced stars $V_1, V_2, \ldots, V_m$. Since the $2m$ vertices $s_1, s_2, \ldots, s_{2m}$ form a clique in G, each V_i, $1 \leq i \leq m$, must include exactly two of these vertices. Moreover, the class that includes s_{m+n+i}, $1 \leq i \leq m - n$, must necessarily include t_i since t_i is only adjacent to s_{m+n+i} in G. Further, since no two of $t_1, t_2, \ldots, t_{m-n}$ are adjacent or have a common neighbour in G, they all must be in $m - n$ different V_i's in the partition. Also no $x \in X \subseteq V(G)$ and a t_j, $1 \le j \le m - n$, are adjacent or have a common neighbour in G. So, no V_i can contain both an $x \in X$ and a t_j. Therefore, there must be $m - n$ sets with exactly two $s_i's$ and one t_j. Without loss of generality, assume that these sets are V_{n+i}, $1 \le i \le m - n$. It then follows that all vertices in $X \subseteq V(G)$ are distributed among $V_1, V_2, \ldots, V_n$. Since each s_i, $1 \le i \le m$, is joined to exactly 3 vertices in X, X is an independent set of $3n$ vertices, and each V_i has to have two s_j's, $1 \le j \le 2m$, it follows that $V_1, V_2, \ldots, V_n$ induce stars that are centered at n of $s_1, s_2, \ldots, s_m$. Without loss of generality, assume that $s_i, s_{m+i} \in V_i$ for $1 \le i \le n$. This means that V_i contains three vertices of X that are connected to s_i in G. Equivalently, this implies that $S_i \subseteq V_i$ for $1 \le i \le n$. It now follows that $S_1, S_2, \ldots, S_n$ is an exact cover of the set cover instance. $\square$

Problem: Determine the time complexity of INDUCED STAR PARTITION for $K_{1,3}$-free split graphs and $K_{1,4}$-free split graphs.

Since INDUCED STAR PARTITION is NP-complete for bipartite graphs and split graphs, we are naturally interested in its status on co-bipartite graphs. Co-bipartite graphs are also important as they are $K_{1,3}$-free. We provide a 1.5-approximation algorithm for MIN INDUCED STAR PARTITION on co-bipartite graphs (Theorem 12). But we have been able to neither find a polynomial time algorithm nor prove that it is NP-complete for this family of graphs. As a natural next step, we consider the class of co-tripartite graphs (that are $K_{1,4}$-free) and prove that INDUCED STAR PARTITION is NP-complete for this graph class.

Theorem 10. INDUCED STAR PARTITION *is NP-complete for co-tripartite graphs.*

Proof. Given a partition of the vertex set of a co-tripartite graph, it can be verified in polynomial time whether each set in the partition induces a star. So, with such a partition as the certificate for a YES instance, the problem can be easily seen to be in NP.

We prove that INDUCED STAR PARTITION for co-tripartite graphs is NP-hard by providing a polynomial reduction from the PARTITIONING INTO TRIANGLES problem for tripartite graphs, which is known to be NP-complete [5]. The PARTITIONING INTOTRIANGLES problem is, given a graph G with $|V(G)| = 3k$, to decide whether $V(G)$ can be partitioned into k sets such that each set induces a K_3 in G. Let $G(A \cup B \cup C, E)$ with $|A| = |B| = |C| = k$ be a tripartite instance of PARTITIONING INTO TRIANGLES. It is mapped to a co-tripartite INDUCED STAR PARTITION instance $G'(A^* \cup B \cup C, E)$ as follows. The graph $G'(A^* \cup B \cup C, E)$ has vertex set $A^* \cup B \cup C$, where $A^* = A \cup A'$ and $|A'| = |A|$, and edge set $E = E(G[A \cup B]) \cup E(G[A \cup C]) \cup E(G^c[B \cup C]) \cup \{xy \mid x, y \in A \cup A'\} \cup \{xy \mid x, y \in B\} \cup \{xy \mid x, y \in C\}$.

We will argue that G is a YES instance of PARTITIONING INTO TRIANGLES if and only if (G', k) is a YES instance of INDUCED STAR PARTITION. Let S_1, $S_2, \ldots, S_k$ be a partition of $V(G)$ into triangles. Since G is a tripartite graph,

$|S_i \cap D| = 1$ for every $1 \leq i \leq k$ and for every $D \in \{A, B, C\}$. Then by construction, $G'[S_i] \cong K_{1,2}$ with center in A and hence $G'[S_i \cup \{x_i\}] \cong K_{1,3}$ where $A' = \{x_1, x_2, \ldots, x_k\}$.

Conversely, assume that $S_1, S_2, \ldots, S_k$ is a partition of $V(G')$ into k induced stars. Then each of these stars must be isomorphic to $K_{1,3}$ since a co-tripartite graph is $K_{1,4}$-free. Since $N[x] = A' \cup A$ is a clique in G' for all $x \in A'$, no $x \in A'$ can be the center of any of the $K_{1,3}$'s in G'. So, $|A' \cap S_i| \leq 1$ and, since $|A'| = k$, $|A' \cap S_i| = 1$ for $1 \leq i \leq k$. W.l.o.g., let $x_i \in S_i$ for all $1 \leq i \leq k$, where $A' = \{x_1, x_2, \ldots, x_k\}$. Thus the center of the $K_{1,3}$ induced by S_i in G' must be in A since $x_i \in A'$ is not adjacent to any vertex in $B \cup C$. In addition, since $\{x_i, x, y\} = S_i \setminus A$ is an independent set, $x \in B$ and $y \in C$ or vice versa. So, $S_i \setminus \{x_i\}$ induces a $K_{1,2}$ in G' and a triangle in G, by construction. Thus $(S_1 \setminus \{x_1\}, S_2 \setminus \{x_2\}, \ldots, S_k \setminus \{x_k\})$ is a partition of $V(G)$ into triangles. $\square$

Problem: Determine the time complexity of INDUCED STAR PARTITION for co-bipartite graphs.

6   Approximation Algorithms

From Theorem 5, it follows that MIN INDUCED STAR PARTITION does not allow any decent approximation algorithms on general graphs. From literature on DOMINATING SET, we have following true for MIN INDUCED STAR PARTITION: (1) It has efficient $O(\log n)$-approximation algorithms for $K_{1,3}$-free graphs; there are no efficient $c \log n$ approximation algorithms for $K_{1,3}$-free graphs for some $c > 0$ [10,17]. (2) For $K_{1,3}$-free graphs of maximum degree d, there are $(d + 1)$-approximation algorithms [10,17]. (3) There are polynomial time algorithms for bipartite permutation graphs [2,8] and trees [3]. In this section, we present an $\frac{r}{2}$-approximation algorithm for $K_{1,r}$-free graphs ($r \geq 2$). This implies an exact algorithm for cluster graphs (or $K_{1,2}$-free graphs) and a 1.5-approximation algorithm for line graphs and co-bipartite graphs as they are $K_{1,3}$-free. We also present a 2-approximation algorithm for split graphs.

We define the MIN K_1-INDUCED STAR PARTITION problem as follows: *Given a graph G, find an induced star partition of G which has a minimum number of stars isomorphic to K_1.* The following theorem follows from the work of [11].

Theorem 11 (Kelmans [11]). *The MIN K_1-INDUCED STAR PARTITION problem has a polynomial time algorithm for $K_{1,r}$-free graphs, where $r \geq 1$.*

The next theorem complements the NP-completeness result of Theorem 8.

Theorem 12. MIN INDUCED STAR PARTITION *has an efficient $\frac{r}{2}$-approximation algorithm for $K_{1,r}$-free graphs, where $r \geq 2$.*

Proof. Let G be a $K_{1,r}$-free graph on n vertices, where $r \geq 2$. Let p^* denote the number of induced stars isomorphic to K_1 in an optimal solution to MIN K_1-INDUCED STAR PARTITION. Then, since each induced star on the remaining $n - p^*$ vertices is isomorphic to one of $K_{1,i}, 1 \leq i < r$, the total number of star in this induced star partition is at most $p^* + \frac{n-p^*}{2}$. We note that such an induced star partition can be computed in polynomial time by Theorem 11. We now prove that $p^* + \frac{n-p^*}{2} \leq \frac{r}{2} \cdot isp(G)$.

Suppose that an optimal solution for MIN INDUCED STAR PARTITION for G has p stars each isomorphic to K_1 and q stars each isomorhic to one of $K_{1,i}$, $1 \leq i < r$ so that $isp(G) = p + q$. We note that $n - p \leq qr$ since each of the q stars on the $n - p$ vertices has at most r vertices. Since this solution is a feasible solution for MIN K_1-INDUCED STAR PARTITION, $p \geq p^*$. Also, $x + \frac{n-x}{2} = \frac{n+x}{2}$ is an increasing function of x. So, we have that $p^* + \frac{n-p^*}{2} \leq p + \frac{n-p}{2} \leq p + \frac{rq}{2} \leq \frac{r}{2}(p+q) = \frac{r}{2} \cdot isp(G)$. $\qquad\square$

The theorem below complements the NP-completeness result of Theorem 9.

Theorem 13. MIN INDUCED STAR PARTITION *has an efficient 2-approximation algorithm for split graphs.*

Proof. Let (C, I) be a clique-independent set partition of the input split graph G. We note that such a partition can be computed in polynomial time. The algorithm below produces an induced star partition of size $|C| \leq \omega(G)$. Since there must be at least $\lceil \frac{\omega}{2} \rceil$ stars in any induced star partition of a graph G, it follows that it is a 2-approximation algorithm.

Input: A split graph $G(C \cup I, E)$, where C is a clique and I is an independent
     set in G with $C = \{x_1, x_2, \ldots, x_k\}$.
Output: An induced star partition $(S_1, S_2, \ldots, S_k)$ of G.
  1: for $i = 1$ to k
  2:    $S_i = \{x_i\} \cup (N(x_i) \cap I)$
  3:    $G = G \setminus S_i$
  4: end for

$\qquad\square$

7 Conclusion

This paper carries out a preliminary study of the induced star partition problem. Although this problem coincides with the dominating set problem on K_3-free graphs, our study reveals that it resembles, in its hardness, the coloring problem on general graphs and that it is actually harder than the coloring problem on many graph classes, e.g., on bipartite graphs and on split graphs. Determining the time complexity of this problem for other graph classes, such as co-bipartite graphs and P_4-free graphs, remains as some interesting open problems.

References

1. van Bevern, R., et al.: Partitioning perfect graphs into stars. J. Graph Theory **85**, 297–335 (2017)
2. Brandstädt, A., Kratsch, D.: On the restriction of some NP-complete graph problems to permutation graphs. In: Budach, L. (ed.) FCT 1985. LNCS, vol. 199, pp. 53–62. Springer, Heidelberg (1985). https://doi.org/10.1007/BFb0028791
3. Cockayne, E.J., Goodman, S., Hedétniemi, S.T.: A linear algorithm for the domination number of a tree. Inf. Process. Lett. **4**, 41–44 (1975)
4. Corneil, D.G., Perl, Y.: Clustering and domination in perfect graphs. Discret. Appl. Math. **9**, 27–39 (1984)
5. Ćustić, A., Klinz, B., Woeginger, G.J.: Geometric versions of the three-dimensional assignment problem under general norms. Discret. Optim. **18**, 38–55 (2015)
6. Dor, D., Tarsi, M.: Graph decomposition is NP-complete: a complete proof of Holyer's conjecture. SIAM J. Comput. **26**(4), 1166–1187 (1997)
7. Duginov, O.: Partitioning the vertex set of a bipartite graph into complete bipartite subgraphs. Discret. Math. Theor. Comput. Sci. **16**(3), 203–214 (2014)
8. Farber, M., Keil, J.M.: Domination in permutation graphs. J. Algorithms **6**, 309–321 (1985)
9. Garey, M.R., Johnson, D.S.: Computers and Intractability; A Guide to the Theory of NP-Completeness. W. H. Freeman & Co., New York (1990)
10. Kann, V.: On the approximability of NP-complete optimization problems. Ph.D. thesis, Department of Numerical Analysis and Computing Science, Royal Institute of Technology, Stockholm (1992)
11. Kelmans, A.K.: Optimal packing of induced stars in a graph. Discret. Math. **173**, 97–127 (1997)
12. Kirkpatrick, D.G., Hell, P.: On the completeness of a generalized matching problem. In: Lipton, R., Burkhard, W., Savitch, W., Friedman, E., Aho, A. (eds.) (STOC 1978) 10th ACM Symposium on Theory of Computing, pp. 240–245 (1978)
13. Maffray, F., Preissmann, M.: On the NP-completeness of the k-colorability problem for triangle-free graphs. Discret. Math. **162**, 313–317 (1996)
14. Monnot, J., Toulouse, S.: The path partition problem and related problems in bipartite graphs. Oper. Res. Lett. **35**, 677–684 (2007)
15. Müller, H., Brandstädt, A.: The NP-completeness of Steiner Tree and Dominating Set for chordal bipartite graphs. Theoret. Comput. Sci. **53**, 257–265 (1987)
16. Pantel, S.: Graph packing problems. M.Sc. Thesis, Simon Fraser University (1998)
17. Vazirani, V.V.: Approximation Algorithms. Springer, Heidelberg (2001). https://doi.org/10.1007/978-3-662-04565-7
18. West, D.B.: Introduction to Graph Theory, 2nd edn. Prentice-Hall, Upper Saddle River (2000)
19. Yuan, J., Wang, Q.: Note: partitioning the vertices of a graph into induced matchings. Discret. Math. **263**, 323–329 (2003)
20. Zuckerman, D.: Linear degree extractors and the inapproximability of Max Clique and Chromatic Number. Theory Comput. **3**, 103–128 (2007)

Fault-Tolerant Additive Weighted Geometric Spanners

Sukanya Bhattacharjee and R. Inkulu[✉]

Department of Computer Science and Engineering, IIT Guwahati, Guwahati, India
{s.bhattacharjee,rinkulu}@iitg.ac.in

Abstract. Let S be a set of n points and let w be a function that assigns non-negative weights to points in S. The additive weighted distance $d_w(p,q)$ between two points $p, q \in S$ is defined as $w(p) + d(p,q) + w(q)$ if $p \neq q$ and it is zero if $p = q$. Here, $d(p,q)$ is the (geodesic) Euclidean distance between p and q. For a real number $t > 1$, a graph $G(S, E)$ is called a *t-spanner* for the weighted set S of points if for any two points p and q in S the distance between p and q in graph G is at most $t.d_w(p,q)$ for a real number $t > 1$. For some integer $k \geq 1$, a t-spanner G for the set S is a *(k, t)-vertex fault-tolerant additive weighted spanner*, denoted with (k,t)-VFTAWS, if for any set $S' \subset S$ with cardinality at most k, the graph $G \setminus S'$ is a t-spanner for the points in $S \setminus S'$. For any given real number $\epsilon > 0$, we present algorithms to compute a $(k, 4 + \epsilon)$-VFTAWS for the metric space (S, d_w) resulting from the points in S belonging to either $\mathbb{R}^d$ or located in the given simple polygon. Note that $d(p,q)$ is the geodesic Euclidean distance between p and q in the case of simple polygons whereas in the case of $\mathbb{R}^d$ it is the Euclidean distance along the line segment joining p and q.

Keywords: Computational geometry · Geometric spanners · Approximation algorithms

1 Introduction

When designing geometric networks on a given set of points in a metric space, it is desirable for the network to have short paths between any pair of nodes while being sparse with respect to the number of edges. Let $G(S, E)$ be an edge-weighted geometric graph on a set S of n points in $\mathbb{R}^d$. The weight of any edge $(p,q) \in E$ is the Euclidean distance $|pq|$ between p and q. The distance in G between any two nodes p and q, denoted by $d_G(p,q)$, is defined as the length of a shortest (that is, minimum-weighted) path between p and q in G. The graph G is called a *t-spanner* for some $t \geq 1$ if for any two points $p, q \in S$ we have $d_G(p,q) \leq t.|pq|$. The smallest t for which G is a t-spanner is called the *stretch*

R. Inkulu—This research is supported in part by NBHM grant 248(17)2014-R&D-II/1049.

S. P. Pal and A. Vijayakumar (Eds.): CALDAM 2019, LNCS 11394, pp. 29–41, 2019.
https://doi.org/10.1007/978-3-030-11509-8_3

factor of G, and the number of edges of G is called its size. Althöfer et al. [4] first attempted to study sparse spanners on edge-weighted graphs that have the triangle-inequality property. The text by Narasimhan and Smid [11] details various results on Euclidean spanners, including a $(1 + \epsilon)$-spanner for the set S of n points in $\mathbb{R}^d$ that has $O(\frac{n}{\epsilon^{d-1}})$ edges for any $\epsilon > 0$.

As mentioned in Abam et al., [2], the cost of traversing a path in a network is not only determined by the lengths of the edges on the path but also by the delays occurring at the nodes on the path. [2] models these delays with the additive weighted metric. Let S be a set of n points in $\mathbb{R}^d$. For every $p \in S$, let $w(p)$ be the non-negative weight associated to p. The following additive weighted distance function d_w on S defining the metric space (S, d_w) is considered in [2]: for any $p, q \in S$, $d_w(p, q)$ equals to 0 if $p = q$; otherwise, it is equal to $w(p) + |pq| + w(q)$. The doubling dimesnion of a metric space (S, d) is defined as follows. If p is a point of S and $R > 0$ is a real number, then the d-ball with center p and radius R is the set $q \in S : d(p, q) \leq R$. The doubling dimension of (S, d) is the smallest real number d such that the following is true: For every real number $R > 0$, every d-ball of radius R can be covered by at most 2^d d-balls of radius $R/2$. As it turns out, results similar to the Euclidean setting are possible when the doubling dimension of the metric space is bounded by a constant d: in this case there is a $(1 + \epsilon)$-spanner with $n/\epsilon^{O(d)}$ edges [8]. However, as pointed in [2], (S, d_w) is not necessarily have bounded doubling dimension, even if the underlying metric (that measures the unweighted distance between any two points) has bounded doubling dimension. For any fixed constant $\epsilon > 0$, [2] gave an algorithm to compute a $(5 + \epsilon)$-spanner with a linear number of edges for (S, d_w). Recently, Abam et al. [3] showed that there exists a $(2+\epsilon)$-spanner with a linear number of edges for the metric space (S, d_w) that has bounded doubling dimension. And, [2] gives a lower bound on the stretch factor, showing that $(2 + \epsilon)$ stretch is nearly optimal. Bose et al. [5] studied the problem of computing spanners for a weighted set of points. They considered the points that lie on the plane to have positive weights associated to them; and defined the distance d_w between any two distinct points $p, q \in S$ as $d(p, q) - w(p) - w(q)$. Under the assumption that the distance between any pair of points is non-negative, they showed the existence of a $(1 + \epsilon)$-spanner with $O(\frac{n}{\epsilon})$ edges.

A simple polygon $P_\mathcal{D}$ containing $h \geq 0$ number of disjoint simple polygons within it is termed the *polygonal domain $\mathcal{D}$*. (When h equals to 0, the polygonal domain $\mathcal{D}$ is a simple polygon.) The free space $\mathcal{F}(\mathcal{D})$ of the given polygonal domain $\mathcal{D}$ is defined as the closure of $P_\mathcal{D}$ excluding the union of the interior of polygons contained in $P_\mathcal{D}$. Essentially, a path between any two given points in $\mathcal{F}(\mathcal{D})$ must be in the free space $\mathcal{F}(\mathcal{D})$ of $\mathcal{D}$. Given a set S of n points in the free space $\mathcal{F}(\mathcal{D})$ defined by the polygonal domain $\mathcal{D}$, computing geodesic spanners in $\mathcal{F}(\mathcal{D})$ is considered in Abam et al. [1]. For any two distinct points $p, q \in S$, $d_\pi(p, q)$ is defined as the geodesic Euclidean distance along a shortest path $\pi(p, q)$ between p and q in $\mathcal{F}(\mathcal{D})$. [1] showed that for the metric space (S, π), for any constant $\epsilon > 0$, there exists a $(5 + \epsilon)$-spanner of size $O(\sqrt{h}n(\lg n)^2)$. Further, for any constant $\epsilon > 0$, [1] gave a $(\sqrt{10} + \epsilon)$-spanner with $O(n(\lg n)^2)$ edges when $h = 0$ i.e., the polygonal domain is a simple polygon with no holes.

For a network to be vertex fault-tolerant, i.e., when a subset of nodes is removed, the induced network on the remaining nodes requires to be connected. Formally, a graph $G(S, E)$ is a *k-vertex fault-tolerant t-spanner*, denoted by (k, t)-VFTS, for a set S of n points in $\mathbb{R}^d$ if for any subset S' of S with size at most k, the graph $G \setminus S'$ is a t-spanner for the points in $S \setminus S'$. Algorithms in Levcopoulos et al., [9], Lukovszki [10], and Czumaj and Zhao [6] compute a (k, t)-VFTS for the set S of points in $\mathbb{R}^d$. These algorithms are also presented in [11]. [9] devised an algorithm to compute a (k, t)-VFTS of size $O(\frac{n}{(t-1)^{(2d-1)(k+1)}})$ in $O(\frac{n \lg n}{(t-1)^{4d-1}} + \frac{n}{(t-1)^{(2d-1)(k+1)}})$ time and another algorithm to compute a (k, t)-VFTS with $O(k^2 n)$ edges in $O(\frac{kn \lg n}{(t-1)^d})$ time. [10] gives an algorithm to compute a (k, t)-VFTS of size $O(\frac{kn}{(t-1)^{d-1}})$ in $O(\frac{1}{(t-1)^d}(n \lg^{d-1} n \lg k + kn \lg \lg n))$ time. The algorithm in [6] computes a (k, t)-VFTS having $O(\frac{kn}{(t-1)^{d-1}})$ edges in $O(\frac{1}{(t-1)^{d-1}}(kn \lg^d n + nk^2 \lg k))$ time with total weight of edges upper bounded by a $O(\frac{k^2 \lg n}{(t-1)^d})$ multiplicative factor of the weight of MST of the given set of points.

Our Results. The spanners computed in this paper are first of their kind as we combine fault-tolerance with the additive weighted set of points. We devise the following algorithms for computing vertex fault-tolerant additive weighted geometric spanners (VFTAWS) for any $\epsilon > 0$ and $k \geq 1$:

* Given a set S of n weighted points in $\mathbb{R}^d$, our first algorithm presented herewith computes a $(k, 4 + \epsilon)$-VFTAWS having $O(kn)$ edges. We incorporate fault-tolerance to the recent results of [3] while retaining the same stretch factor and increasing the number of edges in the spanner by a multiplicative factor of $O(k)$.
* Given a set S of n weighted points in a simple polygon, we present an algorithm to compute a $(k, 4 + \epsilon)$-VFTAWS that has $O(\frac{kn}{\epsilon^2} \lg n)$ edges. Our algorithm combines the clustering based algorithms from [1] and [2], and with the careful addition of more edges, we show that k fault-tolerance is achieved.

Unless specified otherwise, the points are assumed to be in Euclidean space $\mathbb{R}^d$. The Euclidean distance between two points p and q is denoted by $|pq|$. The distance between two points p, q in the metric space X is denoted by $d_X(p, q)$. The length of the shortest path between p and q in a graph G is denoted by $d_G(p, q)$.

Section 2 details the algorithm and its analysis to compute a $(k, 4 + \epsilon)$-VFTAWS when the input weighted points are in $\mathbb{R}^d$. For the input weighted points located in a simple polygon, Sect. 3 describes an algorithm to compute a $(k, 4 + \epsilon)$-VFTAWS.

2   Vertex Fault-Tolerant Additive Weighted Spanner for Points in $\mathbb{R}^d$

In this section, we describe an algorithm to compute a (k, t)-VFTAWS for the set S of n non-negative weighted points in $\mathbb{R}^d$, where $t > 1$ and $k \geq 1$ are real numbers. For any two points $p, q \in S$, the additive weighted distance $d_w(p, q)$ is defined as the $w(p) + |pq| + w(q)$. Following the algorithm mentioned in [3], we first partition all the points belonging to S into at least $k + 1$ clusters. For creating these clusters, the points in set S are sorted in non-decreasing order with respect to their weights. Then the first $k + 1$ points in this sorted list are chosen as the centers of $k + 1$ distinct clusters. As the algorithm progress, more points are added to these clusters as well as more clusters (with cluster centers) may also be created. In any iteration of the algorithm, for any point p in the remaining sorted list, among the current set of cluster centers, we determine the cluster center c_j nearest to p. Let C_j be the cluster to which c_j is the center. It adds p to the cluster C_j if $|pc_j| \leq \epsilon.w(p)$; otherwise, a new cluster C_p with p as its centre is initiated. Let $C = \{c_1, \ldots, c_z\}$ be the final set of cluster centers obtained through this procedure. For every $i \in [1, z]$, the cluster to which c_i is the center is denoted by C_i. Using the algorithm from [12], we compute a $(k, (2 + \epsilon))$-VFTS $\mathcal{B}$ for the set C of cluster centers. We note that the degree of each vertex of $\mathcal{B}$ is $O(k)$. We denote the stretch of $\mathcal{B}$ by $t_\mathcal{B}$. First, the graph $\mathcal{G}$ is initialized to $\mathcal{B}$; further, points in $S \setminus C$ are included in $\mathcal{G}$ as vertices. Our algorithm to compute a $(k, 4 + \epsilon)$-VFTAWS differs from [3] with respect to both the algorithm used in computing $\mathcal{B}$ and the set of edges added to $\mathcal{B}$. The latter part is described now. For every $i \in [1, z]$, let C_i' be the set comprising of $\min\{k + 1, |C_i|\}$ least weighted points of cluster C_i. For each point $p \in S \setminus C$, if p belongs to cluster C_l, then for each $v \in B_l \cup C_l'$, our algorithm introduces an edge between p and v with weight $|pv|$ to $\mathcal{G}$. Here, B_l is the set comprising of all the neighbors of the center (c_l) of cluster C_l in the graph $\mathcal{B}$. In the following theorem, we prove that the graph $\mathcal{G}$ is indeed a $(k, 4 + \epsilon)$-VFTAWS with $O(kn)$ edges.

Theorem 1. *Let S be a set of n weighted points in $\mathbb{R}^d$ with non-negative weights associated to points with weight function w. For any fixed constant $\epsilon > 0$, the graph $\mathcal{G}$ is a $(k, (4 + \epsilon))$-VFTAWS with $O(kn)$ edges for the metric space (S, d_w).*

Proof: From [12], the number of edges in $\mathcal{B}$ is $O(k |C|)$, which is essentially $O(kn)$. Further, the degree of each node in $\mathcal{B}$ is $O(k)$. From each point in $S \setminus C$, we are adding at most $O(k)$ edges. Hence, the number of edges in $\mathcal{G}$ is $O(kn)$.

In proving that $\mathcal{G}$ is a $(k, 4 + \epsilon)$-VFTAWS for the metric space (S, d_w), we show that for any set $S' \subset S$ with $|S'| \leq k$ and for any two points $p, q \in S \setminus S'$ there exists a $(4 + \epsilon)$-spanner path between p and q in $\mathcal{G} \setminus S'$. In the following, we provide the analysis of few cases We subdivide the analysis into cases based on the role p and q play with respect to clusters formed and their centers. In the following we are presenting a few cases.

Case 1: Both p and q are cluster centres of two distinct clusters i.e., $p, q \in C$.

Since $\mathcal{B}$ is a $(k, (2 + \epsilon))$-VFTS for the set C,
$$d_{\mathcal{G} \setminus S'}(p, q) = d_{\mathcal{B} \setminus S'}(p, q)$$
$$\leq t_{\mathcal{B}} \cdot d_w(p, q).$$

Case 2: Both p and q are in the same cluster C_i and one of them, w.l.o.g., say p, is the centre of C_i. Since p is the least weighted point in C_i, there exists an edge joining p and q in $\mathcal{G}$. Hence,
$$d_{\mathcal{G} \setminus S'}(p, q) = d_w(p, q).$$

Case 3: Both p and q are in the same cluster, say C_i; $p \neq c_i$, $q \neq c_i$; and, $c_i \notin S'$. Then,

$$d_{\mathcal{G} \setminus S'}(p, q) = d_w(p, c_i) + d_w(c_i, q)$$
$$= w(p) + |pc_i| + w(c_i) + w(c_i) + |c_i q| + w(q)$$
$$\leq w(p) + \epsilon \cdot w(p) + w(c_i) + w(c_i) + \epsilon \cdot w(q) + w(q)$$
[since a point x is added to cluster C_l only if $|xc_l| \leq \epsilon \cdot w(x)$]
$$\leq w(p) + \epsilon \cdot w(p) + w(p) + w(q) + \epsilon \cdot w(q) + w(q)$$
[since the points are sorted in the non-decreasing order of their
weights and the first point added to any cluster is taken
as its center]
$$= (2 + \epsilon) \cdot [w(p) + w(q)]$$
$$< (2 + \epsilon) \cdot [w(p) + |pq| + w(q)]$$
$$= (2 + \epsilon) \cdot d_w(p, q).$$

Case 4: Both p and q are in the same cluster, say C_i; $p \neq c_i$, $q \neq c_i$; and, $c_i \in S'$. In the case of $|C_i| \leq k$, there exists an edge between p and q in $\mathcal{G}$. Hence, suppose that $|C_i| > k$. Let S'' be the set of $k + 1$ least weighted points from C_i. If $p, q \in S''$ then there exists an edge between p and q in $\mathcal{G}$. If $p \in S''$ and $q \notin S''$ then as well there exists an edge between p and q. (Argument for the other case in which $q \in S''$ and $p \notin S''$ is analogous.) Now consider the case in which both $p, q \notin S''$. Since p and q are connected to every point in S'' and $|S''| = k + 1$, there exists an $r \in S''$ such that $r \notin S'$ and the edges (p, r) and (r, q) belong to $\mathcal{G} \setminus S'$. Therefore,

$$d_{\mathcal{G} \setminus S'}(p, q) = d_w(p, r) + d_w(r, q)$$
$$= w(p) + |pr| + w(r) + w(r) + |rq| + w(q)$$
$$\leq w(p) + |pc_i| + |c_i r| + w(r) + w(r) + |rc_i| + |c_i q| + w(q)$$
[by triangle inequality]

$$\leq w(p) + \epsilon \cdot w(p) + \epsilon \cdot w(r) + w(r) + w(r) + \epsilon \cdot w(r)$$
$$+ \epsilon \cdot w(q) + w(q)$$

[since a point x is added to cluster C_l only if $|xc_l| \leq \epsilon \cdot w(x)$]

$$\leq w(p) + \epsilon \cdot w(p) + \epsilon \cdot w(p) + w(p) + w(q) + \epsilon \cdot w(q)$$
$$+ \epsilon \cdot w(q) + w(q)$$

[since for any point the edges are added to the $k+1$ least weighted points of the cluster to which it belongs]

$$= (2 + 2\epsilon) \cdot [w(p) + w(q)]$$
$$< (2 + 2\epsilon) \cdot [w(p) + |pq| + w(q)]$$
$$= (2 + 2\epsilon) \cdot d_w(p, q).$$

Case 5: Points p and q belong to two distinct clusters, say $p \in C_i$ and $q \in C_j$. In addition, $p \neq c_i$ and $q \neq c_j$, and neither of the cluster centres belong to S'. Then,

$$d_{\mathcal{G} \setminus S'}(p, q) = d_w(p, c_i) + d_{\mathcal{B}}(c_i, c_j) + d_w(c_j, q)$$
$$= w(p) + |pc_i| + w(c_i) + d_{\mathcal{B}}(c_i, c_j) + w(c_j) + |c_j q| + w(q)$$
$$\leq w(p) + \epsilon \cdot w(p) + w(c_i) + d_{\mathcal{B}}(c_i, c_j) + w(c_j) + \epsilon \cdot w(q) + w(q)$$

[since a point x is added to cluster C_l only if $|xc_l| \leq \epsilon \cdot w(x)$]

$$\leq (1 + \epsilon) \cdot [w(p) + w(q)] + w(c_i) + w(c_j) + t_{\mathcal{B}} \cdot d_w(c_i, c_j)$$

[since $\mathcal{B}$ is a $(k, t_{\mathcal{B}})$-vertex fault-tolerant spanner for the set C]

$$\leq (1 + \epsilon) \cdot [w(p) + w(q)] + w(p) + w(q) + t_{\mathcal{B}} \cdot d_w(c_i, c_j)$$

[since the points are sorted in the non-decreasing order of their weights and the first point added to any cluster is taken as center of that cluster]

$$= (2 + \epsilon) \cdot [w(p) + w(q)] + t_{\mathcal{B}} \cdot [w(c_i) + |c_i c_j| + w(c_j)]$$
$$\leq (2 + \epsilon) \cdot [w(p) + w(q)] + t_{\mathcal{B}} \cdot [w(p) + |c_i c_j| + w(q)]$$

[since the points are sorted in the non-decreasing order of their weights and the first point added to any cluster is taken as its center]

$$\leq (2 + \epsilon) \cdot [w(p) + w(q)] + t_{\mathcal{B}} \cdot [w(p) + w(q) + |c_i p| + |pq| + |q c_j|]$$

[by triangle inequality]

$$\leq (2 + \epsilon) \cdot [w(p) + w(q)] + t_B \cdot [w(p) + w(q)$$
$$+ \epsilon \cdot w(p) + |pq| + \epsilon \cdot w(q)]$$

[since a point x is added to cluster C_l only if $|xc_l| \leq \epsilon \cdot w(x)$]

$$= (2 + \epsilon) \cdot [w(p) + w(q)] + t_B \cdot [(1 + \epsilon) \cdot [w(p) + w(q)] + |pq|]$$
$$< (2 + \epsilon) \cdot [w(p) + w(q) + |pq|] + t_B \cdot (1 + \epsilon) \cdot [w(p) + w(q) + |pq|]$$
$$< t_B(2 + \epsilon) \cdot [w(p) + w(q) + |pq|] \quad \text{when } t_B \geq (2 + \epsilon)$$

[since each point has non-negative weight associated with it]

$$= t_B \cdot (2 + \epsilon) \cdot d_w(p, q).$$

The analysis of the following remaining cases will be presented in the full version: *Case 6*: Both the points p and q are in two distinct clusters, w.l.o.g., say $p \in C_i$ and $q \in C_j$, one of them, say p is the centre of C_i (i.e., $p = c_i$), and $c_j \notin S'$. *Case 7*: Both the points p and q are in two distinct clusters, say $p \in C_i$ and $q \in C_j$; $p \neq c_i$, $q \neq c_j$; and, one of these centers, say c_j, belongs to S' and the other center $c_i \notin S'$. *Case 8*: Points p and q are in two distinct clusters, say $p \in C_i$ and $q \in C_j$; $p \neq c_i$, $q \neq c_j$; and both $c_i, c_j \in S'$. Considering the analysis in all these cases proves that $\mathcal{G}$ is a k-VFTAWS with stretch t upper bounded by $t_B \cdot (2+\epsilon)$. We had chosen t_B to be equal to $(2+\epsilon)$, so that it satisfies all the above cases. Since t_B is $(2 + \epsilon)$, $t = (2 + \epsilon)^2 \leq (4 + 5\epsilon)$. Hence, $\mathcal{G}$ is a $(k, 4 + 5\epsilon)$-VFTAWS for the metric space (S, d_w). $\qquad \square$

3 Vertex Fault-Tolerant Additive Weighted Spanner for Points in Simple Polygon

Given a set S of n points in a simple polygon P, for any two points $p, q \in S$, the shortest path between p and q in P is denoted by $\pi(p, q)$, and the length of that path is indicated by $d_\pi(p, q)$. For a $t \geq 1$, a geodesic t-spanner of S is a graph $\mathcal{G}(S, E')$ such that $d_\pi(p, q) \leq d_\mathcal{G}(p, q) \leq t \cdot d_\pi(p, q)$ for every two points $p, q \in S$. We detail an algorithm to compute a geodesic vertex fault-tolerant additive weighted spanner for the set S of n weighted points located in a simple polygon P.

The following definition for the distance function $d_{\pi,w}$ for the set S of points is considered in [2]: For any $p, q \in S$, $d_{\pi,w}(p, q)$ equals to 0 if $p = q$; otherwise, it is equal to $w(p) + d_\pi(p, q) + w(q)$. Further, $(S, d_{\pi,w})$ was shown as a metric space in [2].

We devise a divide-and-conquer based algorithm to compute a $(k, 4 + \epsilon)$-VFTAWS for the metric space $(S, d_{\pi,w})$. Following [1], we define few terms. Let S' be a set of points contained in a simple polygon P'. A vertical line segment that splits P' into two simple sub-polygons of P' such that each sub-polygon contains at most two-thirds of the points in S' is termed a *splitting segment* with respect to S' and P'. (In the following description, S' and P' are not mentioned with the splitting segment whenever they are clear from the context.)

The *geodesic projection* p_l of a point p onto a splitting segment l is a point on l that has the minimum geodesic Euclidean distance from p among all the points of l. By extending [1], we give an algorithm to compute a $(k, 4 + \epsilon)$-VFTAWS $\mathcal{G}$ for the metric space $(S, d_{\pi,w})$.

Our algorithm partitions P containing points in S into two simple sub-polygons P' and P'' with a splitting segment l. For every point $p \in S$, we compute its geodesic projection p_l onto l and assign $w(p) + d_\pi(p, p_l)$ as the weight of p_l. Let S_l be the set comprising of all the geodesic projections of S onto l. Also, let $d_{l,w}$ be the additive weighted metric associated with points in S_l. We use the algorithm from Sect. 2 to compute a $(k, 4 + \epsilon)$-VFTAWS $\mathcal{G}_l$ for the metric space $(S_l, d_{l,w})$. For every edge (r, s) in $\mathcal{G}_l$, we add an edge between p and q to $\mathcal{G}$ with weight $d_\pi(p, q)$, wherein r (resp. s) is the geodesic projection of p (resp. q) onto l. Let S' (resp. S'') be the set of points contained in the sub-polygon P' (resp. P'') of P. We recursively process P' (resp. P'') with points in S' (resp. S'') unless $|S'|$ (resp. $|S''|$) is less than or equal to one. We prove that the graph $\mathcal{G}$ is a $(k, (12 + 15\epsilon))$-VFTAWS for the metric space $(S, d_{\pi,w})$. (Later, with further refinements to this graph, we improve the stretch factor to $(4 + 14\epsilon)$.) We show that by removing any subset S' with $|S'| \leq k$ from $\mathcal{G}$, for any two points p and q in $S \setminus S'$, there exists a path between p and q in $\mathcal{G} \setminus S'$ such that the $d_\mathcal{G}(p, q)$ is at most $(12 + 15\epsilon) d_{\pi,w}(p, q)$. First, we note that there exists a splitting segment l at some iteration of the algorithm so that p and q are on different sides of l. Let r be a point belonging to $l \cap \pi(p, q)$. Let S'_l be the set comprising of geodesic projections of points in S' on l. Since $\mathcal{G}_l$ is a $(k, (4 + 5\epsilon))$-VFTAWS for the metric space $(S_l, d_{l,w})$, there exists a path Q between p_l and q_l in $\mathcal{G}_l \setminus S'_l$ whose length is upper bounded by $(4 + 5\epsilon) \cdot d_{l,w}(p_l, q_l)$. Let Q' be a path between p and q in $\mathcal{G} \setminus S'$ which is obtained by replacing each vertex v_l of Q by v in S such that the point v_l is the geodesic projection of v on l. In the following, we show that the length of Q', which is $d_{\mathcal{G} \setminus S'}(p, q)$, is upper bounded by $(12 + 15\epsilon) \cdot d_{\pi,w}(p, q)$.

For every $x, y \in S$,

$$
\begin{aligned}
d_{\pi,w}(x, y) &= w(x) + d_\pi(x, y) + w(y) \\
&\leq w(x) + d_\pi(x, x_l) + d_\pi(x_l, y_l) + d_\pi(y_l, y) + w(y) \\
&\text{[by triangle inequality]} \\
&= w(x_l) + d_\pi(x_l, y_l) + w(y_l) \\
&\text{[since the weight associated with projection } z_l \text{ of every point} \\
&z \text{ is } w(z) + d_\pi(z, z_l)] \\
&= d_{l,w}(x_l, y_l). \tag{1}
\end{aligned}
$$

This implies,

$$d_{\mathcal{G}\setminus S'}(p,q) \leq \sum_{x_l,y_l \in Q} d_{\pi,w}(x,y)$$

$$\leq \sum_{x_l,y_l \in Q} d_{l,w}(x_l,y_l)$$

[from (1)]

$$\leq (4+5\epsilon) \cdot d_{l,w}(p_l,q_l) \tag{2}$$

[since $\mathcal{G}_l$ is a $(k,(4+5\epsilon))$-vertex fault-tolerant geodesic spanner]

$$= (4+5\epsilon) \cdot [w(p_l) + d_l(p_l,q_l) + w(q_l)]$$

$$= (4+5\epsilon) \cdot [w(p_l) + d_\pi(p_l,q_l) + w(q_l)]$$

[since P contains l, shortest path between p_l and q_l
is same as the geodesic shortest path between p_l and q_l]

$$= (4+5\epsilon) \cdot [w(p) + d_\pi(p,p_l) + d_\pi(p_l,q_l) + d_\pi(q_l,q) + w(q)] \tag{3}$$

[since the weight associated with projection z_l of every point z
is $w(z) + d_\pi(z,z_l)$].

Since r is a point belonging to both l as well as to $\pi(p,q)$,

$$d_\pi(p,p_l) \leq d_\pi(p,r) \text{ and } d_\pi(q,q_l) \leq d_\pi(q,r). \tag{4}$$

Substituting (4) into (3),

$$d_{\mathcal{G}\setminus S'}(p,q) \leq (4+5\epsilon) \cdot [w(p) + d_\pi(p,r) + d_\pi(p_l,q_l) + d_\pi(r,q) + w(q)]$$

$$\leq (4+5\epsilon) \cdot [w(p) + d_\pi(p,r) + w(r) + d_\pi(p_l,q_l)$$

$$+ w(r) + d_\pi(r,q) + w(q)]$$

[since the weight associated with every point is non-negative]

$$= (4+5\epsilon) \cdot [d_{\pi,w}(p,r) + d_\pi(p_l,q_l) + d_{\pi,w}(r,q)]$$

$$= (4+5\epsilon) \cdot [d_{\pi,w}(p,q) + d_\pi(p_l,q_l)] \tag{5}$$

[since $\pi(p,q)$ intersects l at r, by optimal substructure property
of shortest paths, $\pi(p,q) = \pi(p,r) + \pi(r,q)$].

Consider

$$d_\pi(p_l, q_l) \le d_\pi(p_l, p) + d_\pi(p, q) + d_\pi(q, q_l)$$

[since π follows triangle inequality]

$$\le d_\pi(r, p) + d_\pi(p, q) + d_\pi(q, r)$$

[using (4)]

$$\le w(r) + d_\pi(r, p) + w(p) + w(p) + d_\pi(p, q) + w(q)$$
$$+ w(q) + d_\pi(q, r) + w(r)$$

[since weight associated with every point is non-negative]

$$= d_{\pi,w}(p, r) + d_{\pi,w}(p, q) + d_{\pi,w}(r, q)$$
$$= d_{\pi,w}(p, q) + d_{\pi,w}(p, q)$$

[since $\pi(p, q)$ intersects l at r, by optimal substructure property
of shortest paths, $\pi(p, q) = \pi(p, r) + \pi(r, q)$]

$$= 2d_{\pi,w}(p, q). \tag{6}$$

Substituting (6) into (5),
$$d_{G \setminus S'}(p, q) \le 3(4 + 5\epsilon) \cdot d_{\pi,w}(p, q).$$

Hence, the graph $\mathcal{G}$ computed as described above is a $(k, 12 + \epsilon)$-VFTAWS for the metric space $(S, d_{\pi,w})$. We further improve the stretch factor of $\mathcal{G}$ by applying the refinement given in [3] to the above-described algorithm. In doing this, for each point $p \in S$, we compute the geodesic projection p_γ of p on the splitting line γ and we construct a set $S(p, \gamma)$ as defined herewith. Let $\gamma(p) \subseteq \gamma$ be $\{x \in \gamma : d_{\gamma,w}(p_\gamma, x) \le (1 + 2\epsilon) \cdot d_\pi(p, p_\gamma)\}$. Here, for any $p, q \in S$, $d_{\gamma,w}(p, q)$ is equal to 0 if $p = q$; otherwise, equals to $w(p) + d_\gamma(p, q) + w(q)$. We divide $\gamma(p)$ into c pieces with $c \in O(1/\epsilon^2)$: each piece is denoted by $\gamma_j(p)$ for $1 \le j \le c$, and the piece length is at most $\epsilon \cdot d_\pi(p, p_\gamma)$. For every piece j, we compute the point $p_\gamma^{(j)}$ nearest to p in $\gamma_j(p)$. The set $S(p, \gamma)$ is defined as $\{p_\gamma^{(j)} : p_\gamma^{(j)} \in \gamma_j(p)$ and $1 \le j \le c\}$. For every $r \in S(p, \gamma)$, the non-negative weight $w(r)$ of r is set to $w(p) + d_\pi(p, r)$. Let S_γ be $\cup_{p \in S} S(p, \gamma)$.

We replace the set S_l in computing $\mathcal{G}$ with the set S_γ and compute a $(k, (4 + 5\epsilon))$-VFTAWS $\mathcal{G}_l$ using the algorithm from Sect. 2 for the set S_γ instead. Further, for every edge (r, s) in $\mathcal{G}_l$, we add the edge (p, q) to $\mathcal{G}$ with weight $d_\pi(p, q)$ whenever $r \in S(p, l)$ and $s \in S(q, l)$. The rest of the algorithm remains the same.

In the following, we restate a lemma from [3], which is useful for our analysis.

Lemma 1. *Let P be a simple polygon. Consider two points $x, y \in P$. Let r be the point at which shortest path $\pi(x, y)$ between x and y intersects a splitting segment γ. If $r \notin \gamma(x)$, point x'_γ (resp. y'_γ) is set as x_γ (resp. y_γ). Otherwise x'_γ (resp. y'_γ) is set as the point from $S(x, \gamma)$ (resp. $S(y, \gamma)$) which is nearest to x (resp. y). Then $d_\pi(x, x'_\gamma) + d_\gamma(x'_\gamma, r)$ (resp. $d_\gamma(r, y'_\gamma) + d_\pi(y'_\gamma, y)$) is less than or equal to $(1 + \epsilon) \cdot d_\pi(x, r)$ (resp. $(1 + \epsilon) \cdot d_\pi(r, y)$).*

Theorem 2. *Let S be a set of n weighted points in simple polygon P with non-negative weights associated to points with weight function w. For any fixed constant $\epsilon > 0$, there exists a $(k, (4 + \epsilon))$-vertex fault-tolerant additive weighted geodesic spanner with $O(\frac{kn}{\epsilon^2} \lg n)$ edges for the metric space $(S, d_{\pi,w})$.*

Proof: In constructing a $(k, (4+\epsilon))$-VFTAWS $\mathcal{G}_l$ for the set S_γ of $\frac{n}{\epsilon^2}$ points, we add $O(\frac{kn}{\epsilon^2})$ edges to $\mathcal{G}$ in one iteration. Let $S(n)$ be the size of $\mathcal{G}$ when there are n points. Then $S(n) = S(n_1) + S(n_2) + \frac{kn}{\epsilon^2}$ where n_1, n_2 are the number of points in each of the partitions formed by the splitting segment. Since $n_1, n_2 \geq n/3$, $S(n)$ is $O(\frac{kn}{\epsilon^2} \lg n)$.

For proving that $\mathcal{G}$ is a $(k, (4 + \epsilon))$-VFTAWS for the metric space $(S, d_{\pi,w})$, we show that for any set $S' \subset S$ with $|S'| \leq k$ and for any two points $p, q \in S \setminus S'$ there exists a $(4+\epsilon)$-spanner path between p and q in $\mathcal{G} \setminus S'$. First, we note that there exists a splitting segment l at some iteration of the algorithm so that p and q are on different sides of l. Let r be a point belonging to $l \cap \pi(p,q)$. Let S'_l be the set comprising of geodesic projections of points in S' on l. Since $\mathcal{G}_l$ is a $(k, (4 + 5\epsilon))$-VFTAWS for the metric space $(S_l, d_{l,w})$, there exists a path Q between p_l and q_l in $\mathcal{G}_l \setminus S'_l$ whose length is upper bounded by $(4+5\epsilon) \cdot d_{l,w}(p_l, q_l)$. Let Q' be a path between p and q in $\mathcal{G} \setminus S'$ which is obtained by replacing each vertex v_l of Q by v in S such that the point v_l is the geodesic projection of v on l. In the following, we show that the length of Q', which is $d_{\mathcal{G} \setminus S'}(p, q)$, is upper bounded by $(4 + 14\epsilon) \cdot d_{\pi,w}(p, q)$.

Following the Lemma 1, if $r \notin l(p)$, point p'_l (resp. q'_l) is set as p_l (resp. q_l). Otherwise p'_l (resp. q'_l) is set as the point from $S(p, l)$ (resp. $S(q, l)$) which is nearest to p (resp. q).

$$
\begin{aligned}
d_{l,w}(p'_l, q'_l) &= w(p'_l) + d_l(p'_l, q'_l) + w(q'_l) \\
&\leq w(p'_l) + d_l(p'_l, r) + d_l(r, q'_l) + w(q'_l) \\
&\text{[by triangle inequality]} \\
&\leq w(p'_l) + d_l(p'_l, r) + w(r) + w(r) + d_l(r, q'_l) + w(q'_l) \\
&\text{[since the weight associated with each point is non-negative]} \\
&= w(p) + d_\pi(p, p'_l) + d_l(p'_l, r) + w(r) \\
&\quad + w(r) + d_l(r, q'_l) + d_\pi(q'_l, q) + w(q) \qquad (7) \\
&\text{[due to weight assigned to geodesic projections]}.
\end{aligned}
$$

Applying Lemma 1 with p'_l and q'_l,

$$
d_\pi(p, p'_l) + d_l(p'_l, r) \leq (1 + \epsilon) \cdot d_\pi(p, r), \text{ and} \qquad (8)
$$

$$
d_l(r, q'_l) + d_\pi(q'_l, q) \leq (1 + \epsilon) \cdot d_\pi(r, y). \qquad (9)
$$

Substituting (8) and (9) in (7),

$$d_{l,w}(p'_l, q'_l) \leq w(p) + (1 + \epsilon) \cdot d_\pi(p, r) + w(r) + w(r)$$
$$+ (1 + \epsilon) \cdot d_\pi(r, q) + w(q)$$
$$\leq (1 + \epsilon) \cdot [d_{\pi,w}(p, r) + d_{\pi,w}(r, q)]$$
$$= (1 + \epsilon) \cdot d_{\pi,w}(p, q) \qquad (10)$$

[since $r \in l \cap \pi(p, q)$, by the optimal substructure property
of shortest paths, $\pi(p, q) = \pi(p, r) + \pi(r, q)$].

Replacing p_l (resp. q_l) by p_l' (resp. q_l') in inequality (2),

$$d_{\mathcal{G}\backslash S'}(p, q) \leq (4 + 5\epsilon) \cdot d_{l,w}(p_l', q_l')$$
$$\leq (4 + 5\epsilon)(1 + \epsilon) \cdot d_{\pi,w}(p, q)$$

[from (10)]

$$\leq (4 + 14\epsilon) \cdot d_{\pi,w}(p, q).$$

Thus, $\mathcal{G}$ is a $(k, (4 + \epsilon))$-vertex fault-tolerant additive weighted geodesic spanner for the set S of points located in the simple polygon P. □

4   Conclusions

In this paper, we gave algorithms to achieve k vertex fault-tolerance when the metric is additive weighted. We devised algorithms to compute a $(k, 4 + \epsilon)$-VFTAWS when the input points belong to either of the following: $\mathbb{R}^d$ and simple polygon. Apart from the efficient computation, it would be interesting to explore the lower bounds on the number of edges for the fault-tolerant additive weighted spanners. Besides, the future work in the context of additive spanners could include finding the relation between the vertex-fault tolerance and the edge-fault tolerance, and optimizing various spanner parameters, like degree, diameter and weight.

References

1. Abam, M.A., Adeli, M., Homapour, H., Asadollahpoor, P.Z.: Geometric spanners for points inside a polygonal domain. In: Proceedings of Symposium on Computational Geometry, pp. 186–197 (2015)
2. Abam, M.A., de Berg, M., Farshi, M., Gudmundsson, J., Smid, M.H.M.: Geometric spanners for weighted point sets. Algorithmica **61**(1), 207–225 (2011)
3. Abam, M.A., de Berg, M., Seraji, M.J.R.: Geodesic spanners for points on a polyhedral terrain. In: Proceedings of Symposium on Discrete Algorithms, pp. 2434–2442 (2017)
4. Althöfer, I., Das, G., Dobkins, D., Joseph, D., Soares, J.: On sparse spanners of weighted graphs. Discret. Comput. Geom. **9**(1), 81–100 (1993)
5. Bose, P., Carmi, P., Couture, M.: Spanners of additively weighted point sets. In: Proceedings of Scandinavian Workshop on Algorithm Theory, pp. 367–377 (2008)

6. Czumaj, A., Zhao, H.: Fault-tolerant geometric spanners. Discret. Comput. Geom. **32**(2), 207–230 (2004)
7. Har-Peled, S.: Geometric Approximation Algorithms. American Mathematical Society, Providence (2011)
8. Har-Peled, S., Mendel, M.: Fast construction of nets in low-dimensional metrics and their applications. SIAM J. Comput. **35**(5), 1148–1184 (2006)
9. Levcopoulos, C., Narasimhan, G., Smid, M.H.M.: Improved algorithms for constructing fault-tolerant spanners. Algorithmica **32**(1), 144–156 (2002)
10. Lukovszki, T.: New results of fault tolerant geometric spanners. In: Proceedings of Workshop on Algorithms and Data Structures, pp. 193–204 (1999)
11. Narasimhan, G., Smid, M.H.M.: Geometric Spanner Networks. Cambridge University Press, Cambridge (2007)
12. Solomon, S.: From hierarchical partitions to hierarchical covers: optimal fault-tolerant spanners for doubling metrics. In: Proceedings of Symposium on Theory of Computing, pp. 363–372 (2014)

Maintaining the Visibility Graph
of a Dynamic Simple Polygon

Tameem Choudhury and R. Inkulu[✉]

Department of Computer Science and Engineering, IIT Guwahati, Guwahati, India
{t.choudhury,rinkulu}@iitg.ac.in

Abstract. We devise a fully-dynamic algorithm for maintaining the visibility graph of a given simple polygon P amid vertex insertions and deletions to the simple polygon. Our algorithm takes $O(k(\lg n')^2)$ worst-case time to update the visibility graph when a vertex is inserted to the current simple polygon P', or when a vertex is deleted from P'. Here, k is the number of combinatorial changes needed to the visibility graph due to the insertion (resp. deletion) of a vertex v to P', and n' is the number of vertices of P'. This algorithm preprocesses the initial simple polygon P to build few data structures, including the visibility graph of P. Further, as part of efficiently updating the visibility graph, a fully-dynamic algorithm is designed to compute the vertices of the current simple polygon that are visible from a query point.

Keywords: Computational geometry · Visibility · Dynamic algorithms

1 Introduction

Let P be a simple polygon with n vertices. Two points $p, q \in P$ are said to be mutually *visible* to each other whenever the interior of line segment pq does not intersect any edge of P. The *(vertex-vertex) visibility graph* $G(V, E)$ of a simple polygon P is defined to be an undirected graph whose nodes are the vertices of P and whose edges are pairs of mutually visible vertices of P. Computing visibility graphs is a fundamental problem in computational geometry, and it is studied extensively. For the visibility graph of a simple polygon, Lee [24] and Sharir and Schorr [28] gave algorithms that take $O(n^2 \lg n)$ time. Algorithms by Asano et al. [2] and Welzl [30] to compute the visibility graph take $O(n^2)$, which are worst-case optimal since $|E|$ is $\Theta(n^2)$ in the worst-case. For the triangulated simple polygons, Hershberger [17] gave a $O(n + |E|)$ time algorithm. Since any simple polygon can be triangulated in linear time due to an algorithm from Chazelle [5], the algorithm in Hershberger [17] takes $O(n + |E|)$ time, which is optimal for this problem.

R. Inkulu—This research is supported in part by NBHM grant 248(17)2014-R&D-II/1049.

S. P. Pal and A. Vijayakumar (Eds.): CALDAM 2019, LNCS 11394, pp. 42–52, 2019.
https://doi.org/10.1007/978-3-030-11509-8_4

The following algorithms for visibility computation were devised for a polygonal domain defined with n vertices and h holes. Overmars and Welzl [26] gave an output-sensitive algorithm that takes $O(|E|\lg n)$ time and $O(n + |E|)$ space. Ghosh and Mount [12] presented an algorithm with $O(n \lg n + |E|)$ worst-case time that uses $O(|E| + n)$ space. Keeping the same time complexity as in [12], Pocchiola and Vegter [27] improved the space complexity to $O(n)$. Kapoor and Maheshwari [22,23] proposed an algorithm with time complexity $O(T + |E| + h \lg n)$. Here, T is the time for triangulating the free space of the polygonal domain, which is $O(n + h(\lg h)^{1+\delta})$ due to Bar-Yehuda and Chazelle [3] for a small positive constant δ.

For a point $q \in P$, the *visibility polygon* $VP(q)$ of q is the maximal set of points $x \in P$ such that x is visible to q. The problem of computing the visibility polygon of a point in a simple polygon was first attempted in [8], who presented a $O(n^2)$ time algorithm. Then, ElGindy and Avis [9] and Lee [25] presented an $O(n)$ time algorithms for this problem. Joe and Simpson [21] corrected a flaw in [9,25] and devised an $O(n)$ time algorithm that correctly handles winding in the simple polygon. Algorithms that preprocess the given simple polygon P to efficiently compute visibility polygon of any given query point were studied in [1,2,4,6,7,15,18,29,31]. The *complete visibility polygon* of a convex set C located inside a given simple polygon P is defined as the maximal set S of points such that every point $p \in S$ is visible from every point in C. Ghosh [10] devised an algorithm to compute the complete visibility polygon of a given simple polygon P from a convex set located inside P. [10] also devises algorithms for weak visibility polygon of P from a convex set C located inside P. Ghosh [11] details several visibility related algorithms.

Given a simple polygon P and a point p (resp. line segment) interior to P, algorithms devised in Inkulu and Thakur [19] maintain the visibility polygon of p (resp. weak visibility polygon of p) as vertices are added to P. To our knowledge, [19] gave the first dynamic (incremental) algorithms in the context of maintaining the visibility and weak visibility polygons in simple polygons. And, fully-dynamic algorithms for computing visibility polygon as well as several other variants of dynamic visibility were developed in Inkulu et al. [20], which is under review.

In the context of visibility graphs, when a new vertex is added to the current simple polygon or when a vertex of the current simple polygon is deleted, the visibility graph is updated. An algorithm is said to be *incremental* (resp. *decremental*) whenever it updates the visibility graph amid vertex insertions (resp. vertex deletions). If an algorithm is both incremental as well as decremental, then it is termed *fully-dynamic*. Hence, a fully-dynamic algorithm maintains the visibility graph as the simple polygon is updated with both vertex insertions and deletion of vertices. In [19,20], analogous notions of incremental, decremental, and fully-dynamic algorithms are defined for maintaining visibility polygons under the vertex additions and deletions to simple polygons. In this paper, we extend algorithms from [19,20] to devise fully-dynamic algorithms to maintain the visibility graph of a dynamic simple polygon.

Our algorithm for maintaining the visibility graph is fully-dynamic, in the sense that it maintains the visibility graph amid both the vertex insertions and deletions to the simple polygon. We preprocess the initial simple polygon P that has n vertices to build data structures of size $O(n+|E|)$ in $O(n+|E|\lg|E|)$ time. Here, E is the set comprising of edges of the visibility graph of P. When a vertex v is added to (resp. deleted from) the current simple polygon P', we update the visibility graph in $O(k(\lg n')^2)$ time. Here, n' is the number of vertices of P' and k is the number of updates necessary for the visibility graph due to the insertion (resp. deletion) of v to (resp. from) P'. To our knowledge, this is the first dynamic algorithm for maintaining visibility graphs.

We assume that after adding or deleting any vertex of the current simple polygon, the polygon remains simple. Moreover, it is assumed that every new vertex is added between two successive vertices of the current simple polygon. The initial simple polygon is denoted with P. We use P' to denote the simple polygon just before inserting/deleting a vertex. And, the simple polygon after inserting/deleting a vertex is denoted with P''. Further, we assume that P, P', and P'' are respectively defined with n, n', and n'' vertices. The simple polygon P' just before the start of the current iteration of vertex insertion/deletion is termed the current simple polygon. Whenever we delete a vertex v of P', which is adjacent to vertices v_i and v_{i+1} in P', it is assumed that an edge is introduced between v_i and v_{i+1} after deleting v. Similarly, whenever we insert a vertex v between adjacent vertices v_i and v_{i+1} of P', it is assumed that two edges are introduced: one between v_i and v, and the other between v_{i+1} and v. The boundary of a simple polygon P is denoted with $bd(P)$. Any edge of a visibility graph is termed a visible edge. Unless specified otherwise, the boundary of the simple polygon is assumed to be traversed in the counterclockwise direction, and the vertices are listed in the order they encountered while traversing the simple polygon in counterclockwise order.

Section 2 describes an algorithm to compute all the vertices of a simple polygon P that are visible from any query point belonging to P. Section 3 details a fully-dynamic algorithm to maintain the visibility graph of a given simple polygon. The conclusions are in Sect. 4.

2   Fully-Dynamic Algorithm to Compute Vertices Visible from a Query Point

In this Section, we devise an algorithm to compute all the vertices of a dynamic simple polygon that are visible from a query point belonging to that polygon. This algorithm is needed in maintaining the visibility graph of a dynamic simple polygon.

Given a simple polygon P, the *ray-shooting query* of a query ray $\overrightarrow{r} \in \mathbb{R}^2$ determines the first point of intersection of $\overrightarrow{r}$ with $bd(P)$. Given two points p' and p'' in the interior of a simple polygon P, the *shortest-distance query* between p' and p'' outputs the geodesic Euclidean distance between p' and p''. Both the ray-shooting and geodesic two-point distance queries are required in updating the visibility graph whenever a vertex is inserted or deleted. The following theorem from Goodrich and Tamassia [13] helps in answering ray-shooting as well as the geodesic distance queries between any two points located in the current simple polygon P'.

Proposition 1 ([13] **Theorem 6.3**). *Let T be a planar connected subdivision with n vertices. With $O(n)$-time preprocessing, a fully dynamic data structure of size $O(n)$-space is computed for T that supports point-location, ray-shooting, and shortest-distance queries in $O((\lg n)^2)$ time, and operations InsertVertex, RemoveVertex, InsertEdge, RemoveEdge, AttachVertex, and DetachVertex in $O((\lg n)^2)$ time, all bounds being worst-case.*

Note that the data structures in [13] accommodate the dynamic nature of our simple polygon. In addition, for any two query points q', q'' belonging to simple polygon P, we note that the fully-dynamic data structures in [13] support outputting the first line segment in the geodesic shortest path from q' to q'' in $O((\lg n)^2)$ time.

Given a ray $\overrightarrow{r}$ whose origin q belongs to simple polygon P', the *ray-rotating query* (defined in Chen and Wang [7]) with clockwise (resp. counterclockwise) orientation seeks the first vertex of P' visible to q that will be hit by $\overrightarrow{r}$ when we rotate $\overrightarrow{r}$ by a minimum non-negative angle in clockwise (resp. counterclockwise) direction. The first parameter to ray-rotating query algorithm is the ray and the second parameter determines whether to rotate the input ray by a non-negative angle in clockwise or counterclockwise direction.

Proposition 2 ([7] **Lemma 1**). *By preprocessing a simple polygon having n vertices, a data structure can be built in $O(n)$ time and $O(n)$ space such that each ray-rotating query can be answered in $O(\lg n)$ time.*

To support ray-rotating queries, [7] uses the ray-shooting data structure and two-point shortest distance query data structures from [14]. But for supporting dynamic insertion (resp. deletion) of vertices to (resp. from) the given simple polygon, we use data structures from [13] in place of the ones from [14]. This leads to preprocessing simple polygon P defined with n vertices in $O(n)$ time to compute data structures of $O(n)$ space so that ray-shooting, ray-rotating, and two-point distance queries are answered in $O((\lg n)^2)$ worst-case time.

We preprocess the given initial simple polygon P to answer queries of the following form, amid vertex insertions and deletions to the simple polygon: given a point q in the current simple polygon P' and a cone C with apex at q, our query algorithm outputs all the vertices of P' that belong to the interior of C and are visible from q. This subprocedure is required to update the visibility graph when a vertex is inserted to (resp. deleted from) the simple polygon.

For any two arbitrary rays r_1 and r_2 with their origin at q, the $cone(r_1, r_2)$ comprises of a set S of points in $\mathbb{R}^2$ such that $x \in S$ whenever a ray r with origin at q is rotated with center at q from the direction of r_1 to the direction of r_2 in counterclockwise direction, the ray $\overrightarrow{qx}$ occurs. The $opencone(r_1, r_2)$ is the $cone(r_1, r_2) \setminus \{r_1, r_2\}$.

Let P' be a simple polygon and let q be a point located in $\mathbb{R}^2$. Also, let r_1 and r_2 be two rays with origin at q. The *visvert-inopencone* algorithm listed underneath outputs all the vertices of P' that are visible from q in the $opencone(r_1, r_2)$. This is accomplished by issuing a series of ray-rotating queries in the $opencone(r_1, r_2)$ while essentially sweeping the $opencone(r_1, r_2)$ region to find all the vertices of P' that are visible from q.

We can compute all the vertices that are visible from q by invoking Algorithm 1 (listed underneath) with ray r_1 as both the first parameter as well as the second parameter. This invocation yields all the vertices of P' that are visible from q except those that may lie along the ray r_1. And, one invocation of the ray-shooting query with ray r_1 outputs any vertex along the ray r_1 that is visible from q (Fig. 1).

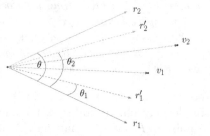

Fig. 1. Illustrating three cones, each correspond to a recursive call in *visvert-inopencone* algorithm

---

Algorithm 1. *visvert-inopencone*(r_1, r_2)

Input : A simple polygon P' and two rays r_1 and r_2 with their origin at a point $q \in P'$

Output: vertices in $P' \cap opencone(r_1, r_2)$ that are visible from q

1: $\theta := \cos^{-1}(\frac{r_1.r_2}{|r_1||r_2|})$
2: Let r_1' and r_2' be the rays with origin at q that respectively make θ_1 and θ_2 counterclockwise angles with ray r_1 such that $\theta_1 < \theta_2 < \theta$.
3: $v_1 :=$ ray-rotating-query$(r_1', counterclockwise)$
4: $v_2 :=$ ray-rotating-query$(r_2', clockwise)$
5: output v_1, v_2
6: *visvert-inopencone*(r_1, r_1')
7: *visvert-inopencone*(r_2', r_2)
8: If $v_1 \neq v_2$ then *visvert-inopencone*$(\overrightarrow{qv_1}, \overrightarrow{qv_2})$

We note that the ray-rotating-query from [7] works only if the input ray does not pass through a vertex of the simple polygon visible from that ray's origin. To take this into account, in Algorithm 1, we perturb r_1 (resp. r_2) to obtain another ray r_1' (resp. r_2') such that $r_1' \in opencone(r_1, r_2)$ (resp. $r_2' \in opencone(r_1, r_2)$) and q is the origin of r_1' (resp. r_2'). After obtaining visible vertices v_1 and v_2 in steps (3) and (4) via ray-rotating queries, in steps (6), (7), and (8) of the Algorithm, we recursively compute vertices visible in three cones.

Lemma 1. *The visvert-inopencone(r_1, r_2) algorithm outputs every vertex of P' in opencone(r_1, r_2) that is visible from q.*

Proof: Consider any non-leaf node b' of the recursion tree. The opencone corresponding to b' is divided into three open cones. Assuming that the set S of vertices of P' within these three open cones are computed correctly, the vertices in S together with the vertices computed at node b' of the recursion tree along the rays that separate these three cones ensure the correctness of the algorithm.

$\square$

Theorem 1. *Our algorithm preprocesses the given simple polygon P having n vertices in $O(n)$ time and computes $O(n)$ sized data structures to facilitate in answering any query of the following form in $O((k+1)(\lg n)^2)$ time: given any point $p \in P'$ and a cone C with apex at p, output all the vertices of P' that lie in C and are visible from p. Here, k is the number of vertices that are visible to v in C and P' is the current simple polygon.*

Note that when there are no vertices to output, our algorithm takes $O((\lg n)^2)$. (Hence, $k + 1$ in the query time complexity.)

3   Fully-Dynamic Algorithm to Maintain the Visibility Graph

In this Section, we devise a dynamic algorithm to maintain the visibility graph of a simple polygon amid insertion and deletion of vertices of the simple polygon. In every iteration, the simple polygon is modified either due to the insertion of a vertex or due to the deletion of a vertex. When a new vertex v is added to the current simple polygon P' or an existing vertex is deleted from P', it results in a new simple polygon P''. When the simple polygon is changed from P' to P'', the visibility graph G' of P' is updated to G'' so that G'' is the visibility graph of P''.

The visibility graph G' of P' is saved with the vertices of P'. In specific, for every vertex v of P', all the edges of G' that incident to P' are stored in a balanced binary search tree (BBST), say B_v. Each leaf of B_v stores a unique visible edge that incident to v. The key for inserting, deleting, and searching for an edge e in B_v is the angle e makes with the positive x-axis. The ordered listing of edges at left to right leaf nodes of B_v is the sorted order with respect to angle visible edges make with the x-axis. Mainly, these ordered lists of visible edges at the nodes of P' are updated in maintaining the visibility graph.

In the preprocessing phase of our algorithm, using [16], we compute the visibility graph G of the initial simple polygon P. For each vertex v of P, we insert the visible edges that incident to v into B_v with the angle each visible edge makes with x-axis as its key in B_v. We also compute data structures that are required for the preprocessing phase of the algorithm in [13] as well as the ones that are necessary for Theorem 1. The space required for the preprocessed data structures include the space for the BBSTs and the space needed for the data structures from [13] and [7].

Lemma 2. *Given a simple polygon P with n vertices, the preprocessing algorithm computes data structures of size $O(n + |E|)$ are computed in $O((n + |E| \lg |E|)$ worst-case time. Here, E is the set comprising of edges of the visibility graph of P.*

Proof: Using the algorithm from Hershberger [16], we compute the visibility graph G of P in $O(n + |E|)$ time. The cost of inserting all the visible edges into BBSTs at vertices of P takes $O(|E| \lg |E|)$ time. The data structures from [13] and [7] are computed in $O(n)$ time.   □

Inserting a Vertex

In this Section, we describe an algorithm to update the visibility graph G' of P' when a vertex v is inserted between vertices v_i and v_{i+1} of P'. We observe that for any edge $e \in G'$, edge e belongs to updated visibility graph G'' if and only if e does not intersect the triangle vv_iv_{i+1}. The following lemma helps in efficiently finding all the edges in G' that intersect with the triangle vv_iv_{i+1}.

Lemma 3. *For every edge ab of $G'(V', E')$ that intersects the triangle vv_iv_{i+1}, either at least one of a or b is visible from v, or there exists a path R in G' from v' to either a or b such that R comprises of a subset of edges that intersect the triangle vv_iv_{i+1} with v' being a vertex visible from v.*

Proof: If either a or b is visible to v, there is nothing to prove. Otherwise, among all the edges of G that intersect the triangles vv_iv_{i+1} and abv, let $a'b' \in E'$ with either $a' = a$ or $b' = b$ is the one that makes the least angle with ab. Refer to Fig. 2. Without loss of generality, we suppose $b' = b$. Note that both a and b are reachable from a' in G' and $a'b$ intersects the triangle vv_iv_{i+1}. Further, a' is closer to v as compared to a, with respect to Euclidean distance. Inductively, a' is reachable in G' from v as described in the lemma statement. Since there are finite vertices, there exists some edge $a''b''$ such that $a''b''$ intersects vv_iv_{i+1} and either a'' or b'' is visible from v.   □

Let I be the set of vertices such that for every vertex r in I, there exists at least one edge of G' that incident to r intersects the triangle vv_iv_{i+1}. Essentially, the above lemma claims that we can find all the edges that need to be removed from G'

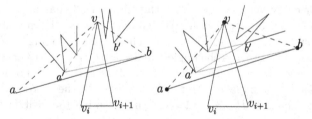

Fig. 2. Illustration for the Lemma 3.

by doing binary searches at the vertices that are visible from v as well as at the vertices in I. By exploiting Lemma 3, our algorithm traverses at most $O(1)$ vertices that are not visible to v and are not in I. In our algorithm, we start with the set of vertices that are visible from v, and from those vertices with a constrained breadth-first traversal of G' we find all the vertices in I. During this traversal, whenever an untraversed vertex v' is encountered, we do the binary search at v' to find all the visible edges in G' that are needed to be removed from G'. We detail these steps in the following.

By using Theorem 1, we compute all the vertices of P' that are visible to v. Let r be a ray with origin at v and oriented in any arbitrary direction. We invoke *visvert-inopencone* procedure (listed in Sect. 2) with r as both the first and second arguments. This invocation outputs all the vertices that are visible to v except for the one that may be located on the ray r. Hence, we do the ray-shooting with ray r to determine whether there is any vertex that lies on ray r and is visible to v. A new BBST B_v to store visible edges that incident to v is created, and all visible edges that incident to v are saved in B_v. Further, for every vertex v' that is determined to be visible from v, we add the visible edge $v'v$ to BBST at v'. This completes introducing new visible edges into G''.

Next, our algorithm removes edges in G' whose endpoints are not mutually visible in P'. As mentioned, we use Lemma 3 in efficiently determining all the edges that needed to be removed. Let S_1 be the set of vertices that are visible from v. For each vertex v' in S_1, we find all the visible edges intersecting the triangle vv_iv_{i+1} by searching BBST at v'. W.l.o.g., suppose the ray $\overrightarrow{v'v_i}$ makes larger angle with ray $\overrightarrow{v'v}$ as compared to ray $\overrightarrow{v'v_{i+1}}$. For any vertex v', let set $S_{v'}$ be the set of vertices of P' that are in the cone formed by rays $\overrightarrow{v'v}$ and $\overrightarrow{v'v_i}$. The set $S_{v'}$ is determined by searching the BBST $B_{v'}$. Further, for every vertex $v'' \in S_{v'}$, we remove the visible edge $v'v''$ from both the BBSTs at v' and v''. For $i \geq 1$, let S_{i+1} be the $\bigcup_{v' \in S_i} S_{v'}$. Analogously, for $i > 1$, for every vertex $v' \in S_i$ and $v' \notin S_j$ with $j < i$, we find the set S_{i+1} comprising of all the visible edges that intersect vv_iv_{i+1}. Our algorithm terminates whenever for some k every vertex in S_k belongs to some S_j for $j < k$. From the Lemma 3, every edge that is required to be removed from G' is guaranteed to be found in the mentioned searches in BBSTs.

Lemma 4. *For every two vertex $v_i, v_j \in P''$, the line segment v_iv_j is an edge in the visibility graph of P'' if and only if v_i and v_j are mutually visible in P''.*

Lemma 5. *When any vertex v is inserted to the current simple polygon P', the insertion algorithm updates the visibility graph of P' in $O(k(\lg n')^2)$ time. Here, n' be the number of vertices of P' and k is the number of combinatorial changes required to the visibility graph G' of P' upon the insertion of a vertex v to P'.*

Proof: From Theorem 1, our algorithm computes all the vertices that are visible from v in $O((k'+1)(\lg n')^2)$ time, where k' is the number of vertices that are visible from v. Every edge that intersects the triangle vv_iv_{i+1} is computed with a search in a BBST. There are at most $O(1)$ searches in BBSTs which do not contribute to removing any visible edge. If there are k'' number of edges that intersect the triangle vv_iv_{i+1}, it takes $O(k'' \lg n')$ time to search and delete nodes from BBSTs. We note that k is the sum of k' and k''. □

Deleting a Vertex

Let v be the vertex that is removed from the current simple polygon. Then for every pair of vertices v_i, v_j of P', if the line segment v_iv_j intersects the triangle

vv_iv_{i+1}, the line segment v_iv_j needs to be introduced as a visible edge in G''. Further, edges in G' that incident to v are needed to be removed from G'.

Let I be the set of vertices such that for every vertex r in I, there exists at least one edge of G' that incident to r intersects the triangle vv_iv_{i+1}. The Lemma 3 requires us to find all the edges that need to be added to G' by exploring only those vertices that are visible from v as well as those in I. Like in the insertion algorithm, we iteratively compute the vertices in I. As the algorithm progress, vertices are incrementally added to I while invoking the *visvert-inopencone* procedure at the vertices of I as well as at $O(1)$ vertices which do not contribute to I. Whenever a vertex v' is encountered that was not traversed in the update procedure, we invoke the *visvert-inopencone* procedure to find all the visible edges in G'' that incident to v''. The deletion algorithm is detailed below.

Let S_1 be the endpoints of visible edges in G' that incident to v. For each vertex v' in S_1, we find all the visible edges in G'' that intersect the triangle vv_iv_{i+1}. W.l.o.g., suppose the ray $\overrightarrow{v'v_i}$ makes larger angle with the ray $\overrightarrow{v'v}$ as compared to ray $\overrightarrow{v'v_{i+1}}$. For any vertex v', let set $S_{v'}$ be the set of vertices of P' that are in the cone formed by rays $\overrightarrow{v'v}$ and $\overrightarrow{v'v_i}$. The set $S_{v'}$ is computed by invoking the *visvert-inopencone* procedure listed in Sect. 2. For every vertex $v'' \in S_2$, we introduce the visible edge $v'v''$ into BBSTs at v' and v''. For $i \geq 1$, let S_{i+1} be the $\bigcup_{v' \in S_i} S_{v'}$. Analogously, for $i > 1$, for every vertex $v' \in S_i$ and $v' \notin S_j$ with $j < i$, we find the set S_{i+1} of vertices that are visible from v' in the cone defined by rays $\overrightarrow{v'v}$ and $\overrightarrow{v'v_i}$ (or, $\overrightarrow{v'v_{i+1}}$). Our algorithm terminates whenever for some k every vertex in S_k belongs to some S_j for $j < k$.

Lemma 6. *For every two vertices $v_i, v_j \in P''$, the line segment v_iv_j is an edge in the visibility graph of P'' if and only if v_i and v_j are mutually visible in P''.*

Lemma 7. *When any vertex v is deleted from the current simple polygon P', the deletion algorithm updates the visibility graph of P' in $O(k(\lg n')^2)$ time. Here, n' be the number of vertices of P', and k is the number of combinatorial changes required in the visibility graph G' of P' upon the deletion of a vertex v from P'.*

Proof: Our algorithm removes all the visible edges that incident to v in $O(k'(\lg n'))$ time, where k' is the number of visible edges in G' that incident at v. Let k'' be the number of visible edges that need to be introduced into G'. From Theorem 1, we find all these edges in $O((k'' + 1)(\lg n')^2)$ time. Further, new visible edges are inserted into respective BBSTs in $O((k'' + 1)(\lg n'))$ time. We note that k is the sum of k' and k''. □

Theorem 2. *Given a simple polygon P with n vertices, we compute data structures of size $O(n+|E|)$ in $O(n+|E| \lg |E|)$ time to support the following. Here, E is the set comprising of edges in the visibility graph G of P. Let P' be the current simple polygon defined with n' vertices. Let G' be the visibility graph of P'. For any vertex v inserted to (resp. deleted from) P', updating G' takes $O(k(\log n')^2)$*

time, where k is the number of updates required to the visibility graph due to the insertion (resp. deletion) of v.

4 Conclusions

We have presented algorithms to maintain the visibility graph of a dynamic simple polygon. To our knowledge, this is the first result that gives an algorithm for maintaining the visibility graph of a dynamic simple polygon. The worst-case time complexity to update the visibility graph of a simple polygon is just a poly-log multiplicative factor to the number of combinatorial changes required to the visibility graph. In addition, as part of the algorithm for updating the visibility graph, we detailed a subprocedure to compute the vertices that are visible from any given query point belonging to a dynamic simple polygon. This algorithm could be of independent interest. We see lots of scope for future work in devising dynamic algorithms in the areas of computing visibility, art gallery, minimum link path, and geometric shortest path problems.

References

1. Aronov, B., Guibas, L.J., Teichmann, M., Zhang, L.: Visibility queries and maintenance in simple polygons. Discret. Comput. Geom. **27**(4), 461–483 (2002)
2. Asano, T., Asano, T., Guibas, L.J., Hershberger, J., Imai, H.: Visibility of disjoint polygons. Algorithmica **1**(1), 49–63 (1986)
3. Bar-Yehuda, R., Chazelle, B.: Triangulating disjoint Jordan chains. Int. J. Comput. Geom. Appl. **4**(4), 475–481 (1994)
4. Bose, P., Lubiw, A., Munro, J.I.: Efficient visibility queries in simple polygons. Comput. Geom. **23**(3), 313–335 (2002)
5. Chazelle, B.: Triangulating a simple polygon in linear time. Discret. Comput. Geom. **6**, 485–524 (1991)
6. Chen, D.Z., Wang, H.: Visibility and ray shooting queries in polygonal domains. Comput. Geom. **48**(2), 31–41 (2015)
7. Chen, D.Z., Wang, H.: Weak visibility queries of line segments in simple polygons. Comput. Geom. **48**(6), 443–452 (2015)
8. Davis, L.S., Benedikt, M.L.: Computational models of space: Isovists and Isovist fields. Comput. Graph. Image Process. **11**(1), 49–72 (1979)
9. ElGindy, H.A., Avis, D.: A linear algorithm for computing the visibility polygon from a point. J. Algorithms **2**(2), 186–197 (1981)
10. Ghosh, S.K.: Computing the visibility polygon from a convex set and related problems. J. Algorithms **12**(1), 75–95 (1991)
11. Ghosh, S.K.: Visibility Algorithms in the Plane. Cambridge University Press, New York (2007)
12. Ghosh, S.K., Mount, D.M.: An output-sensitive algorithm for computing visibility graphs. SIAM J. Comput. **20**(5), 888–910 (1991)
13. Goodrich, M.T., Tamassia, R.: Dynamic ray shooting and shortest paths in planar subdivisions via balanced geodesic triangulations. J. Algorithms **23**(1), 51–73 (1997)

14. Guibas, L.J., Hershberger, J.: Optimal shortest path queries in a simple polygon. J. Comput. Syst. Sci. **39**(2), 126–152 (1989)
15. Guibas, L.J., Motwani, R., Raghavan, P.: The robot localization problem. SIAM J. Comput. **26**(4), 1120–1138 (1997)
16. Hershberger. J.: Finding the visibility graph of a simple polygon in time proportional to its size. In: Proceedings of the Third Annual Symposium on Computational Geometry, pp. 11–20 (1987)
17. Hershberger, J.: An optimal visibility graph algorithm for triangulated simple polygons. Algorithmica **4**, 141–155 (1989)
18. Inkulu, R., Kapoor, S.: Visibility queries in a polygonal region. Comput. Geom. **42**(9), 852–864 (2009)
19. Inkulu, R., Thakur, N.: Incremental algorithms to update visibility polygons. In: Proceedings of Conference on Algorithms and Discrete Applied Mathematics, pp. 205–218 (2017)
20. Inkulu, R., Sowmya, K.: Dynamic algorithms for visibility polygons. CoRR, abs/1704.08219 (2017)
21. Joe, B., Simpson, R.: Corrections to Lee's visibility polygon algorithm. BIT Numer. Math. **27**(4), 458–473 (1987)
22. Kapoor, S., Maheshwari, S.N.: Efficient algorithms for Euclidean shortest path and visibility problems with polygonal obstacles. In: Proceedings of Symposium on Computational Geometry, pp. 172–182 (1988)
23. Kapoor, S., Maheshwari, S.N.: Efficiently constructing the visibility graph of a simple polygon with obstacles. SIAM J. Comput. **30**(3), 847–871 (2000)
24. Lee, D.T.: Proximity and reachability in the plane. Ph.D. thesis, University of Illinois at Urbana-Champaign (1978). Ph.D. thesis and Technical report ACT-12
25. Lee, D.T.: Visibility of a simple polygon. Comput. Vis. Graph. Image Process. **22**(2), 207–221 (1983)
26. Overmars, M.H., Welzl, E.: New methods for computing visibility graphs. In: Proceedings of the Fourth Annual Symposium on Computational Geometry, pp. 164–171 (1988)
27. Pocchiola, M., Vegter, G.: Topologically sweeping visibility complexes via pseudo-triangulations. Discret. Comput. Geom. **16**(4), 419–453 (1996)
28. Sharir, M., Schorr, A.: On shortest paths in polyhedral spaces. SIAM J. Comput. **15**(1), 193–215 (1986)
29. Vegter, G.: The visibility diagram: a data structure for visibility problems and motion planning. In: Proceedings of Scandinavian Workshop on Algorithm Theory, pp. 97–110 (1990)
30. Welzl, E.: Constructing the visibility graph for n-line segments in $O(n^2)$ time. Inf. Process. Lett. **20**(4), 167–171 (1985)
31. Zarei, A., Ghodsi, M.: Efficient computation of query point visibility in polygons with holes. In: Proceedings of the Symposium on Computational Geometry, pp. 314–320 (2005)

On the Bitprobe Complexity of Two Probe Adaptive Schemes Storing Two Elements

Deepanjan Kesh[1(✉)] and Vidya Sagar Sharma[2]

[1] Department of Computer Science and Engineering,
Indian Institute of Technology Guwahati, Guwahati 781039, Assam, India
deepkesh@iitg.ac.in
[2] School of Technology and Computer Science,
Tata Institute of Fundamental Research, Homi Bhabha Road, Navy Nagar, Colaba,
Mumbai 400005, India
vidya.sagar@tifr.res.in

Abstract. In the adaptive bitprobe model, consider the following set membership problem – store subsets of size at most two from an universe of size m, and answer membership queries using two bitprobes. Radhakrishnan *et al.* [3] proposed a scheme for the problem which takes $\mathcal{O}(m^{2/3})$ amount of space, and conjectured that this is also the lower bound for the problem. We propose a proof of the lower bound for the problem, but for a restricted class of schemes. This proof hopefully makes progress over the ideas proposed by Radhakrishnan *et al.* [3] and [4] towards the conjecture.

Keywords: Data structures · Bitprobe model · Adaptive scheme · Lower bound

1 Introduction

The static membership problem in the bitprobe model asks the following question – given a subset $\mathcal{S}$ from a universe $\mathcal{U}$ of elements, how efficiently can it be stored such that membership queries can be answered correctly if only a limited number of bits of our data structure can be probed. The space requirement can be parameterised by the size of our subset $\mathcal{S}$ and the number of probes allowed. If the size of the universe $\mathcal{U}$ is m, the size of our subset is at most n, and the number of bit probes allowed is at most t, Radhakrishnan *et al.* [4] introduced the following notation to denote the amount of space required by the data structure – $s(n, m, t)$. The schemes in this model can be further divided into two distinct classes depending on whether or not the location of a probe depends on the answer received in the previous probes. If the location of the probes is independent of the answers we get, it is called a *non-adaptive scheme*, otherwise it is called an *adaptive scheme*. The notations for the space required in the two cases are $s_N(n, m, t)$ and $s_A(n, m, t)$, respectively.

© Springer Nature Switzerland AG 2019
S. P. Pal and A. Vijayakumar (Eds.): CALDAM 2019, LNCS 11394, pp. 53–64, 2019.
https://doi.org/10.1007/978-3-030-11509-8_5

1.1   The Bitprobe Model

We define the bitprobe model in the context of two probe adaptive data structure where the size of the subset is at most two. We store the elements in our universe in a hierarchical data structure, as depicted in the diagram above (Fig. 1). It consists of three tables – $\mathcal{A}, \mathcal{B}$, and $\mathcal{C}$. Any element, say x, of $\mathcal{U}$ has a location in each of the three tables, which will be denoted by $\mathcal{A}(x), \mathcal{B}(x)$, and $\mathcal{C}(x)$. There are two aspects to the bitprobe model – storage scheme and query scheme.

Given a subset $\mathcal{S}$ of size at most two, the storage scheme sets the bits in the three tables such that the query scheme, described below, can correctly identify the members of $\mathcal{S}$.

The query scheme works as follows. If we want to determine whether an element x is a member of $\mathcal{S}$, we first query in table $\mathcal{A}$ at the bit location $\mathcal{A}(x)$. If the bit stored is 0, then we query table $\mathcal{B}$, else we query table $\mathcal{C}$. If the next query corresponding to the element's location returns a 1, we declare that the element belongs to $\mathcal{S}$. We declare otherwise if the bit stored is 0. This tree is also known as the *decision tree* of an element (Fig. 1).

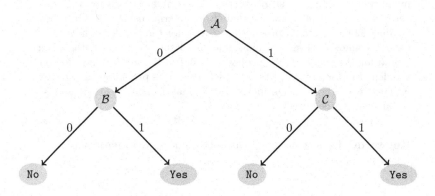

Fig. 1. The decision tree of an element.

1.2   Problem Statement

In this paper, we attempt to prove the following conjecture, albeit for a restricted class of schemes.

Conjecture 1. $s_A(2, m, 2) = \Omega(m^{2/3})$ [Radhakrishnan *et al.* [3]].

Buhrman *et al.* [1] showed, using probabilistic arguments, the existence of an adaptive scheme for $s_A(2, m, 2)$ that uses $\mathcal{O}(m^{3/4})$ amount of space. Radhakrishnan *et al.* [3] greatly improved the result when they proposed a scheme that was not only deterministic but also used just $\mathcal{O}(m^{2/3})$ amount of space. In the same paper, they proved that $\Omega(m^{2/3})$ is indeed the lower bound for a restricted class of schemes, and conjectured that the same holds true for the general class. A further attempt at the lower bound was made by Radhakrishnan et al. [4] where they proved that $\Omega(m^{4/7})$ space was necessary for any such scheme.

1.3 Terminologies

We now establish certain notations that we are going to use throughout the rest of the paper. Elements that query the same bit location in table $\mathcal{A}$ are said to belong to the same *block*. So, each bit location in $\mathcal{A}$ corresponds to a block of elements of our universe. Alternatively, we will say for elements that belong to the same block as *mapping* to the same block, or mapping to the same bit location.

All elements within a block are numbered $1, 2, 3, \ldots$, which we will refer to as the *index* of an element. So, any element of the universe $\mathcal{U}$ can be referred to by a tuple (u, v) where u is the block number to which it belongs, and v is its index within that block.

In tables $\mathcal{B}$ and $\mathcal{C}$, the collection of elements mapping to the same location are said to belong to the same *set*. Alternatively, two elements that query the same bit location are said to belong to the same set. So, there is a bit corresponding to every set in tables $\mathcal{B}$ and $\mathcal{C}$.

1.4 Defining the Class of Schemes

We now state the one restriction that defines the class of schemes our proof addresses.

Restrictions 1. *If two elements belong to the same set either in table $\mathcal{B}$ or in table $\mathcal{C}$, then their indices are the same.*

We also make the following assumptions about our class of schemes for the sake of simplicity based on the ideas found in the lower bound proofs in Sect. 4 of Radhakrishnan et al. [3] and Sects. 3 and 4 of Radhakrishnan et al. [4]. These assumptions do not affect the final result.

Assumptions 2. *1. All elements of our universe have the same decision tree.*
2. The three tables $\mathcal{A}, \mathcal{B},$ and $\mathcal{C}$ do not share any bit.
3. All the blocks in table $\mathcal{A}$ are of equal size. Let that size be b.

1.5 Motivation

The motivation for Restriction 1 comes from the design patterns of the various explicit schemes proposed in the literature for adaptive bitprobe data structures. Though the notion of blocks and indices, as defined in Subsect. 1.3, may be not be explicit in those schemes, yet the way the mapping functions have been defined in each of those cases suggest that they respect Restriction 1.

To take an example, in Radhakrishnan et al. [3], we have the notion of blocks and reserving bit vectors for multiple blocks in the scheme for

$s_A(n, m, \lceil \log n+1 \rceil +1)$ in Theorem 1, characteristic vectors for multiple blocks and superblocks in Theorem 2 for $s_A(2, m, 2)$, for $s_A(n, m, 1 + \lceil \log(\lfloor n/2 \rfloor + 2) \rceil)$ in Theorem 3, and in Theorem 4 where the relationship between the parameters n, m, and t are slightly more involved. These notions of blocks, superblocks, bit vectors, and characteristic vectors for blocks are exactly the notion of blocks and indices as defined in Sect. 1.3, but in a different guise. Furthermore, the schemes, as stated in [3], also implies that all of them satisfy Restriction 1. Table 1 lists a few more schemes in the literature satisfying the restriction.

Table 1. A few more schemes satisfying Restriction 1.

Source	n	t	Theorems
Lewenstein *al.* [2]	2	3	Theorem 1
	2	$t \geq 3$	Theorem 2
	$n > 2$	$n + 1$	Theorem 3
Radhakrishnan *et al.* [4]	2	$t > 2$	Theorem 1
	$n \geq 2$	$t > n$	Theorem 2

1.6 Our Results

Our result, discussed in the following section, is that any scheme so designed that it satisfies Restriction 1 must use $\Omega(m^{2/3})$ space (Theorem 2). The proof relies on the notion of *bad elements* (Definition 1) and *good locations* (Definition 2), and how frequently they occur in any given scheme. It is our hope that a more general definition of the aforementioned terms would settle the conjecture for the general class of schemes.

2 Lower Bound

In this section we proceed to prove the main result of this paper – establishing the lower bound for the class of all schemes satisfying Restriction 1.

2.1 Notations

By Restriction 1, elements belonging to the same set in tables $\mathcal{B}$ and $\mathcal{C}$ have the same indices. We use $\mathcal{B}_i$ to denote the family of all sets formed by elements whose indices are i. As there are b elements in each block, so the indices range from 1 to b, inclusive. So, there will be b such families in each of the tables $\mathcal{B}$ and $\mathcal{C}$. So, the various families range from $\mathcal{B}_1, \mathcal{B}_2, \ldots, \mathcal{B}_b$ and $\mathcal{C}_1, \mathcal{C}_2, \ldots, \mathcal{C}_b$, one family each in tables $\mathcal{B}$ and $\mathcal{C}$ for each index.

If S is a set in the family $\mathcal{B}_i$, then the elements in S will look like $\{(u_1, i), (u_2, i), \dots\}$, i.e. all members of S will have index i. Since the second component is common for all elements in the set, we will omit it and refer to an element only by its block number. As as example, if the set $S \in \mathcal{B}_1$ has the members $\{a, b, c, \dots\}$, then the actual members of S are $\{(a, 1), (b, 1), (c, 1), \dots\}$. The index can be deduced from the family to which the set belongs.

Observation 3. $\bigcup_{S \in \mathcal{B}_i} S = \bigcup_{S \in \mathcal{C}_i} S = \mathcal{A}, \text{for } 1 \leq i \leq b.$

Proof. Every block of table $\mathcal{A}$ contains exactly one element with index i, and all elements with with index i must belong to some set in the families $\mathcal{B}_i$ and $\mathcal{C}_i$, hence the equality.

Without loss of generality, let the families in tables $\mathcal{B}$ and $\mathcal{C}$ corresponding to the indices 1 and 2 be such that

$$|\mathcal{B}_i| + |\mathcal{C}_i| \leq |\mathcal{B}_j| + |\mathcal{C}_j|, \text{ where } i = 1, 2 \text{ and } j \geq 3. \tag{1}$$

So the space corresponding to the first two families is minimal among all the families. Among the first two families, we also have

$$|\mathcal{C}_1| \leq |\mathcal{C}_2|. \tag{2}$$

For quick and easy reference, we will use separate letters to denote the members of these families.

$$\mathcal{B}_1 = \{W_1, W_2, \dots\};$$
$$\mathcal{B}_2 = \{X_1, X_2, \dots\};$$
$$\mathcal{C}_1 = \{Y_1, Y_2, \dots\}; \text{and}$$
$$\mathcal{C}_2 = \{Z_1, Z_2, \dots\}.$$

2.2   Bad Elements and Good Locations

Let us now focus on the classes corresponding to the first and second indices, namely $\mathcal{B}_1, \mathcal{B}_2, \mathcal{C}_1$, and $\mathcal{C}_2$. We define the notion of a bad element of $\mathcal{C}_1$ as follows (Fig. 2).

Definition 1. *An element e of a set Y in $\mathcal{C}_1$ is a bad element for Y if there exists distinct elements a and b, different from e, such that :*

1. *Elements e, a, and b belong to the same set in $\mathcal{B}_1$.*
2. *Elements e and a belong to the same set in $\mathcal{C}_1$, namely Y.*
3. *Elements e and b belong to the same set in $\mathcal{C}_2$.*

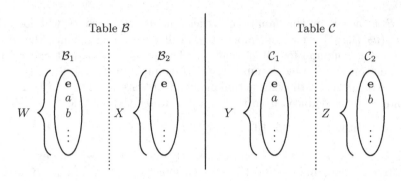

Fig. 2. In this arrangement, e is a bad element for the set Y.

Why is e bad for $Y \in C_1$? Consider the set $X \in B_2$ to which e belongs. If e is bad, then we will establish in Lemma 1 that $X \cap Y = \{e\}$. Why is this situation not agreeable? To come up with a data structure that packs in more and more elements in less and less bits, the natural situation would be to have large intersections between sets. Whereas if there are 10 bad elements in $Y \in C_1$, then those 10 elements must be in separate sets of B_2, and hence $|B_2| \geq 10$. So, intuitively, if there are too many bad elements in Y, then the cardinality of B_2 would be large. As each set corresponds to one bit in our data structure, the size of the data structure would grow.

Lemma 1. *Let e be an element which belongs to the sets $X \in B_2$ and $Y \in C_1$. If e is bad for Y in C_1, then $X \cap Y = \{e\}$.*

Proof. As e is a bad element for the set Y in C_1, a set $W \in B_1$ contains the three elements e, a, and b. Without loss of generality, let the set $Y \in C_1$ contain the elements e and a. Furthermore, a set $Z \in C_2$ contains the elements e and b. We will prove the statement of the lemma by contradiction.

Let f be different from e, a, and b such that $X \cap Y = \{e, f\}$. Recall that $X \in B_2$ implies $(e, 2), (f, 2) \in X$. Similarly, $Y \in C_1$ implies $(e, 1), (f, 1) \in Y$. We will show that under the current arrangement, we cannot store the set $S = \{(a, 1), (e, 2)\}$.

Suppose we set the block a in table A to 0. So, when we query about $(a, 1)$, according to the decision tree (Fig. 1) the second query would be in table B at location W in B_1. So, the bit corresponding to W must be set to 1.

This implies that we cannot set the blocks e and b in table A to 0. If we do set block e to 0, the queries corresponding to $(e, 1)$ will get 0 against the first query, consequently it will query table B at location W and get a 1. So, we would incorrectly deduce that $(e, 1)$ is a member of S. We can similarly argue about block b.

So, the blocks e and b are set to 1, and hence, the second query for the element $(e, 2)$ will be made in table C. Recall that $(e, 2)$ belongs to set Z in C_2. As $(e, 2)$ is an element we want to store, so the bit corresponding to the set Z must be set to 1.

Under this assignment of bits, let us consider the membership query for the element $(b, 2)$. We first query block b in table $\mathcal{A}$ and we will get a 1, and hence we will next query table $\mathcal{C}$. The element $(b, 2)$ belongs to the set Z in this table, and the bit corresponding to Z has been set to 1. Thus the element $(b, 2)$ would be incorrectly deduced to be a member of $\mathcal{S}$.

We, thus, deduce that our initial supposition of setting the block a in table $\mathcal{A}$ to 0 is wrong. Similarly, we can prove that one cannot set the block a to 1 either, because if we do set it to 1, then we can infer that the element $(f, 2)$ will create a problem.

To avoid this situation, $X \cap Y$ cannot contain anything else besides e.

We now define the notion of a *good location*, which is set against the concept of *bad elements*.

Definition 2. *A location or a set of $\mathcal{B}_2$, say X, is said to be a* good *location for a set Y of $\mathcal{C}_1$ if X does not contain any bad element of Y. Good locations of $\mathcal{C}_1$ is the union of the good locations of all of its sets.*

2.3   Counting the Bad Elements

In this section, we determine the number of bad elements that can be created due to the members of some set $W \in \mathcal{B}_1$.

Lemma 2. *If $|W| < |\mathcal{C}_1| + |\mathcal{C}_2|$, then there is a way of arranging the elements of W in $\mathcal{C}_1$ and $\mathcal{C}_2$ such that no bad elements are formed.*

Proof. Let us recall that $|\mathcal{C}_1| \leq |\mathcal{C}_2|$ (Eq. 2).

We will calculate the maximum number of elements of W one can accommodate in the sets of $\mathcal{C}_1$ and $\mathcal{C}_2$ without creating a bad element. We pick two elements of W and put them together in a set of $\mathcal{C}_1$ and in a set of $\mathcal{C}_2$, and we do this $|\mathcal{C}_1| - 1$ times. So, we have engaged $2(|\mathcal{C}_1| - 1)$ elements of W, and $|\mathcal{C}_1| - 1$ members each of $\mathcal{C}_1$ and $\mathcal{C}_2$.

We now put one element of W in each of the remaining sets of $\mathcal{C}_2$, and put all of these elements together in the remaining set of $\mathcal{C}_1$. So, the total number of elements of W we have utilised so far is $2(|\mathcal{C}_1| - 1) + \{|\mathcal{C}_2| - (|\mathcal{C}_1| - 1)\} = |\mathcal{C}_1| + |\mathcal{C}_2| - 1$. One can easily verify that this arrangement does not create any bad element for any of the sets of $\mathcal{C}_1$ due to $W \in \mathcal{B}_1$.

Lemma 3. *If $|W| = |\mathcal{C}_1| + |\mathcal{C}_2| + 1$, then there is no way of arranging the elements of W in $\mathcal{C}_1$ and $\mathcal{C}_2$ such that no bad elements are formed.*

Proof. We will prove the above statement by contradiction. Let us assume that there exists some way of arranging the elements of W over the sets $\mathcal{C}_1$ and $\mathcal{C}_2$ such that no bad elements are created. Let us consider the distribution of the elements of W in the various sets of $\mathcal{C}_1$. We capture it using the following notations –

n_0 – the number of sets of $\mathcal{C}_1$ which contain no element of W;

n_1 – the number of sets of $\mathcal{C}_1$ which contain exactly one element of W;

n_2 – the number of sets of $\mathcal{C}_1$ which contain exactly two elements of W; and

$n_{\geq 3}$ – the number of sets of $\mathcal{C}_1$ which contain three or more elements of W.

So, we have $n_0 + n_1 + n_2 + n_{\geq 3} = |\mathcal{C}_1|$. Moreover, due to the size of W, there has to be at least one set in $\mathcal{C}_1$ that contains three elements or more. Consequently, $n_{\geq 3} > 0$, and $n_0, n_1, n_2 \geq 0$.

The sets containing 0 elements of W cannot give rise to any bad element. The elements that are alone in the sets of $\mathcal{C}_1$ also cannot give rise to any bad elements, no matter how they are placed in $\mathcal{C}_2$. We can assume that they are all together in a single set of $\mathcal{C}_2$.

Consider the sets of $\mathcal{C}_1$ that have pairs of elements of W. The only way to avoid being a bad element for a pair is to be together or individually in the sets of $\mathcal{C}_2$ with no other element of W being present. So, the number of sets these elements would occupy in $\mathcal{C}_2$ is at least n_2. The number of elements of W considered so far is $n_1 + 2n_2$. The number of sets of $\mathcal{C}_2$ occupied so far is at least $1 + n_2$ if $n_1 > 0$, else it is just n_2.

Any element that is part of a set of $\mathcal{C}_1$ that contains three or more elements of W must be in such a set of $\mathcal{C}_2$ which does not contain any other element of W. The number of such elements are $|W| - (n_1 + 2n_2)$, and the number of sets they occupy of $\mathcal{C}_2$ is also $|W| - (n_1 + 2n_2)$.

So, the number of sets of $\mathcal{C}_2$ must satisfy the following inequality.

$$|\mathcal{C}_2| \geq 1 + n_2 + |W| - (n_1 + 2n_2), \text{ if } n_1 > 0;$$
$$|\mathcal{C}_2| \geq n_2 + |W| - 2n_2, \qquad \text{otherwise .}$$

Substituting the values of $|W|$ and $|\mathcal{C}_1|$ in the above inequalities and simplifying, we get the following absurdities –

$$2 + n_0 + n_{\geq 3} \leq 0, \text{ if } n_1 > 0;$$
$$1 + n_0 + n_{\geq 3} \leq 0, \text{ otherwise,}$$

which concludes the proof.

Lemma 4. *If $|W| = |\mathcal{C}_1| + |\mathcal{C}_2|$, then depending on the sizes of $\mathcal{C}_1$ and $\mathcal{C}_2$ a bad element may or may not be created.*

Proof. The proof will follow along the lines of the proof of Lemma 3. We will borrow the notations used in that proof.

If we proceed according to the previous proof, the number of sets of $\mathcal{C}_2$ must satisfy the same inequalities.

$$|\mathcal{C}_2| \geq 1 + n_2 + |W| - (n_1 + 2n_2), \text{ if } n_1 > 0;$$
$$|\mathcal{C}_2| \geq n_2 + |W| - 2n_2, \qquad \text{otherwise.}$$

Substituting the current values of $|W|$ and $|\mathcal{C}_1|$ in the above inequalities and simplifying, we get the following inequalities.

$$1 + n_0 + n_{\geq 3} \leq 0, \text{ if } n_1 > 0;$$
$$n_0 + n_{\geq 3} \leq 0, \text{ otherwise.}$$

One can observe that if $|\mathcal{C}_1| = |\mathcal{C}_2|$, then $n_{\geq 3} \geq 0$. Now, if both n_0 and $n_{\geq 3}$ are 0, the second inequality will be satisfied and no bad elements will be created.

If $|\mathcal{C}_1| < |\mathcal{C}_2|$, then one can easily see that $n_{\geq 3} > 0$. So, neither of the inequalities can be satisfied, implying that one or more bad elements will be created.

Theorem 1. *If $|W| \geq |\mathcal{C}_1| + |\mathcal{C}_2|$, then at least $|W| - (|\mathcal{C}_1| + |\mathcal{C}_2|)$ bad elements will be created in $\mathcal{C}_1$ due to $W \in \mathcal{B}_1$.*

Proof. We will prove by contradiction. Accordingly, let us assume that the number of bad elements created is x and $x < |W| - (|\mathcal{C}_1| + |\mathcal{C}_2|)$.

We know that each member of W is a number denoting some block of table $\mathcal{A}$. Let us remove the blocks corresponding to the x bad elements from our universe. In this new universe, let the corresponding notations be $W', \mathcal{B}_1', \mathcal{C}_1'$, and $\mathcal{C}_2'$. The new $W' \in \mathcal{B}_1'$ will now contain $|W| - x$ elements, and none of these elements are bad for the sets of $\mathcal{C}_1'$. But, $|W'| = |W| - x > |\mathcal{C}_1| + |\mathcal{C}_2| = |\mathcal{C}_1'| + |\mathcal{C}_2'|$, and we have managed to arrange them in the sets of $\mathcal{C}_1'$ and $\mathcal{C}_2'$ without creating any bad element. This is absurd due to Lemma 3.

2.4   Counting the Good Locations

Let the elements of our universe be so distributed among the sets of $\mathcal{B}_1$ that there are g sets whose cardinality is $< |\mathcal{C}_1| + |\mathcal{C}_2|$. These sets do not necessarily give rise to bad elements (Lemma 2). For the sake of simplicity, let these sets be $W_1, W_2, \ldots, W_g$. On the other hand, if $W_i(i > g)$ be such a set of $\mathcal{B}_1$ whose cardinality is $\geq |\mathcal{C}_1| + |\mathcal{C}_2|$, then at least $|W_i| - (|\mathcal{C}_1| + |\mathcal{C}_2|)$ bad elements are created (Theorem 1).

We calculate the total number of bad elements created due to the first $|\mathcal{B}_1| - 1$ members of $\mathcal{B}_1$. That number is

$$\geq \sum_{i=g+1}^{|\mathcal{B}_1|-1} \left\{ |W_i| - \Big(|\mathcal{C}_1| + |\mathcal{C}_2| \Big) \right\} = \left\{ \sum_{i=g+1}^{|\mathcal{B}_1|-1} |W_i| \right\} - \Big(|\mathcal{B}_1| - 1 - g \Big) \Big(|\mathcal{C}_1| + |\mathcal{C}_2| \Big).$$

Observation 3 tells us that

$$|\mathcal{A}| = \sum_{i=1}^{i=|\mathcal{B}_1|} |W_i| = \sum_{i=1}^{i=g} |W_i| + \left(\sum_{i=g+1}^{i=|\mathcal{B}_1|-1} |W_i| \right) + |W_{|\mathcal{B}_1|}|$$

$$\leq g\Big(|\mathcal{C}_1| + |\mathcal{C}_2| - 1 \Big) + \left(\sum_{i=g+1}^{i=|\mathcal{B}_1|-1} |W_i| \right) + |W_{|\mathcal{B}_1|}|$$

$$\implies \sum_{i=g+1}^{i=|\mathcal{B}_1|-1} |W_i| \geq |\mathcal{A}| - g\Big(|\mathcal{C}_1| + |\mathcal{C}_2| - 1 \Big) - |W_{|\mathcal{B}_1|}|.$$

Substituting it in the inequality for the number of bad elements, we get

$$\geq |\mathcal{A}| - g\Big(|\mathcal{C}_1| + |\mathcal{C}_2| - 1 \Big) - |W_{|\mathcal{B}_1|}| - \Big(|\mathcal{B}_1| - 1 - g \Big) \Big(|\mathcal{C}_1| + |\mathcal{C}_2| \Big)$$

$$= |\mathcal{A}| - |W_{|\mathcal{B}_1|}| + g - \Big(|\mathcal{B}_1| - 1 \Big) \Big(|\mathcal{C}_1| + |\mathcal{C}_2| \Big).$$

The total number of locations of $\mathcal{B}_2$ available to the sets of $\mathcal{C}_1$ is $|\mathcal{B}_2||\mathcal{C}_1|$. So, the number of good locations available to $\mathcal{C}_1$ before populating $W_{|\mathcal{B}_1|}$ is

$$\leq |\mathcal{B}_2||\mathcal{C}_1| - |\mathcal{A}| + |W_{|\mathcal{B}_1|}| - g + \Big(|\mathcal{B}_1| - 1 \Big) \Big(|\mathcal{C}_1| + |\mathcal{C}_2| \Big). \tag{3}$$

2.5   Size of $W_{|\mathcal{B}_1|}$

We now define a notion of a bad element of $W_{|\mathcal{B}_1|}$, which is analogous to the definition of a bad element of $\mathcal{C}_1$ (Definition 1) (Fig. 3).

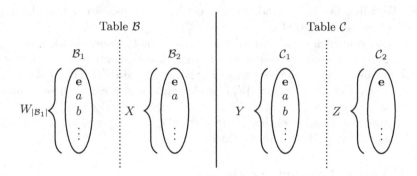

Fig. 3. In this arrangement, e is a bad element for the set $W_{|\mathcal{B}_1|}$.

Definition 3. *An element e of $W_{|\mathcal{B}_1|}$ is a bad element for $W_{|\mathcal{B}_1|}$ if the following happens.*

1. *The set, say Y, of $\mathcal{C}_1$ which contains e also contains at least two more elements of $W_{|\mathcal{B}_1|}$.*
2. *The set in $\mathcal{B}_2$ which contains e also contains at least one more element of $W_{|\mathcal{B}_1|} \bigcap Y$.*

Lemma 5. *If e is bad for $W_{|\mathcal{B}_1|}$, then the set containing e in $\mathcal{C}_2$ cannot contain any other element of $W_{|\mathcal{B}_1|}$.*

Proof. Let e belong to the sets X in $\mathcal{B}_2$, Y in $\mathcal{C}_1$, and Z in $\mathcal{C}_2$. As e is bad for $W_{|\mathcal{B}_1|}$, let e, a, and b belong to both $W_{|\mathcal{B}_1|}$ and Y. Moreover, let e and a belong to X in $\mathcal{B}_2$.

The proof of the lemma follows along the lines of the proof of Lemma 1, and is proved by contradiction. If indeed $Z \in \mathcal{C}_2$ contains one more element of $W_{|\mathcal{B}_1|}$, then one can show that we cannot store the set $S = \{(b, 1), (e, 2)\}$ in our data structure.

Similar to the implication of Lemmas 1, Lemma 5 implies that the bad elements of $W_{|\mathcal{B}_1|}$ must be kept in separate sets of $\mathcal{C}_2$. We now estimate the size of $W_{|\mathcal{B}_1|}$.

Let the members of $\mathcal{C}_1$ be denoted by $Y_1, Y_2, \ldots, Y_{|\mathcal{C}_1|}$. Let the number of good locations for each of those members be $g_1, g_2, \ldots, g_{|\mathcal{C}_1|}$, respectively. We now distribute the elements of $W_{|\mathcal{B}_1|}$ as follows.

We distribute g_1 elements of $W_{|\mathcal{B}_1|}$ in the set Y_1. The set Y_1 has g_1 good locations in $\mathcal{B}_2$; we put one of these elements in each of the good locations of

Y_1 in $\mathcal{B}_2$. We next distribute g_2 elements of $W_{|\mathcal{B}_1|}$ in the set Y_2, and put these elements in each of the good locagtions of Y_2. We do the same for each of the sets of $\mathcal{C}_1$. The number of elements of $W_{|\mathcal{B}_1|}$ this accounts for is given by Eq. 3.

For some j, if $g_j = 1$, we put one more element of $W_{|\mathcal{B}_1|}$ in Y_j. This achieves the following – every set of $\mathcal{C}_1$ which has non-zero good locations has at least two elements of $W_{|\mathcal{B}_1|}$. The number of such sets could be as high as $|\mathcal{C}_1|$. The number of elements of $W_{|\mathcal{B}_1|}$ accounted for so far is

$$\leq \left(|\mathcal{B}_2||\mathcal{C}_1| - |\mathcal{A}| + |W_{|\mathcal{B}_1|}| - g + \Big(|\mathcal{B}_1| - 1\Big)\Big(|\mathcal{C}_1| + |\mathcal{C}_2|\Big) \right) + |\mathcal{C}_1|. \tag{4}$$

from Equation 3

The situation so far is as follows. Every set of $\mathcal{C}_1$ that has some good locations in $\mathcal{B}_2$ has at least two elements of $W_{|\mathcal{B}_1|}$. Every good location in $\mathcal{B}_2$ corresponding to every set of $\mathcal{C}_1$ has at least one element of $W_{|\mathcal{B}_1|}$. If we put any more elements of $W_{|\mathcal{B}_1|}$ in any of the relevant sets of $\mathcal{C}_1$, we would start creating bad elements (Definition 3).

So, the total number of bad elements created is at least the size of the set $W_{|\mathcal{B}_1|}$ minus the elements of $W_{|\mathcal{B}_1|}$ accounted for in the previous equation, i.e.

$$|W_{|\mathcal{B}_1|}| - \left(|\mathcal{B}_2||\mathcal{C}_1| - |\mathcal{A}| + |W_{|\mathcal{B}_1|}| - g + \Big(|\mathcal{B}_1| - 1\Big)\Big(|\mathcal{C}_1| + |\mathcal{C}_2|\Big) \right) - |\mathcal{C}_1|.$$

These bad elements must be kept separate in $\mathcal{C}_2$. So, the cardinality of $W_{|\mathcal{B}_1|}$ must satisfy

$$|\mathcal{C}_2| \geq |W_{|\mathcal{B}_1|}| - \left(|\mathcal{B}_2||\mathcal{C}_1| - |\mathcal{A}| + |W_{|\mathcal{B}_1|}| - g + \Big(|\mathcal{B}_1| - 1\Big)\Big(|\mathcal{C}_1| + |\mathcal{C}_2|\Big) \right) - |\mathcal{C}_1|. \tag{5}$$

2.6 Size of Our Data Structure

Simplifying Eq. 5, we get the following inequality –

$$|\mathcal{A}| \leq |\mathcal{B}_1|\Big(|\mathcal{C}_1| + |\mathcal{C}_2|\Big) + |\mathcal{B}_2||\mathcal{C}_1| - g.$$

So, we can have the following inequality for $|\mathcal{B}_1| + |\mathcal{C}_1| + |\mathcal{B}_2| + |\mathcal{C}_2|$ –

$$\left(\frac{|\mathcal{B}_1| + |\mathcal{C}_1| + |\mathcal{B}_2| + |\mathcal{C}_2|}{2} \right)^2 \geq \Big(|\mathcal{B}_1| + |\mathcal{B}_2|\Big)\Big(|\mathcal{C}_1| + |\mathcal{C}_2|\Big)$$

$$\geq |\mathcal{B}_1|\Big(|\mathcal{C}_1| + |\mathcal{C}_2|\Big) + |\mathcal{B}_2||\mathcal{C}_1| - g \geq |\mathcal{A}|$$

$$|\mathcal{B}_1| + |\mathcal{C}_1| + |\mathcal{B}_2| + |\mathcal{C}_2| \geq 2|\mathcal{A}|^{\frac{1}{2}}.$$

As indices 1 and 2 satisfy Eq. 1, we have –

$$|\mathcal{B}| + |\mathcal{C}| = \sum_{i=1}^{b} \Big(|\mathcal{B}_i| + |\mathcal{C}_i|\Big) \geq \tfrac{b}{2}\Big(|\mathcal{B}_1| + |\mathcal{C}_1| + |\mathcal{B}_2| + |\mathcal{C}_2|\Big)$$

$$\geq b|\mathcal{A}|^{\frac{1}{2}}$$

Finally, the size of our data structure, s, is

$$s = |\mathcal{A}| + |\mathcal{B}| + |\mathcal{C}| \geq |\mathcal{A}| + b|\mathcal{A}|^{\frac{1}{2}}$$
$$= \frac{m}{b} + b\frac{m^{\frac{1}{2}}}{b^{\frac{1}{2}}} = \frac{m}{b} + b^{\frac{1}{2}}m^{\frac{1}{2}}.$$

The right hand side is minimised at $b = 2^{\frac{2}{3}}m^{\frac{1}{3}}$. So, after substitution we have

$$s \geq \frac{3}{2^{\frac{2}{3}}}m^{\frac{2}{3}},$$

which gives us the minimum amount of space that is required by any scheme that satisfy Restriction 1. We summarise the result in the following theorem.

Theorem 2. *The minimum amount of space required by any adaptive two element two proble scheme satisfying Restriction 1 is $3/2^{\frac{2}{3}}\ m^{\frac{2}{3}}$.*

3   Conclusion

We have presented a proof for the conjecture $s_A(2, m, 2) = \Omega(m^{2/3})$ for a restricted class of schemes. We hope that the notions of *bad elements* and *good locations* can be generalised so that it applies to all schemes, and consequently resolve the conjecture.

Acknowledgement. The authors would like to thank Protyai Ghosal for useful insights and discussions throughout the duration of the work.

References

1. Buhrman, H., Miltersen, P.B., Radhakrishnan, J., Venkatesh, S.: Are bitvectors optimal? SIAM J. Comput. **31**(6), 1723–1744 (2002)
2. Lewenstein, M., Munro, J.I., Nicholson, P.K., Raman, V.: Improved explicit data structures in the bitprobe model. In: Schulz, A.S., Wagner, D. (eds.) ESA 2014. LNCS, vol. 8737, pp. 630–641. Springer, Heidelberg (2014). https://doi.org/10.1007/978-3-662-44777-2_52
3. Radhakrishnan, J., Raman, V., Srinivasa Rao, S.: Explicit deterministic constructions for membership in the bitprobe model. In: auf der Heide, F.M. (ed.) ESA 2001. LNCS, vol. 2161, pp. 290–299. Springer, Heidelberg (2001). https://doi.org/10.1007/3-540-44676-1_24
4. Radhakrishnan, J., Shah, S., Shannigrahi, S.: Data structures for storing small sets in the bitprobe model. In: de Berg, M., Meyer, U. (eds.) ESA 2010. LNCS, vol. 6347, pp. 159–170. Springer, Heidelberg (2010). https://doi.org/10.1007/978-3-642-15781-3_14

On m-Bonacci-Sum Graphs

Kalpana Mahalingam[✉] and Helda Princy Rajendran[✉]

Department of Mathematics, Indian Institute of Technology, Chennai,
Chennai 600036, India
kmahalingam@iitm.ac.in, heldaprincy96@gmail.com

Abstract. We introduce the notion of *m-bonacci-sum graphs* denoted by $G_{m,n}$ for positive integers m, n. The vertices of $G_{m,n}$ are $1, 2, \ldots, n$ and any two vertices are adjacent if and only if their sum is an *m-bonacci number*. We show that $G_{m,n}$ is bipartite and for $n \geq 2^{m-2}$, $G_{m,n}$ has exactly $(m-1)$ components. We also find the values of n such that $G_{m,n}$ contains cycles as subgraphs. We also use this graph to partition the set $\{1, 2, \ldots, n\}$ into $m - 1$ subsets such that each subset is ordered in such a way that sum of any 2 consecutive terms is an *m-bonacci number*.

1 Introduction

In [7], Barwell posed the problem of an ordering of the set $\{1, 2, 3, \ldots, 34\}$ such that the sum of the consecutive pairs of numbers is a Fibonacci number. Motivated by this problem, in 2014, Fox et al. defined the notion *Fibonacci-sum graph* [1] and found values of n for which such an ordering exists. Some properties of *Fibonacci-sum graph* are discussed in [5]. An extension of *Fibonacci-sum graphs* called as the *Fibonacci-sum set graphs* was introduced in [11]. Several basic structural properties of these graphs were also discussed [11].

The concept of Fibonacci numbers was generalized and was called *m-bonacci numbers* [4]. Blackmore et al. [8] have associated *m-bonacci numbers* to the class of energy minimizing n-dimensional lattices and area preserving transformations on a torus. Other generalizations of Fibonacci numbers were given in [2,3].

In this paper, we consider an extension of the question posed by Barwell: Given any set $\{1, 2, \cdots, n\}$, does there exist a partition of it into $m - 1$ sets (say $\{A_i\}_{i=1}^{m-1}$) and an ordering of elements in each A_i such that the sum of any two consecutive numbers in the set is an *m-bonacci number*. We address the problem by introducing the notion of *m-bonacci-sum graph*.

We use the definition of *m-bonacci number* given in [4]. For $m \geq 2$, $n \geq 1$, we define *m-bonacci-sum graph* $G_{m,n}$ as the graph on vertex set $\{1, 2, \ldots, n\}$ and any two vertices are adjacent if and only if their sum is an *m-bonacci number*. Using such a graph we try to partition the given vertex set based on the connected components of the graph. We also find values for n for which the vertices of each component can be ordered so that any two consecutive vertices sum up to an *m-bonacci number*.

This paper is arranged as follows. In Sect. 2 we recall some facts about *m-bonacci numbers*. In Sect. 3, we define *m-bonacci-sum graph* and we discuss some

© Springer Nature Switzerland AG 2019
S. P. Pal and A. Vijayakumar (Eds.): CALDAM 2019, LNCS 11394, pp. 65–76, 2019.
https://doi.org/10.1007/978-3-030-11509-8_6

basic properties of such graphs. In Sect. 4, we show that the *m-bonacci-sum graph* is bipartite and is not connected. In fact, we show precisely that an *m-bonacci-sum graph* has exactly $m - 1$ components. In Sect. 5, we give answers to the extension of the question posed by Barwell for all $m \geq 3$. We end with a few concluding remarks.

2   Preliminaries

We begin the section by recalling the notion of *m-bonacci numbers* (See [4]).

Definition 1. *The m-bonacci sequence* $\{Z_{n,m}\}_{n \geq 1}$ *is defined by*

$$Z_{i,m} = 0, \ 1 \leq i \leq m - 1, \ Z_{m,m} = 1$$

and for $n \geq m + 1$,

$$Z_{n,m} = \sum_{i=n-m}^{n-1} Z_{i,m}$$

Each $Z_{i,m}$ *is called a m-bonacci number.*

For example, when $m = 5$, the sequence is

$$\{Z_{n,5}\}_{n=1}^{\infty} = \{0, 0, 0, 0, 1, 1, 2, 4, 8, 16, 31 \ldots\}$$

We also recall some known facts about *m-bonacci numbers* which will be used later.

Lemma 1. *1. For each* $m \geq 2$, $2Z_{k,m} \geq Z_{k+1,m}$ *for all* $k \geq 1$.
2. If a sum of m m-bonacci numbers equals another m-bonacci number, then those $m + 1$ numbers must be consecutive.
3. For each $m \geq 2$, *the first $2m + 1$ terms of the m-bonacci sequence are* $\underbrace{0, 0, \ldots, 0}_{m-1 \ times}, 1, 1, 2, 2^2, \ldots, 2^{m-1}, 2^m - 1$.

3   *m*-Bonacci-Sum Graph

We begin the section by introducing the notion of *m-bonacci-sum graph* formally in the following way. For basic concepts on Graph theory, we refer the reader to [9,10].

Definition 2. *For fixed* $m \geq 2$, *for each* $n \geq 1$, *m-bonacci-sum graph denoted by* $G_{m,n} = (V, E)$ *is the graph defined on the vertex set* $V = [n] = \{1, 2, \ldots, n\}$ *and with edge set*

$$E = \{\{i, j\} : i, j \in V, \ i \neq j, \ i + j \text{ is an m-bonacci number}\}.$$

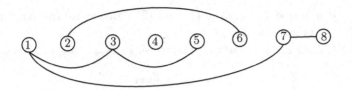

Fig. 1. *4-bonacci graph* $G_{4,8}$

This shows that *m-bonacci-sum graph* is a simple graph. When $m = 2$, $G_{m,n}$ is *Fibonacci-sum graph*. We illustrate an example of $G_{4,8}$ in Fig. 1.

We show that for any $m \geq 2$, the *m-bonacci-sum graph* with n vertices will be a subgraph of the *m-bonacci-sum graph* with vertices greater than n.

Lemma 2. *For fixed m, for all n, $G_{m,n}$ is a subgraph of $G_{m,n+i}$, $i \geq 0$.*

In the following lemma we see that the *m-bonacci numbers* $1, 2, 2^2, \ldots, 2^{m-2}$ are not adjacent to any vertex less than the respective *m-bonacci number*.

Lemma 3. *If $2^i \leq n$ for some $i \in \{1, 2, \ldots, m-2\}$, in $G_{m,n}$, 2^i is not adjacent to any $k < 2^i$.*

3.1 Degrees

In this section we discuss about the adjacency and degrees of each vertex in *m-bonacci-sum graph*. The following Lemma talks about the vertices to which n is adjacent in $G_{m,n}$.

Lemma 4. *Let $n \geq 2$, and let k be so that $Z_{k,m} \leq n < Z_{k+1,m}$. In $G_{m,n}$, vertex n is adjacent to only*

$$
\begin{array}{ll}
Z_{k+1,m} - n & \text{if } n \leq \frac{Z_{k+2,m}}{2}, 2n \neq Z_{k+1,m} \\
Z_{k+1,m} - n \text{ and } Z_{k+2,m} - n & \text{if } n > \frac{Z_{k+2,m}}{2}
\end{array}.
$$

Proof. Let $x \in [n]$ be adjacent to n in $G_{m,n}$; i.e., $x + n = Z_{i,m}$ for some i. Since, $G_{m,n}$ is a simple graph and $1 \leq x \leq n$, $x + n$ can be at most $2n - 1$ Therefore, we have,

$$x + n \leq 2n - 1 < 2Z_{k+1,m}$$

$$= Z_{k+1,m} + \sum_{i=k-(m-1)}^{k} Z_{i,m}$$

$$= Z_{k+2,m} + Z_{k-(m-1),m}$$

$$< Z_{k+3,m} \tag{1}$$

Also we have $Z_{k,m} \leq n$. Therefore,

$$Z_{k,m} < 1 + Z_{k,m} \leq x + n \tag{2}$$

Thus, $Z_{k,m} < x + n < Z_{k+3,m}$ (by (1) and (2)). Hence, we can conclude that $x + n = Z_{k+1,m}$ or $x + n = Z_{k+2,m}$.

If $n > \frac{Z_{k+2,m}}{2}$, then $Z_{k+2,m} - n < n$. Therefore, $Z_{k+2,m} - n \in [n]$. Also,

$$
\begin{aligned}
Z_{k+1,m} - n &\leq Z_{k+1,m} - Z_{k,m} \\
&= Z_{k-1,m} + \cdots + Z_{k-(m-1),m} \\
&= Z_{k,m} - Z_{k-m,m} \\
&\leq Z_{k,m}.
\end{aligned}
\tag{3}
$$

Hence $Z_{k+1,m} - n \in [n]$ (by (3) and $Z_{k,m} \leq n$). Thus, in this case n is adjacent to both $Z_{k+1,m} - n$ and $Z_{k+2,m} - n$ in $G_{m,n}$.

If $n \leq \frac{Z_{k+2,m}}{2}$, then

$$
x + n < 2n \leq Z_{k+2,m}
\tag{4}
$$

So $i \neq k+2$. Thus, in this case if $2n \neq Z_{k+1,m}$, n is adjacent to only $Z_{k+1,m} - n$ in $G_{m,n}$.   □

In $G_{m,n}$ if n is a m-bonacci number, then degree of n must be at most one. This will be proved in the following lemma by using Lemma 4.

Lemma 5. *Let $n \geq 2$, and k be so that $Z_{k,m} \leq n < Z_{k+1,m}$. Then in $G_{m,n}$ if $Z_{k,m} \neq 2Z_{k+1,m}$, the vertex $Z_{k,m}$ has only one neighbour, namely $Z_{k+1,m} - n$.*

Proof. If $n = Z_{k,m}$, then by Lemma 4, in $G_{m,n}$, if $Z_{k+1,m} \neq 2Z_{k,m}$, n is adjacent to only $Z_{k+1,m} - n$ ($Z_{k,m} \leq \frac{Z_{k+2,m}}{2}$ for all k).

Now, assume that $Z_{k,m} < n < Z_{k+1,m}$. In $G_{m,n}$, the only possible adjacent vertex less than $Z_{k,m}$ is $Z_{k+1,m} - n$. It is enough to prove that $Z_{k,m}$ has no adjacent vertices larger than $Z_{k,m}$ in $G_{m,n}$. Let v be any vertex in $G_{m,n}$ such that $Z_{k,m} < v \leq n$. Then

$$
\begin{aligned}
Z_{k+1,m} = Z_{k,m} + \sum_{i=k-(m-1)}^{k-1} Z_{i,m} \\
\leq Z_{k,m} + Z_{k,m} \\
< Z_{k,m} + v \\
\leq Z_{k,m} + n \\
< Z_{k,m} + Z_{k+1,m} \\
\leq Z_{k+2,m}
\end{aligned}
\tag{5}
$$

This shows that $Z_{k,m} + v$ is not a m-bonacci number (since $Z_{k+1,m} < Z_{k,m} + v < Z_{k+2,m}$ by (5)), and hence v is not adjacent to $Z_{k,m}$. Therefore, in $G_{m,n}$, $Z_{k,m}$ has only neighbour $Z_{k+1,m} - n$ provided $Z_{k+1,m} \neq 2Z_{k,m}$.   □

In previous two lemmas we discussed about adjacency of vertices in $G_{m,n}$. Now, it is left to find the degree of each vertex in m-bonacci-sum graph. The following theorem gives the formula to find the degree of each vertex in any m-bonacci-sum graph.

Theorem 1. *Let $n \geq 1$ and let $x \in [n]$. Let $k \geq m+1$ satisfy $Z_{k,m} \leq x < Z_{k+1,m}$ and $l \geq k$ satisfy $Z_{l,m} \leq x + n < Z_{l+1,m}$. Then the degree of x in $G_{m,n}$ is*

$$deg_{G_{m,n}}(x) = \begin{cases} l - k & \text{if } 2x \text{ is not an } m - bonacci \text{ number} \\ l - k - 1 & \text{if } 2x \text{ is an } m - bonacci \text{ number} \end{cases}$$

Proof. Let $s \in [n]$. Since

$$Z_{k,m} < x + s \leq x + n < Z_{l+1,m}$$

the possible choices for $x+s$ to be an m-*bonacci number* is $Z_{k+1,m}, \ldots, Z_{l,m}$. If $2x$ is not an m-*bonacci number*, then all choices are possible and hence $deg_{G_{m,n}}(x) = l - k$. Otherwise, only one choice is not possible and hence $deg_{G_{m,n}}(x) = l - k - 1$. $\qquad\square$

The following corollary can be proved directly by applying Lemma 1 and Theorem 1. By above Theorem we can find at least few possible pendant vertices for any m and n. Apart from the vertices listed below there may be other pendant vertices too.

Corollary 1. *Let $k \geq m + 1$ and n satisfy $Z_{k,m} \leq n < Z_{k+1,m}$. If $n < \frac{Z_{k+2,m}}{2}$, then degree of the vertices $Z_{k,m}, Z_{k+1,m}, \ldots, n$ are at most one. If $n \geq \frac{Z_{k+2,m}}{2}$, then degree of the vertices $Z_{k,m}, Z_{k,m} + 1, \ldots, Z_{k+2,m} - n - 1$ are at most one.* $\square$

4 Properties of m-Bonacci-Sum Graph

In [5], Arman et al. gave a proof for *Fibonacci-sum graph* to be bipartite. We show that the property can be extended also to m-*bonacci-sum graph*.

Theorem 2. *For $n \geq 2$, $G_{m,n}$ is bipartite.*

As an example, one can see in Fig. 1, $G_{4,8}$ is bipartite with partite sets $V_1 = \{1, 2, 4, 5, 8\}$ and $V_2 = \{3, 6, 7\}$.

In [6], Costain proved that *Fibonacci-sum graph* is connected. But this is not the case for any m. In fact, we show that for all $m \geq 2$ and $n \geq 2^{m-2}$, $G_{m,n}$ has exactly $(m - 1)$ components.

Theorem 3. *For each $n \geq 2^{m-2}$, $G_{m,n}$ has exactly $(m - 1)$ components.*

Proof. Let $m \geq 2$ and $n \geq 2^{m-2}$. For each m and n, we define the partition of vertex set with $(m - 1)$ partitions in the following way. For each i, $1 \leq i \leq m - 1$, we define $A_i = B_{i,n}$ where, $B_{i,0} = \{2^{i-1}\}$ and for $1 \leq j \leq n$,

$$B_{i,j} = \{k : k \leq n, \ k + c \text{ is an } m\text{-bonacci number } for \text{ some } c < k, \ c \in B_{i,j-1}\} \cup B_{i,j-1}. \tag{6}$$

Note that $A_i \neq \emptyset$ as $2^{i-1} \in A_i$ for $1 \leq i \leq m - 1$, since $n \geq 2^{m-2}$. Let H_i be the subgraph induced by vertices of A_i. We now show that each H_i is connected.

Let $x \neq y$ and $x, y \in A_i$. If x and y are adjacent then we are done. Otherwise, we have the following cases.

<u>Case 1</u>: Either x or y is 2^{i-1}. Without loss of generality assume that $x = 2^{i-1}$. Now, by definition of A_i there exists $u_1 < y$, $u_1 \in A_i$ such that u_1 and y are adjacent. If u_1 is adjacent to x we are done. Otherwise, again by definition of A_i, there exists $u_2 < u_1$ in A_i such that u_1 and u_2 are adjacent. If $u_2 = x$ or u_2 is adjacent to x we are done. Otherwise we proceed like above. Since $x = 2^{i-1} < \cdots < u_2 < u_1$, this process stops in a finite number of steps and we get a u_j such that $u_j = x$ or it is adjacent to x. Thus, we have a path between x and y.

<u>Case 2</u>: $x, y \neq 2^{i-1}$. Without loss of generality we can assume that $x < y$. Now, for x we can find a sequence of vertices $v_1 = 2^{i-1} < v_2 < \cdots < v_k < x$ in A_i such that v_i and v_{i+1} are adjacent and v_k and x are adjacent. Now, by case 1, there exists a path between v_1 and y. Thus, we get a walk between x and y from which we can get a path.

Thus, in either case there exists a path between x and y. Hence, each H_i is connected. It is clear from Lemma 4 and definition of A_i, that, $[n] = \bigcup\limits_{i=1}^{i=m-1} A_i$. Thus, for $m = 2$, the graph is connected.

Let $m \geq 3$. We now prove that each H_i is a component in $G_{m,n}$ i.e. to prove that there is no edge between $a_{ik} \in A_i$ and $a_{jl} \in A_j$ for any $i \neq j$. By definition of A_i, it is enough to prove that $\{A_i\}_{i=1}^{m-1}$ forms a partition of $[n]$. Suppose not, assume that there exists an $x \in A_i \cap A_j$ for some i, j such that $i \neq j$ and x is the least such positive integer. Then, by definition of A_i and A_j, x must be adjacent to some $a_{il} \in A_i$ and $a_{jk} \in A_j$ for some l, k such that $a_{il} < x$ and $a_{ik} < x$. This x should be greater than 3 (since $m \geq 3$, 3 is not adjacent to 2). Then, by Lemma 4, $a_{jk} = Z_{s+1,m} - x$ and $a_{il} = Z_{s+2,m} - x$ (or $a_{jk} = Z_{s+2,m} - x$ and $a_{il} = Z_{s+1,m} - x$) where $Z_{s,m} \leq x < Z_{s+1,m}$, $x > \frac{Z_{s+2,m}}{2}$.

Let $b = Z_{s+1,m} + x - Z_{s+2,m}$. Then,

$$b = Z_{s+1,m} + x - Z_{s+2,m} > x + x - Z_{s+2,m} > 0 \ (since \ x > \frac{Z_{s+2,m}}{2}) \quad (7)$$

Also we have,

$$b = Z_{s+1,m} + x - Z_{s+2,m} = x - \sum_{i=s-(m-2)}^{s} Z_{i,m} \quad (8)$$

Since, $x > 3$, $Z_{s,m} > 0$ and hence $b < x$ (by Eq. 8). Also, $b + a_{il} = Z_{s+1,m}$. This implies that $b \in A_i$. But, $b + a_{jk} = 2Z_{s+1,m} - Z_{s+2,m} = Z_{s-(m-1),m}$ and hence b and a_{jk} are adjacent. Thus, $b \in A_i \cap A_j$ and $b < x$. This is a contradiction to our choice of x such that x is minimal. Hence, $A_i \cap A_j = \emptyset$ for $i \neq j$. Therefore, each H_i forms a connected component of $G_{m,n}$. $\qquad\square$

In Fig. 1, one can see that $G_{4,8}$ has 3 components.

Corollary 2. *If $x \in A_k$ for some k in $G_{m,r}$, then $x \in A_k$ in $G_{m,s}$ for all $s \geq r$.*

Proof. Let $x \in A_k$ for some k in $G_{m,r}$. Then, we can find a path from 2^{k-1} to x in $G_{m,r}$. By Lemma 2, this path will be there in $G_{m,s}$, for all $s \geq r$. Hence, in $G_{m,s}$, $x \in A_k$. Hence the result.   □

We now investigate the isolated vertices in $G_{m,n}$ where n can be one of the $m+1, m+2, \ldots, 2m-1^{th}$ term of m-bonacci sequence.

Proposition 1. *Let $m \geq 3$. In $G_{m,2^i}$, $1 \leq i \leq m-2$, 2^i and 2^{i-1} are the only isolated vertices. In $G_{m,2^{m-1}}$, 2^{m-2} is the only isolated vertex.*

Proof. By Lemma 3, it is clear that 2^i is an isolated vertex in $G_{m,2^i}$. We know from Lemma 1 that 2^i and 2^{i+1} are consecutive m-*bonacci numbers*. Let x be such that $2^{i-1} < x \leq 2^i$. Then,

$$2^{i-1} + x \leq 2^{i-1} + 2^i = 3 \cdot 2^{i-1} < 2^{i+1}$$

Thus, $2^i < 2^{i-1} + x < 2^{i+1}$ (i.e.), $2^{i-1} + x$ lies between two consecutive m-*bonacci numbers*. Hence, 2^{i-1} is not adjacent to any x such that $2^{i-1} < x \leq 2^i$. Also, by Lemma 3, 2^{i-1} is not adjacent to any $x < 2^{i-1}$. Hence, 2^{i-1} is also an isolated vertex.

We now show that there are no other isolated vertices in $G_{m,2^{i-1}}$. Let u be any other vertex in $G_{m,2^i}$. Clearly $u < 2^i$. We now have the following cases.

    <u>Case 1</u>: $u \neq 2^j$ for any $j < i$. We know by Lemmas 1 and 4, that $2u$ is not an m-*bonacci number* and u is adjacent to at least one vertex in $G_{m,u}$. This adjacency is preserved in all $G_{m,n}$, $n \geq u$ and hence in $G_{m,2^i}$. Hence, u is not isolated in this case.

    <u>Case 2</u>: $u = 2^j$ for some $j < i-1$. Since $j < i-1$, we have that $2^i - 2^j \neq 2^j$ and hence, u is adjacent to $2^i - 2^j$ (since 2^i is an m-*bonacci number*). Thus, u is not isolated in this case too.

Therefore, in $G_{m,2^i}$, 2^i and 2^{i-1} are the only isolated vertices. Similarly one can show that, 2^{m-2} is the only isolated vertex in $G_{m,2^{m-1}}$.   □

We now give a lower bound on the number of vertices n so that the m-*bonacci-sum graph* $G_{m,n}$ does not contain any isolated vertices. We omit the proof as it is similar.

Proposition 2. *Let $m \geq 2$. $G_{m,n}$ has at least one isolated vertex if $n < N$, $N = 3 \cdot 2^{m-2} - 1$. For all $n \geq N$, $G_{m,n}$ has no isolated vertex.*

4.1   Cycles in m-Bonacci-Sum Graph

In this section, we discuss the existence of cycles in any given *m-bonacci sum graph*. In Theorem 4, we find the least value of n such that $G_{m,n}$ has a cycle in at least one of its components. We use *m-bonacci numbers* $Z_{2(m+1)+i,m}$, for $0 \leq i \leq m-1$ to calculate this least value.

Using Lemma 1 and Definition 1 and a direct calculation, we have the following,

$$Z_{i,m} = 0, \ 1 \leq i \leq m-1 \tag{9}$$

$$Z_{m,m} = 1, \ Z_{i,m} = 2^{i-(m+1)}, \ m+1 \leq i \leq 2m \tag{10}$$

$$Z_{2(m+1)+i,m} = 2^{m+(i+1)} - ((i+1)2^i + 2^{i+1}), \ -1 \leq i \leq m-1 \tag{11}$$

By Theorem 3, we know that if $n \geq 2^{m-2}$, $G_{m,n}$ has $(m-1)$ components. In Theorem 3, we have denoted the vertex sets of different components as A_i. In the following theorem we follow the same notation. We denote the component with vertex set A_i as H_i for all $1 \leq i \leq m-1$. In the following theorem, we discuss the occurrence of a cycle in each component of $G_{m,n}$ and their relation with the number of vertices (i.e.), n. We use the following lemma.

Lemma 6. *Let $Z_{2(m+1)+i,m} - 2^{i-1} \leq x < Z_{2(m+1)+(i+1),m} - 2^i$, $i < m$. If degree of x is two in $G_{m,x}$, then x must be in some H_t, $t \leq i$.*

Proof. Given that $Z_{2(m+1)+i,m} - 2^{i-1} \leq x < Z_{2(m+1)+(i+1),m} - 2^i$, $i < m$. If degree of x is at most 1 in $G_{m,x}$, then there is nothing to prove. Now assume that degree of x is 2 in $G_{m,x}$. We have two cases. We only prove for the case when $Z_{2(m+1)+i,m} - 2^{i-1} \leq x < Z_{2(m+1)+i,m}$, as the proof is similar when $Z_{2(m+1)+i,m} \leq x < Z_{2(m+1)+(i+1),m} - 2^i$.

Let $Z_{2(m+1)+i,m} - 2^{i-1} \leq x < Z_{2(m+1)+i,m}$. Now, the possibilities for x are $Z_{2(m+1)+i} - j, 1 \leq j \leq 2^{i-1}$. Then, for any j, the vertex $Z_{2(m+1)+i} - j$ is adjacent to j. By Lemma 3, the vertex 2^t, $t \geq i$ is not adjacent to any of $1, 2, \ldots, 2^{i-1}$ and by the construction of A_t, the vertices $1, 2, \ldots, 2^{i-1}$ do not belong to A_t, for all $t > i$ and hence not in the component H_t, $t > i$. Since $Z_{2(m+1)+i,m} - j$ is adjacent to j, for each j, $1 \leq j \leq 2^{i-1}$, we have that, $Z_{2(m+1)+i,m} - j$ also does not belong to H_t, $t > i$. Hence the result. □

Theorem 4. *Let $m \geq 2$. Let $G_{m,i}$ be the m-bonacci-sum graph with i vertices. Then*

1. *$G_{m,i}$ has no cycles for $1 \leq i \leq Z_{2(m+1)+1,m} - 2$.*
2. *$G_{m,j}$ has at least one cycle in each of the components $H_1, H_2, \ldots, H_i$, where H_i is the component containing the vertex 2^{i-1} and the components $H_j, j > i$ contains no cycles for all $Z_{2(m+1)+i,m} - 2^{i-1} \leq j < Z_{2(m+1)+(i+1),m} - 2^i$, $i \leq (m-1)$.*

Proof. 1. Let $G_{m,i}$ be the *m-bonacci-sum graph* with i vertices and $i \in \{1, 2, \ldots, Z_{2(m+1)+1,m} - 2\}$. By Eq. 11, we get that,

$$Z_{2(m+1)+1,m} - 2 = 2^{m+2} - 8 - 2 = 2^{m+2} - 10$$

Let $1 \leq i \leq 2^{m+2} - 10$. Then, there exists a '$k$' such that $Z_{k,m} \leq i < Z_{k+1,m}$ and $m + 1 \leq k \leq 2(m+1) + 1$. Then, i can take the value at most $Z_{k+1,m} - 1$ and hence $2i \leq Z_{k+2,m}$ (since $1 \leq i \leq 2^{m+2} - 10$ and by Eqs. 9, 10 and 11). Thus, by Lemma 4, degree of i in $G_{m,i}$ will be at most 1 for all i. Also, for any n, $G_{m,n+1}$ can be obtained from $G_{m,n}$ by adding the new vertex $n + 1$ and the possible new edges. This concludes that

$$G_{m,i}, 1 \leq i \leq Z_{2(m+1)+1,m} - 2 = 2^{m+2} - 10$$

contains no cycle (since degree of i in $G_{m,i}$ can be at most 1, adding an isolated or pendant vertex will not create any cycle).

2. One can easily verify by direct calculation that, for $1 \leq i \leq m - 2$

$$Z_{2(m+1)+i,m} - 2^{i-1} + Z_{2(m+1)+i,m} - 3 \cdot 2^{i-1} = Z_{2(m+1)+(i+1),m} \qquad (12)$$

Now, for $i = m - 1$ we have, $Z_{2(m+1)+(m-1),m} = Z_{3m+1,m}$ and,

$$Z_{3m+1,m} - 2^{m-2} + Z_{3m+1,m} - (3 \cdot 2^{m-2} - 1) = Z_{2(m+1)+m,m} \qquad (13)$$

Also, one can easily verify that for each i, $1 \leq i \leq m - 2$, the vertices

$$Z_{2(m+1)+i,m} - 2^{i-1}, \ 2^{i-1}, \ 3 \cdot 2^{i-1}, \ Z_{2(m+1)+i,m} - (3 \cdot 2^{i-1}) \qquad (14)$$

form a cycle in the respective component H_i (by Eq. 12). For $m - 1$, the vertices

$$Z_{2(m+1)+(m-1),m} - 2^{m-2}, \ 2^{m-2}, \ 3 \cdot 2^{m-2} - 1, \ Z_{2(m+1)+(m-1),m} - (3 \cdot 2^{m-2} - 1) \qquad (15)$$

form a cycle in the component H_{m-1} (by Eq. 13). Since, $Z_{2(m+1)+i} - 2^{i-1} \leq j < Z_{2(m+1)+(i+1)} - 2^i$, $i \leq (m - 1)$ we can conclude that, the components $H_1, H_2, \ldots, H_i$ of $G_{m,j}$, contains a cycle (by Eqs. 14 and 15).

Consider the component H_t, $t > i$. We show that H_t does not contain a cycle. Let $t > i$. Arrange the vertices of H_t in an increasing order (i.e.), $A_t = \{a_{t1}, a_{t2}, \ldots, a_{tk}\}$ where $a_{t1} < a_{t2} < \cdots < a_{tk} \leq j$. Since $j < Z_{2(m+1)+(i+1)} - 2^i$, by Lemma 6 we can conclude that degree of a_{ts} in $G_{m,a_{ts}}$, $1 \leq s \leq l$ is at most 1. Thus, H_t must be a tree and has no cycle. Hence the result. $\square$

The following corollary follows directly from the above theorem.

Corollary 3. *In G_{m,n_i}, where $n_i = Z_{2(m+1)+i} - 2^{i-1}$, the component H_i has exactly one cycle.*

5   Extension of Barwell's Problem

In this section, we consider an extension of Barwell's problem [7]: Given a set $\{1, 2, 3, \cdots, n\}$, does there exist a partition of the set in to say k sets and an ordering of each of these partition such that, sum of any two consecutive numbers is a $k + 1$-*bonacci number*. Existence of such a partition is equivalent to the existence of paths in each of the component of the graph $G_{k+1,n}$ that spans the vertices of the respective components. By direct verification, it is easy to observe that each component of $G_{3,n}$, $1 \leq n \leq 9$ has a path which spans all the vertices of the corresponding component.

Throughout this section we follow the notation used in Theorem 3 (i.e.), for a given graph $G_{m,n}$, the vertex set of $G_{m,n}$ is partitioned into A_i and H_i represents the component corresponding to A_i. We first prove the following.

Proposition 3. *Let $m \geq 4$. In $G_{m,n}$, at least one number out of four consecutive numbers will lie in A_1.*

Proof. We prove this result by induction on n. Since $m \geq 4$, we have 1,1,2,4,8 as consecutive *m-bonacci numbers*. This implies that $1, 3, 5, 7 \in A_1$. Thus for $1 \leq n \leq 8$, among any four consecutive numbers at least one number is in A_1. Let $k \geq 8$ and assume that for all $k < n + 1$, the result is true. Now to prove that among $n + 1, n + 2, n + 3, n + 4$ at least one number is in A_1. We have the following cases.

Case 1: $Z_{k,m} < n + 1, n + 2, n + 3, n + 4 < Z_{k+1,m}$ for some k. Then, by Lemma 1, $2(n + i) > Z_{k+1,m}$ for $1 \leq i \leq 4$. Therefore, $n + i$ is adjacent to $Z_{k+1,m} - (n + i)$ for $1 \leq i \leq 4$. Here, $Z_{k+1,m} - (n + 1), Z_{k+1,m} - (n + 2), Z_{k+1,m} - (n + 3)$ and $Z_{k+1,m} - (n + 4)$ are consecutive numbers and are less than $n + 1$ (by Lemma 4). Therefore, by induction hypothesis, at least one of them will be in A_1. This implies that at least one of the numbers $n + 1, n + 2, n + 3, n + 4$ must be in A_1.

Case 2: $n + i = Z_{k,m}$ for some $i \in \{1, 2, 3, 4\}$. If $n + i = Z_{k,m}$, $i = 2, 3, 4$, then $n + (i - 1) \in A_1$. If $n + 1 = Z_{k,m}$, then $2(n + 1) \geq Z_{k+1,m}$ (by Lemma 1) and $n \in A_1$ (since $1 \in A_1$). If $2(n + 1) = Z_{k+1,m}$, then $n + 2 \in A_1$ (since $n \in A_1$ and $n + (n + 2) = 2(n + 1) = Z_{k+1,m}$). If $2(n + 1) > Z_{k+1,m}$, then each $n + i$ is adjacent to $Z_{k+1,m} - (n + i)$ which are consecutive numbers. Hence by induction assumption one of them will be in A_i. This implies that one of the numbers $n + 1, n + 2, n + 3, n + 4$ lie in A_1.

Hence the result.   □

By direct verification, it is easy to observe that for $m \geq 5$ and $1 \leq n \leq 12$, each component of $G_{m,n}$ has a path which spans all the vertices of the corresponding component. We now show that for any $n > 12$, in $G_{m,n}$ H_1 has no path which spans all the vertices of H_1.

Theorem 5. *Let $m \geq 5$. Then, in $G_{m,n}$, $n > 12$ the component H_1 does not contain a path that spans H_1.*

Proof. We prove this result by induction on n. When $n = 13$, $G_{m,13}$ for all $m \geq 5$ is given in Fig. 2

From Fig. 2, one can observe that H_1 is the component containing vertex 1 and has no path which spans A_1. Thus, the result holds for $n = 13$. Let $l \geq 13$ and assume that the result is true for all $l < n$. Now, we prove for n. If $n \notin A_1$,

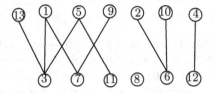

Fig. 2. $G_{m,13}$, $m \geq 5$

then by induction the result holds for $G_{m,n}$. Assume that $n \in A_1$. If degree of n is at most one, then the result holds by induction assumption (since, by adding a vertex of degree at most one will not create any new path which spans the whole component). Let the degree of n be 2 in $G_{m,n}$. Then, by the proof of the first statement of Theorem 4, $n \geq Z_{2(m+1)+1,m} - 1$. By Lemma 4, $n > \frac{Z_{k+2,m}}{2}$. Let k be such that $Z_{k,m} \leq n < Z_{k+1,m}$. We have $k \geq 2(m + 1)$. By Corollary 1, the vertices $Z_{k,m}, Z_{k,m} + 1, \ldots, Z_{k+2,m} - n - 1$ have degree at most 1. Since $k \geq 2(m + 1)$, by Proposition 2, the vertices $Z_{k,m}, Z_{k,m} + 1, \ldots, Z_{k+2,m} - n - 1$ are pendant vertices in $G_{m,n}$. Now,

$$Z_{k+2,m} - n - 1 > Z_{k+2,m} - Z_{k+1,m} - 1 \text{ (since } n < Z_{k+1,m})$$
$$= Z_{k,m} + Z_{k-1,m} + \cdots + Z_{k-(m-2),m} - 1 \qquad (16)$$

Since $k \geq 2(m+1)$ and $m \geq 5$, we have $Z_{k,m} + Z_{k-1,m} + \cdots + Z_{k-(m-2),m} - 1 > 16$. Thus, the number of pendant vertices is more than 16 and they are consecutive numbers. By Proposition 3, at least three from these consecutive 16 pendant vertices must be in A_1. That is, the component H_1 contains at least 3 pendant vertices. This implies that H_1 has no path which spans A_1. Hence, the result. $\square$

Similarly, one can show that when $m = 3$ and $m = 4$ we need at most 9 and 11 vertices for a path that spans the vertices in their respective components to exist. Thus, we conclude that the set $\{1, 2, \cdots n\}$ can be partitioned in to k subsets and ordered so that any two consecutive numbers in the respective subsets adds up to a $(k + 1)$-*bonacci number* for the following values of k and n. It was shown in [1], that for $m = 2$ (i.e.), using *Fibonaaci-sum graphs*, n can take the value $9, 11, f_t$ or $f_t - 1$ where f_t is the tth Fibonacci number.

	$k + 1$	n
Fibonacci-sum graph	2	$9, 11, f_t, f_t - 1$
Tribonacci-sum graph	3	≤ 9
4-bonacci-sum graph	4	≤ 11
m-bonacci-sum graph	≥ 5	≤ 12

6 Conclusion

In this paper we extended the notion of *Fibonacci-sum graph* to *m-bonacci- sum graph*. We also, considered a generalization of the question posed by Barwell and found the answer using *m-bonacci-sum graph* for all values of m. It is easy to verify that in a *Tribonacci-sum graph*, H_1 does not contain such a path for $n \geq 10$, however there does exist a path in H_2 that spans A_2 for $n \leq 18$. It will be interesting to see for what values of n and m a particular component H_i has a path that spans A_i. It will also be interesting to look into automorphisms of such graphs.

Acknowledgement. The second author wishes to acknowledge the fellowship received from Department of Science and Technology under INSPIRE fellowship (IF170077).

References

1. Fox, K., Kinnersely, W.B., McDonald, D., Orflow, N., Puleo, G.J.: Spanning paths in Fibonacci-sum graphs. Fiboncci Quart **52**, 46–49 (2014)
2. Chamberlain, J., Higgins, N., Yürekli, O.: M-bonacci numbers and their finite sums. Int. J. Math. Educ. Sci. Technol. **34**(6), 935–940 (2003)
3. Asiru, M.A.: Sequences with M-bonacci property and their finite sums. Int. J. Math. Educ. Sci. Technol. **39**(6), 819–829 (2008)
4. Kappraff, J.: Beyond Measure: A Guided Tour Through Nature, Myth and Number (Chap. 21). World Scientific, Singapore (2002)
5. Arman, A., Gunderson, D.S., Li, P.C.: Properties of the Fibonacci-sum graph. Preprint, arXiv:1710.10303v1 [math CO] (2017)
6. Costain, G.: On the additive graph generated by a subset of the natural numbers. Master's thesis, McGill University (2008)
7. Barwell, B.: Problem 2732, problems and conjectures. J. Recreational Math. **34**, 220–223 (2006)
8. Blackmore, D., Kappraff, J.: Phyllotaxis and toral dynamical systems. ZAMM (1996)
9. Bondy, J.A., Murty, U.S.R.: Graph Theory. Springer, New York (2008)
10. West, D.B.: Introduction to Graph Theory. Prentice-Hall of India, New Delhi
11. Mphako-Banda, E.G., Kok, J., Naduvath, S.: Some properties of Fibonacci-sum set-graphs. Preprint, arXiv:1802.02452v1, [math.GM] (2018)

Linear Time Algorithm to Check the Singularity of Block Graphs

Ranveer Singh[1]([⊠]), Naomi Shaked-Monderer[1,2], and Avi Berman[1]

[1] Technion-Israel Institute of Technology, 32000 Haifa, Israel
{singh,nomi,berman}@technion.ac.il
[2] The Max Stern Yezreel Valley College, 19300 Yezreel Valley, Israel

Abstract. A block graph is a graph in which every block is a complete graph. Let G be a block graph and let $A(G)$ be its $(0,1)$-adjacency matrix. Graph G is called nonsingular (singular) if $A(G)$ is nonsingular (singular). Characterizing nonsingular block graphs is an interesting open problem proposed by Bapat and Roy in 2013. In this article, we give a linear time algorithm to check whether a given block graph is singular or not.

Keywords: Block · Block graph · Nonsingular graph · Nullity

AMS Subject Classifications: 15A15 · 05C05

1 Introduction

Let $G = (V(G), E(G))$ be a undirected graph with vertex set $V(G)$ and edge set $E(G)$. Let the cardinality $|V(G)|$ (also called *the order of G*) be equal to n. The adjacency matrix $A(G) = (a_{ij})$ of G is the square matrix of order n defined by

$$a_{ij} = \begin{cases} 1 & \text{if the vertices } i,j \text{ are connected by an edge,} \\ 0 & \text{if } i = j \text{ or } i,j \text{ are not connected by an edge,} \end{cases}$$

where $1 \leq i,j \leq n$. A graph G is called *nonsingular (singular)* if $A(G)$ is *nonsingular (singular)*, that is, the determinant of $A(G)$ is nonzero (zero). The *rank* of G, denoted by $r(G)$, is the rank of the adjacency matrix $A(G)$. If G has full rank, that is, $r(G) = n$, then G is nonsingular, otherwise, it is singular. The *nullity* of G, denoted by $\eta(G)$, is equal to the number of zero eigenvalues of $A(G)$. By the rank-nullity theorem $r(G) + \eta(G) = n$. Thus a zero nullity of a graph G implies that it is nonsingular while a positive nullity implies that it is singular. A *cut-vertex* of G is a vertex whose removal results in an increase in the number of connected components. Let $G_1 = (V_1, E_1)$ and $G_2 = (V_2, E_2)$ be

This work is supported by the Joint NSFC-ISF Research Program (jointly funded by the National Natural Science Foundation of China and the Israel Science Foundation (Nos. 11561141001, 2219/15).

S. P. Pal and A. Vijayakumar (Eds.): CALDAM 2019, LNCS 11394, pp. 77–90, 2019.
https://doi.org/10.1007/978-3-030-11509-8_7

graphs on disjoint sets of vertices. Their *disjoint union* $G_1 + G_2$ is the graph $G_1 + G_2 = (V_1 \cup V_2, E_1 \cup E_2)$. A *coalescence* of graphs G_1 and G_2 is any graph obtained from the disjoint union $G_1 + G_2$ by identifying a vertex v_1 of G_1 with a vertex v_2 of G_2, that is, merging v_1, v_2 into a single vertex v. If v_1, and v_2 have loops of weights α_1, α_2, respectively, then in the coalescence v will have a loop of weight $(\alpha_1 + \alpha_2)$.

A *block* in a graph is a maximal connected subgraph that has no cut-vertex ([21], p.15). Note that if a connected graph has no cut-vertex, then it itself is a block. The blocks and cut-vertices in any graph can found in linear time [11]. A block in a graph G is called a *pendant block* if it has one cut-vertex of G. Two blocks are called *adjacent blocks* if they share a cut-vertex of G. A complete graph on n vertices is denoted by K_n. If every block of a graph is a complete graph then it is called a *block graph*. An example of block graph is given in Fig. 1(a), where the blocks are the induced subgraph on the vertex-sets $\{3, 5, 6, 7\}, \{1, 4, 3, 2\}, \{2, 11, 12\}$, and $\{1, 10, 8, 9\}$. For more details on the block graphs see ([2], Chap. 7).

A well-known problem, proposed in 1957, by Collatz and Sinogowitz, is to characterize graphs with positive nullity [20]. Nullity of graphs is applicable in various branches of science, in particular, quantum chemistry, Hückel molecular orbital theory [9,14] and social networks theory [15]. There has been significant work on the nullity of undirected graphs like trees [6,8,10], unicyclic graphs [12,16], and bicyclic graphs [3,5,23]. It is well-known that a tree is nonsingular if and only if it has a perfect matching. Since a tree is a block graph, it is natural to investigate in general which block graphs are nonsingular. The combinatorial formulae for the determinant of block graphs are known in terms of size and arrangements of its blocks [4,17,18]. But, due to a myriad of the possibility of sizes and arrangements of the blocks it is difficult in general to classify nonsingular or singular block graphs. In [19], some classes of singular and nonsingular block graphs are given. However, it is still an open problem. In this paper, we give a linear time algorithm to determine whether a given graph is singular or not.

Thanks to the eminent computer scientist Alan Turing, the matrix determinant can be computed in the polynomial time using the LUP decomposition of the matrix, where L is a lower triangular matrix, U is an upper triangular matrix, P is a permutation matrix. Interestingly, the asymptotic complexity of the matrix determinant is the same as that of matrix multiplication of two matrices of the same order. The theorem which relates the complexity of matrix product and the matrix determinant is as follows.

Theorem 1 [1]. *Let $M(n)$ be the time required to multiply two $n \times n$ matrices over some ring, and A be an $n \times n$ matrix. Then, we can compute the determinant of A in $O(M(n))$ steps.*

Rank, nullity is also calculated using LUP decomposition, hence their complexities are also the same as that of the matrix determinant or product of two matrices of the same order. The fastest matrix product takes time of $O(n^{2.373})$

[7,13,22]. In this article, we give a linear time algorithm to check whether a given block graph G of order n, is singular or nonsingular (full rank). The algorithm is linear in n. We will not make use of matrix product or elementary row/column operations during any stage of the computation, instead, we prove first results using elementary row/column operation on matrices and use them to treat pendant blocks of the block graphs.

The rest of the article is organised as follows. In Sect. 2, we mention some notations and preliminaries used in the article. In Sect. 3, we give necessary and sufficient condition for any block graph to be nonsingular by using prior results of elementary row/column operation given in Sect. 2. In Sect. 4, we provide an outline for Algorithm 1 and provide some examples in support of it. A proof of the correctness of Algorithm 1 is given in Sect. 5.

2   Notations and Preliminary Results

If Q is a subgraph of graph G, then $G \setminus Q$ denotes the induced subgraph of G on the vertex subset $V(G) \setminus V(Q)$. Here, $V(G) \setminus V(Q)$ is the standard set-theoretic subtraction of vertex sets. If Q consists of a single vertex v, then we write $G \setminus v$ for $G \setminus Q$. J, j, O, o denote the all-one matrix, all-one column vector, zero matrix, zero column vector of suitable order, respectively. w denotes a $(0, 1)$-column vector of suitable order. $diag(x_1, \ldots, x_n)$ denotes the diagonal matrix of order n, where the i-th diagonal entry is $x_i, i = 1, \ldots, n$. If a graph has no vertices (hence no edges) we call it a *void* graph.

Theorem 2. *Consider a matrix*

$$M = \begin{bmatrix} x_1 & 1 & \cdots & 1 \\ 1 & x_2 & \ddots & \vdots \\ \vdots & \ddots & \ddots & 1 \\ 1 & \cdots & 1 & x_n \end{bmatrix} = J - D, \tag{1}$$

where $D = diag(1 - x_1, \ldots, 1 - x_n)$.

1. *If exactly one of* $x_1, \ldots, x_n$ *is equal to 1, then* M *is nonsingular.*
2. *If any two (or more) of* $x_1, \ldots, x_n$ *are equal to 1, then* M *is singular.*
3. *If* $x_i \neq 1, i = 1, \ldots, n$. *Let*

$$S = \sum_{i=1}^{n} \frac{1}{1 - x_i}.$$

Then
(a) M *is nonsingular if and only if* $S \neq 1$.
(b) *if* $S = 1$ *and* x_{n+1} *is any real number, then the matrix*

$$\begin{bmatrix} M & j \\ j^T & x_{n+1} \end{bmatrix}$$

is nonsingular.

4. *If M is nonsingular, then the matrix*

$$\tilde{M} = \begin{bmatrix} M & j \\ j^T & \alpha \end{bmatrix}$$

can be transformed to the following matrix using elementary row and column operations

$$\begin{bmatrix} M & o \\ o^T & \alpha + \gamma \end{bmatrix},$$

where

$$\gamma = \begin{cases} -\frac{S}{S-1} & \text{if } x_i \neq 1, i = 1, \dots n, \\ -1 & \text{if exactly one of } x_1, \dots, x_n \text{ is equal to 1.} \end{cases} \tag{2}$$

Proof. 1. Without loss of generality, let $x_1 = 1$. By subtracting the 1-st row from the rest of the rows, M can be transformed to the matrix,

$$\begin{bmatrix} 1 & 1 & \cdots & 1 \\ 0 & (x_2 - 1) & \ddots & \vdots \\ \vdots & \ddots & \ddots & 1 \\ 0 & \cdots & 0 & (x_n - 1) \end{bmatrix},$$

whose determinant $\prod_{i=2}^{n}(x_i - 1)$ is nonzero.

2. In this case two rows (or columns) are the same.

3. (a) We have

$$D - J = D^{1/2}(I - D^{-1/2}JD^{-1/2})D^{1/2},$$

$$D^{-1/2}JD^{-1/2} = D^{-1/2}jj^T D^{-1/2} = (D^{-1/2}j)(D^{-1/2}j)^T.$$

Let $y = D^{-1/2}j$. The $J - D = M$ is nonsingular if and only if $I - yy^T$ is nonsingular. Since the eigenvalues of yy^T are $\|y\|^2 = y^T y$ and 0, the eigenvalues of the matrix $I - yy^T$ are $1 - \|y\|^2$ and 1 hence it is nonsingular if and only if $\|y\|^2 \neq 1$. That is M is nonsingular if and only if

$$\sum_{i=1}^{n} \frac{1}{1 - x_i} \neq 1.$$

(b)   i. If $x_{n+1} \neq 1$: as $S = 1$, this implies

$$S + \frac{1}{1 - x_{n+1}} \neq 1.$$

Hence the result follows by 3(a).

ii. If $x_{n+1} = 1$: then result follows by 1.

4. (a) If $x_i \neq 1, i = 1, \dots n$. Let $y = D^{-1/2}j$. We have,

$$(I - yy^T)(I + tyy^T) = I + (t - 1 - t\|y\|^2)yy^T,$$

and therefore if $\|y\|^2 \neq 1$,

$$(I - yy^T)^{-1} = I + \frac{1}{1 - \|y\|^2} yy^T.$$

Thus if $J - D$ is invertible, where $d_i = 1 - x_i$, then, by the above,

$$(J - D)^{-1} = -(D - J)^{-1} = -D^{-1/2}\left(I + \frac{1}{1 - \|y\|^2} yy^T\right) D^{-1/2},$$

where $y = D^{-1/2}j$.

$$j^T M^{-1} j = -j^T D^{-1/2}\left(I + \frac{1}{1 - \|y\|^2} yy^T\right) D^{-1/2} j$$

$$= -y^T\left(I + \frac{1}{1 - \|y\|^2} yy^T\right) y = \frac{\|y\|^2}{\|y\|^2 - 1}.$$

Let

$$P = \begin{bmatrix} I & -M^{-1}j \\ o^T & 1 \end{bmatrix}.$$

Then

$$P^T \tilde{M} P = \begin{bmatrix} M & o \\ o^T & \alpha + \gamma \end{bmatrix},$$

where

$$\gamma = -j^T M^{-1} j = -\frac{\|y\|^2}{\|y\|^2 - 1} = -\frac{S}{S - 1}.$$

(b) When exactly one of $x_1, \ldots, x_n$ is equal to 1, without loss of generality, $x_1 = 1$, we can write

$$\tilde{M} = \begin{bmatrix} 1 & 1 & \ldots & 1 & 1 \\ 1 & x_2 & \ddots & \vdots & \vdots \\ \vdots & \ddots & \ddots & 1 & 1 \\ 1 & \ldots & 1 & x_n & 1 \\ 1 & \ldots & 1 & 1 & \alpha \end{bmatrix}$$

On subtracting the first column from the last column, and subsequently subtracting the first row from the last row. We get the following matrix,

$$\tilde{M} = \begin{bmatrix} 1 & 1 & \ldots & 1 & 0 \\ 1 & x_2 & \ddots & \vdots & \vdots \\ \vdots & \ddots & \ddots & 1 & 0 \\ 1 & \ldots & 1 & x_n & 0 \\ 0 & \ldots & 0 & 0 & \alpha - 1 \end{bmatrix},$$

hence $\gamma = -1$.

Lemma 1 [18, Lemma 2.3]. *If G is a coalescence of G_1 and G_2 at a vertex v having loop of weight α, then*

$$\det(G) = \det(G_1)\det(G_2 \setminus v) + \det(G_1 \setminus v)\det(G_2) - \alpha \det(G_1 \setminus v)\det(G_2 \setminus v).$$

3   Transformation of the Adjacency Matrix

Elementary row/column operations transform the adjacency matrix $A(G)$ into another matrix $EA(G)$, which may have nonzero diagonal entries. The graph corresponding to $EA(G)$ can be obtained by a change of weights on the edges and addition of loops on the vertices in G according to the elementary row/column operations on $A(G)$. This can be seen as follows. Let B be a pendant block of a block graph G, having the cut-vertex v of G. As reordering the vertices of G does not change the rank of G, we can write the adjacency matrix $A(G)$ as the follows.

$$A(G) = \begin{bmatrix} A(B \setminus v) & j & O^T \\ j^T & 0 & w^T \\ O & w & A(G \setminus B) \end{bmatrix}.$$

Let the pendant block B have m vertices.

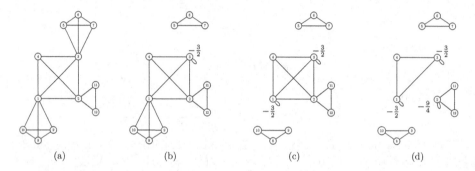

(a)   (b)   (c)   (d)

Fig. 1. Results of elementary row/column operations on the successive pendant blocks of the block graph in (a). In (d) all the four weighted induced subgraphs are nonsingular, hence, the block graph in (a) is nonsingular.

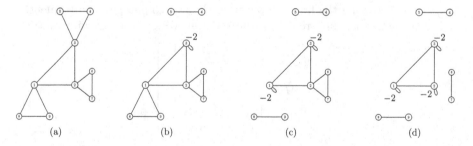

(a)   (b)   (c)   (d)

Fig. 2. Results of elementary row/column operations on the successive pendant blocks of the block graph in (a). In (d) the weighted induced subgraph on vertices $1, 2, 3$ is singular, hence, the block graph in (a) is singular.

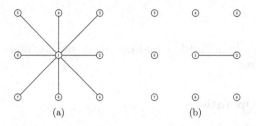

Fig. 3. Results of elementary row/column operations on the successive pendant blocks of the block graph in (a). In (d) there are isolated vertices with zero weights, hence, the block graph in (a) is singular.

1. If $m \geq 3$: as $A(B \setminus v)$ is the complete graph on $m - 1 \geq 2$ vertices, it is non-singular. All the diagonal elements in $A(B \setminus v)$ are zero. Using Theorem 2(4), $A(G)$ is transformed to the following matrix.

$$M_1 = \begin{bmatrix} A(B \setminus v) & o & o^T \\ o^T & -\frac{m-1}{m-2} & w^T \\ O & w & A(G \setminus B) \end{bmatrix}.$$

Let

$$M_{12} = \begin{bmatrix} -\frac{m-1}{m-2} & w^T \\ w & A(G \setminus B) \end{bmatrix}.$$

We have, $r(A(G)) = r(M_1) = r(A(B \setminus v)) + r(M_{12})$. As $A(B \setminus v)$ is nonsingular, $A(G)$ is nonsingular if and only if M_{12} is nonsingular.

2. If $m = 2$: In this case G is a coalescence of $B = K_2$ with $G \setminus (B \setminus v)$ at the vertex v. As K_2 is nonsingular and $K_2 \setminus v$ is singular, using Lemma 1, G is nonsingular if and only if $A(G \setminus B)$ is nonsingular.

In view of the above discussion, in order to know whether block graph G is singular or nonsingular, we need to further examine the matrices M_{12} for $m \geq 3$, $A(G \setminus B)$, for $m = 2$. Note that, M_{12}, is the matrix corresponding to the induced subgraph on the vertex-set $V(G) \setminus V(B \setminus v)$ of G, with a loop on the vertex v having weight $-\frac{m-1}{m-2}$. By selecting a pendant block from the graph $G \setminus B$ (for $m = 2$) or the graph corresponding to M_{12} (for $m \geq 3$) we can further investigate the rank of G. This process will continue until we cover all the blocks of G. The matrix corresponding to the pendant block at any step is of the form M in Theorem 2. Hence we use the prior results from Sect. 2 to check whether G is singular or not. Examples are given in Fig. 1 (nonsingular), Fig. 2 (singular) and Fig. 3 (singular), where at each step a pendant block is chosen for the elementary row/column operations. The detailed algorithm is given in the next section.

4   Algorithm

For the purpose of the algorithm we first define two auxiliary operations on a set of sets of integers.

4.1   Auxiliary Operations

Let $U = \{S_1, \ldots, S_k\}$, where, $S_i, i = 1, \ldots, k$, is a set of positive integers. We define two operations on U:

1. $U - S_i = \{S_j \mid j \neq i\}$.
2. $U -^{\star} S_i = \{S_j \setminus S_i \mid j \neq i\}$.

Example 1. Let $S_1 = \{2, 3, 4, 5\}, S_2 = \{2, 7, 6, 5\}, S_3 = \{1, 3, 8, 5\}$. Consider $U = \{S_1, S_2, S_3\}$. Then,

1. $U - S_2 = \{\{2, 3, 4, 5\}, \{1, 3, 8, 5\}\}$.
2. $U -^{\star} S_2 = \{\{3, 4\}, \{1, 3, 8\}\}$.

The cardinality of a set $\mathcal{U}$ is denoted by $|\mathcal{U}|$. In Example 1, the cardinality of U is 3.

4.2   Algorithm

Given a block graph G, with blocks $B_1, \ldots, B_k$. We have a set BV which is the set of vertex-sets of the blocks $B_1, \ldots, B_k$, in G. CV is the set of cut-vertex-sets of the blocks $B_1, \ldots, B_k$. Let $V(B_i)$ denote the vertex set of the block B_i and $C(B_i)$ be the set of cut-vertices in it. That is, we have

$$BV = \{V(B_1), \ldots, V(B_k)\},$$

$$CV = \{C(B_1), \ldots, C(B_k)\}.$$

A vertex-set $V(B_i)$ in BV is *pendant vertex-set* if $C(B_i)$ contains exactly one cut-vertex, that is, $|C(B_i)| = 1$. Let $f(p)$ be the number of times the cut-vetex p appears in CV. Let $w(i)$ be the weight assigned to the vertex i, $i = 1, \ldots, n$ of G. We first give an outline of the procedure to check whether a given block graph G is singular or nonsingular. We start with the sets BV, CV. Initially, $w(i) = 0, i = 1, \ldots, n$.

BEGIN- If BV has no pendant vertex-set, go to **END**. Otherwise, pick a pendant vertex-set V_p, where the cut-vertex is p.
Check if $w(i) \neq 1, \forall i \in (V_p \setminus p)$.

1. If false:
   (a) If for more than one i, $w(i)$ is 1, then G is **singular**. **STOP**.
   (b) If for exactly one i, $w(i)$ is 1, then induced subgraph on $V_p \setminus p$ is nonsingular. **DO**
       i. $w(p) := w(p) - 1$.

Algorithm 1: Algorithm to check whether a block graph G is singular or not.

Result: G is **singular** or G is **nonsingular**.
$BV = \{V(B_1), \ldots, V(B_k)\}$,
$CV = \{C(B_1), \ldots, C(B_k)\}$,
$W = \{w(1), \ldots, w(n)\}$, $w(i) = 0, i = 1, \ldots, n$,
$f(i)$ is the number of times vertex i appears in CV.

```
 1  GSING (BV, CV);
 2  m := 0;                                          ▷ m is a variable to check the existence of a pendant vertex-set.
 3  for i = 1 : |CV| do
 4      if (|CV(i)| = 1) then
 5          SBV := BV(i);
 6          p := CV(i);
 7          V := SBV \ p;
 8          m := 1;
 9          break;
10      end
11  end
12  if (m = 1) then
13      t := 0;                                      ▷ t is a variable to count the number of nonzero weights.
14      for i = 1 : |V| do
15          if (w(V(i)) ~= 1) then
16              t := t + 1;
17          end
18      end
19      if (t ~= |V|) then
20          if (t < |V| - 1) then
21              G is singular ;
22              exit;
23          else
24              w(p) := w(p) - 1;
25              BV = BV - SBV;
26              if (f(p) == 2) then
27                  CV := CV -* p;
28              else
29                  CV := CV - p;
30                  f(p) := f(p) - 1;
31              end
32          end
33          GSING (BV, CV)
34      else
35          S := 0;                                   ▷ S is a variable to calculate the sum as in Theorem 2.3.
36          for i = 1 : |V| do
37              S := S + 1/(1-w(V(i)));
38          end
39          if S ≠ 1 then
40              w(p) := w(p) - S/(S-1);
41              BV := BV - SBV;
42              if (f(p) == 2) then
43                  CV := CV -* p;
44              else
45                  CV := CV - p;
46                  f(p) := f(p) - 1;
47              end
48          else
49              BV := BV -* SBV;
50              CV := CV -* p;
51          end
52          GSING (BV, CV)
53      end
54  else
55      for i = 1 : |BV| do
56          if BV(i) is singular then
57              G is singular;
58              break;
59          end
60      end
61      G is nonsingular;
62  end
```

ii. $BV := BV - V_p$.

iii. If $f(p) = 2$, then $CV := CV -^* p$, else, $CV := CV - p, f(p) := f(p) - 1$.

BEGIN the algorithm for BV, CV.

2. If true: Let $S = \sum_{i \in (V_p \backslash p)} \frac{1}{1-w(i)}$. **Check If** $S \neq 1$.

   (a) If true: Then the induced subgraph on $V_p \backslash p$ is nonsingular. **DO**

      i. $w(p) := w(p) - \frac{S}{S-1}$.

      ii. $BV := BV - V_p$.

      iii. If $f(p) = 2$, $CV := CV -^* p$, else, $CV := CV - p, f(p) := f(p) - 1$.

     BEGIN the algorithm for BV, CV.

   (b) If false: then induced subgraph on $V_p \backslash p$ is singular. **DO**

      i. $BV := BV -^* V_p$.

      ii. $CV := CV -^* p$.

     BEGIN the algorithm for BV, CV.

END: Let $H_1, \ldots, H_t$ be the induced subgraphs of G on the vertex sets in BV with the addition of weighted loops added in the process. If H_i is nonsingular for all $i = 1, \ldots, t$, then G is **nonsingular** else G is **singular**. Note that, each of H_i is of form M in Theorem 2, hence, using Theorem 2, it is checked in linear time whether H_i is singular or nonsingular.

The algorithm of the above outline is given in Algorithm 1. Where the function **GSING** takes BV, CV as input, and makes recursive use of basic arithmetic, auxiliary operations which take at the most linear time, and check whether the block graph G is singular or not. Now we provide some examples in support of Algorithm 1.

Example 2. Consider the block graph given in Fig. 1(a).

$$BV = \{\{3,5,6,7\}, \{1,10,8,9\}, \{3,4,1,2\}, \{2,11,12\}\},$$

$$CV = \{\{3\}, \{1\}, \{3,1,2\}, \{2\}\},$$

$f(3) = f(2) = f(1) = 2$.

1. 1st call to $GSING(BV, CV)$. As 3 is the only vertex in $\{3\}$ in CV, the corresponding vertex-set $\{3,5,6,7\}$ in BV is selected as pendant vertex-set. Next, induced graph on vertex set $\{3,5,6,7\} \backslash 3$ is investigated. $S = 3$, as $w(5) = w(6) = w(7) = 0$. Then $w(3)$ will be updated to $w(3) - \frac{3}{2}$, that is vertex 3 will now have a loop of weight $-\frac{3}{2}$. Now,

$$BV = \{\{1,10,8,9\}, \{3,4,1,2\}, \{2,11,12\}\},$$

As, $f(3) = 2$,

$$CV = \{\{1\}, \{1,2\}, \{2\}\}.$$

2. 2nd call to $GSING(BV, CV)$. As 1 is the only vertex in $\{1\}$ in CV, the corresponding vertex-set $\{1,10,8,9\}$ in BV is selected as pendant vertex-set. Next, induced graph on vertex set $\{1,10,8,9\} \backslash 1$ is investigated. As

$w(10) = w(8) = w(9) = 0$, $S = 3$. Then $w(1)$ will be updated to $w(1) - \frac{3}{2}$, that is vertex 1 will now have a loop of weight $-\frac{3}{2}$.

$$BV = \{\{3, 4, 1, 2\}, \{2, 11, 12\}\},$$

As, $f(1) = 2$,

$$CV = \{\{2\}, \{2\}\}.$$

3. 3rd call to $GSING(BV, CV)$. As 2 is the only vertex in $\{2\}$ in CV, the corresponding vertex-set $\{3, 4, 1, 2\}$ in BV is selected as pendant vertex-set. Next, induced graph on vertex set $\{3, 4, 1, 2\} \setminus 2$ is investigated. As $w(1) = -\frac{3}{2}, w(3) = -\frac{3}{2}, w(4) = 0$, $S = \frac{9}{5}$. Then $w(2)$ will be updated to $w(2) - \frac{9}{4}$, that is vertex 2 will now have a loop of weight $-\frac{9}{4}$.

$$BV = \{\{2, 11, 12\}\},$$

As, $f(1) = 2$,

$$CV = \{\}.$$

4. 4-th call to $GSING(BV, CV)$. As CV is empty, there is no pendant vertex-set. We need to investigate the induced graph H_1 on vertex set $\{2, 11, 12\}$ (with loop on 2). As $w(2) = -\frac{9}{4}, w(11) = 0, w(12) = 0$, and hence $S = \frac{30}{13} \neq 1$. Which means H_1 is nonsingular, hence G is nonsingular.

Example 3. The steps for block graph in Fig. 2(a) are similar to the steps in Example 2 for block graph in Fig. 1(a). In the 4-th call to $GSING(BV, CV)$, we need to investigate the induced graph H_1 on vertex set $\{1, 2, 3\}$ (with loops). As $w(1) = w(2) = w(3) = -2$, and hence $S = 1$. Which means H_1 is singular, hence G is singular.

Example 4. Consider block graph in Fig. 3(a). We have,

$$BV = \{\{1, 2\}, \{1, 3\}, \{1, 4\}, \{1, 5\}, \{1, 6\}, \{1, 7\}, \{1, 8\}, \{1, 9\}\},$$

$$CV = \{\{1\}, \{1\}, \{1\}, \{1\}, \{1\}, \{1\}, \{1\}, \{1\}\},$$

$f(1) = 2$.

1. 1st call to $GSING(BV, CV)$. As 1 is the only vertex in $\{1\}$ in CV, the corresponding vertex-set $\{1, 2\}$ in BV is selected as pendant vertex-set. Next, induced graph on vertex set $\{1, 2\} \setminus 1$ is investigated. As $w(2) = 0$, $S = 1$. Thus

$$BV = \{\{3\}, \{4\}, \{5\}, \{6\}, \{7\}, \{8\}, \{9\}\},$$

$$CV = \{\{\}, \{\}, \{\}, \{\}, \{\}, \{\}, \{\}\},$$

2. 2nd call to $GSING(BV, CV)$. As CV is empty, there is no pendant vertex-set. We need to investigate the induced graphs on vertex-sets in BV. But as all are singleton vertices without loops. Hence the block graph is singular.

5 Proof of Correctness

Just before (i)-th calling of function $GSING(BV, CV)$ in Algorithm 1, the adjacency matrix $A(G)$ is transformed to the matrix M_i which is of the form

$$
M_i = \begin{bmatrix}
A(G_1) & & & & & \\
& \ddots & & & & \\
& & A(G_{i-1}) & & & \\
& & & A(G_i \setminus i) & j & \\
& & & j^T & \alpha & w^T \\
& & & & w & M^\star
\end{bmatrix}.
$$

where each G_j, $j = 1, \ldots, i-1$, are some induced subgraph of block B (possibly with loops) whose vertex-set selected as pendant vertex-set during j-th calling of $GSING(BV, CV)$. It is clear that if any one of $G_1, \ldots, G_{i-1}$ is singular, then G is singular. Note that if any of $G_1, \ldots, G_{i-1}$ is empty then it is to be considered as nonsingular by convention. Now, G_i is the induced subgraph (possibly with loops) whose vertex-set is selected as pendant vertex-set during i-th calling of $GSING(BV, CV)$. Let i be the cut-vertex in the selected pendant vertex-set. Let α be the weight of loop at i, if there is no loop at i then $\alpha = 0$. $M^\star$ is the matrix corresponding to induced subgraph (with possible loops) of G, on the vertex-set $V(G) \setminus (V(G_1) \cup \ldots \cup V(G_i))$. Thus, if $G_1, \ldots, G_{i-1}$ are nonsingular, then G is nonsingular if and only if the following submatrix is nonsingular,

$$
\tilde{M} = \begin{bmatrix}
A(G_i \setminus i) & j & O \\
j^T & \alpha & w^T \\
O^T & w & M^\star
\end{bmatrix}. \tag{3}
$$

$A(G_i)$ is of form $J - D$, where $D = diag(d_1, d_2, \ldots, d_{|V(G_i)|})$ is diagonal matrix, $d_i = 1 - w(i), i = 1, 2, \ldots, |V(G_i)|$.

1. If more than one diagonal entries of $A(G_i \setminus i)$ are equal to 1, then as, at least two rows and two columns of $A(G_i \setminus i)$ are same, it is obvious that G is singular.
2. If exactly one diagonal entry of $A(G_i \setminus i)$ is 1, the using Theorem 2.1, $A(G_i \setminus i)$ is nonsingular. By Theorem 2.4, $\tilde{M}$ can be transformed to the following matrix,

$$
\begin{bmatrix}
A(G_i \setminus i) & o & O \\
o^T & \alpha - 1 & w^T \\
O^T & w & M^\star
\end{bmatrix}.
$$

Thus, $\tilde{M}$ is nonsingular if and only if the following nonsingular matrix is nonsingular,

$$
\begin{bmatrix}
\alpha - 1 & w^T \\
w & M^\star
\end{bmatrix}.
$$

3. If $w(i) \neq 1, i = 1, \ldots, |V(G_i)-1|$ and $S = \sum_{i=1}^{|V(G_i)-1|} \frac{1}{1-w(i)} \neq 1$. Then $A(G_i \setminus i)$ is nonsingular. By Theorem 2.4, $\tilde{M}$ can be transformed to the following matrix,

$$\begin{bmatrix} A(G_i \setminus i) & o & O \\ o^T & \alpha - \frac{S}{S-1} & w^T \\ O^T & w & M^\star \end{bmatrix}.$$

Thus, $\tilde{M}$ is nonsingular if and only if the following nonsingular matrix is nonsingular,

$$\begin{bmatrix} \alpha - \frac{S}{S-1} & w^T \\ w & M^\star \end{bmatrix}.$$

4. If $A(G_i \setminus i)$ is singular. Then by Theorem 2.3(b), $A(G_i)$ is nonsingular. The graph corresponding to $\tilde{M}$ is a coalescence of G_i with the induced graph (possibly with loop) of $M^\star$.

   As, $A(G_i \setminus i)$ is singular and $A(G_i)$ is nonsingular, by Lemma 1, $\tilde{M}$ is nonsingular if and only if $M^\star$ is nonsingular.

This completes the proof.

Acknowledgments. The authors are grateful to Dr. Cheng Zheng for his valuable comments and suggestions.

References

1. Aho, A.V., Hopcroft, J.E.: The Design and Analysis of Computer Algorithms. Pearson Education India, New Delhi (1974)
2. Bapat, R.B.: Graphs and Matrices. Springer, London (2014). https://doi.org/10.1007/978-1-4471-6569-9
3. Bapat, R.: A note on singular line graphs. Bull. Kerala Math. Assoc. **8**(2), 207–209 (2011)
4. Bapat, R., Roy, S.: On the adjacency matrix of a block graph. Linear Multilinear Algebra **62**(3), 406–418 (2014)
5. Berman, A., Friedland, S., Hogben, L., Rothblum, U.G., Shader, B.: An upper bound for the minimum rank of a graph. Linear Algebra Appl. **429**(7), 1629–1638 (2008)
6. Cvetkovic, D.M., Doob, M., Sachs, H.: Spectra of Graphs. Pure Applied Mathematics, vol. 87. Academic Press, New York (1980)
7. Davie, A.M., Stothers, A.J.: Improved bound for complexity of matrix multiplication. Pro. R. Soc. Edinb.: Sect. A Math. **143**(02), 351–369 (2013)
8. Fiorini, S., Gutman, I., Sciriha, I.: Trees with maximum nullity. Linear Algebra Appl. **397**, 245–251 (2005)
9. Gutman, I., Borovicanin, B.: Nullity of graphs: an updated survey. In: Selected Topics on Applications of Graph Spectra, pp. 137–154. Mathematical Institute SANU, Belgrade (2011)
10. Gutman, I., Sciriha, I.: On the nullity of line graphs of trees. Discret. Math. **232**(1–3), 35–45 (2001)
11. Hopcroft, J.E., Tarjan, R.E.: Efficient algorithms for graph manipulation (1971)

12. Hu, S., Xuezhong, T., Liu, B.: On the nullity of bicyclic graphs. Linear Algebra Appl. **429**(7), 1387–1391 (2008)
13. Le Gall, F.: Powers of tensors and fast matrix multiplication. In: Proceedings of the 39th International Symposium on Symbolic and Algebraic Computation, pp. 296–303. ACM (2014)
14. Lee, S.L., Li, C.: Chemical signed graph theory. Int. J. Quantum Chem. **49**(5), 639–648 (1994)
15. Leskovec, J., Huttenlocher, D., Kleinberg, J.: Signed networks in social media. In: Proceedings of the SIGCHI Conference on Human Factors in Computing Systems, pp. 1361–1370. ACM (2010)
16. Nath, M., Sarma, B.K.: On the null-spaces of acyclic and unicyclic singular graphs. Linear Algebra Appl. **427**(1), 42–54 (2007)
17. Singh, R., Bapat, R.B.: B-partitions, application to determinant and permanent of graphs. Trans. Comb. **7**(3), 29–47 (2018)
18. Singh, R., Bapat, R.: On characteristic and permanent polynomials of a matrix. Spec. Matrices **5**, 97–112 (2017)
19. Singh, R., Zheng, C., Shaked-Monderer, N., Berman, A.: Nonsingular block graphs: an open problem. arXiv preprint arXiv:1803.03947 (2018)
20. Von Collatz, L., Sinogowitz, U.: Spektren endlicher grafen. Abhandlungen aus dem Mathematischen Seminar der Universität Hamburg **21**, 63–77 (1957)
21. West, D.B.: Introduction to Graph Theory, vol. 2. Prentice Hall, Upper Saddle River (2001)
22. Williams, V.V.: Breaking the Coppersmith-Winograd barrier (2011)
23. Xuezhong, T., Liu, B.: On the nullity of unicyclic graphs. Linear Algebra Appl. **408**, 212–220 (2005)

b-Coloring of the Mycielskian
of Regular Graphs

S. Francis Raj$^{(\boxtimes)}$ and M. Gokulnath

Department of Mathematics, Pondicherry University, Puducherry 605014, India
francisraj_s@yahoo.com, gokulnath.math@gmail.com

Abstract. The b-chromatic number $b(G)$ of a graph G is the maximum k for which G has a proper vertex coloring using k colors such that each color class contains at least one vertex adjacent to a vertex of every other color class. In this paper, we have mainly investigated on the b-chromatic number of the Mycielskian of regular graphs. In particular, we have obtained the exact value of the b-chromatic number of the Myciel-skian of some classes of graphs. This includes a few families of regular graphs, graphs with $b(G) = 2$ and split graphs. In addition, we have found bounds for the b-chromatic number of the Mycielskian of some more families of regular graphs in terms of the b-chromatic number of their original graphs.

Keywords: b-coloring · b-chromatic number · Mycielskian of graphs · Regular graphs

2000 AMS Subject Classification: 05C15

1 Introduction

All graphs considered in this paper are simple, finite and undirected. Let G be a graph with vertex set $V(G)$ and edge set $E(G)$. A b-coloring of a graph G using k colors is a proper coloring of the vertices of G using k colors in which each color class has a color dominating vertex (c.d.v.), that is, a vertex that has a neighbor in each of the other color classes. The b-chromatic number, $b(G)$ of G is the largest k such that G has a b-coloring using k colors. For a given b-coloring of a graph, a set of c.d.v.'s, one from each class, is known as a color dominating system (c.d.s.) of that b-coloring. The concept of b-coloring was introduced by Irving and Manlove [6] in analogy to the achromatic number of a graph G (which gives the maximum number of color classes in a complete coloring of G). Clearly, $b(G) \leq \Delta(G) + 1$, where $\Delta(G)$ is the maximum degree of G.

It is clear from the definition of $b(G)$ that the chromatic number, $\chi(G)$ of G is the least k for which G admits a b-coloring using k colors and hence $\chi(G) \leq b(G)$. Graphs for which there exists a b-coloring using k colors for every integer k such that $\chi(G) \leq k \leq b(G)$ are known as b-continuous graphs. It can be observed that all graphs are not b-continuous. For instance, Q_3 has a b-coloring using 2

© Springer Nature Switzerland AG 2019
S. P. Pal and A. Vijayakumar (Eds.): CALDAM 2019, LNCS 11394, pp. 91–96, 2019.
https://doi.org/10.1007/978-3-030-11509-8_8

colors and 4 colors but none with 3 colors, and therefore Q_3 is not b-continuous. Hence the natural question that arises is to characterize graphs which are b-continuous. There are a few papers in this direction. For instance, see [1,3,4]. The b-spectrum of a graph G is the set of positive integers k for which G has a b-coloring using k colors and is denoted by $S_b(G)$, that is, $S_b(G) = \{k : G$ has a b-coloring using k colors$\}$. Clearly, $\{\chi(G), b(G)\} \subseteq S_b(G)$ and G is b-continuous if and only if $S_b(G) = \{\chi(G), \chi(G) + 1, \ldots, b(G)\}$.

Let the vertices of a graph G be ordered as $v_1, v_2, \ldots, v_n$ with $d(v_1) \geq d(v_2) \geq \ldots \geq d(v_n)$. Then the m-degree, $m(G)$ of G is defined by $m(G) = \max\{i : d(v_i) \geq i - 1, 1 \leq i \leq n\}$. For any graph G, $b(G) \leq m(G) \leq \Delta(G) + 1$. Also for any regular graph, $m(G) = \Delta(G) + 1$.

In a search for triangle-free graphs with arbitrary large chromatic numbers, Mycielski [10] developed an interesting graph transformation as follows. For a graph $G = (V, E)$, the Mycielskian of G, denoted by $\mu(G)$, is the graph with vertex set $V(\mu(G)) = V \cup V' \cup \{u\}$ where $V' = \{x' : x \in V\}$ and the edge set $E(\mu(G)) = E \cup \{xy' : xy \in E\} \cup \{y'u : y' \in V'\}$. The vertex x' is called the twin of the vertex x (and x the twin of x') and the vertex u is called the root of $\mu(G)$. In $\mu(G)$, if $A \subseteq V$, let A' denotes the set of twin vertices of A in $\mu(G)$ and for every $x \in V$ and any non-negative integer i, define $N_i(x) = \{y \in V : d_G(x, y) = i\}$.

For notation and terminologies not mentioned in this paper, see [11].

2   b-Coloring of the Mycielskian of Regular Graphs

In [2], it has been shown that if G is a graph with b-chromatic number b and for which the number of vertices of degree at least b is at most $2b - 2$, then $b(\mu(G))$ lies in the interval $[b + 1, 2b - 1]$. While considering regular graphs G, in [7,8] it has been shown that $b(G) = \Delta(G) + 1$, when the girth of G is at least 6 or when the girth is at least 5 with no induced C_6. For these regular graphs, the number of vertices of degree at least b is 0 and hence $b(\mu(G))$ lies in the interval $[b + 1, 2b - 1]$. What we intend to do in Sect. 2 is to find the exact value of $b(\mu(G))$ or at least find some better bounds. Also, we would like to investigate on the Mycielskian of k-regular graphs which are b-continuous.

Let us start with the following observations on k-regular graphs with girth at least 7.

Observation 1. Let G be a k-regular graph with girth at least 7. For $v \in V(G)$,

  (i) $N_1(v)$ and $N_2(v)$ are independent sets.
  (ii) For $y, z \in N_2(v)$, $[N_1(y) \cap N_1(z)] \cap N_3(v) = \emptyset$ and there exists at most one edge between $N_1(y)$ and $N_1(z)$ (otherwise, we will get a C_6 or a C_4).
  (iii) For $w \in N_1(v)$ and $x \in N_3(v)$, there exists at most one edge between x and $N_2(w)$.

Theorem 1. For $k \geq 3$, if G is a k-regular graph with girth at least 7, then $b(\mu(G)) = 2k + 1 = 2b(G) - 1$.

Proof. Let G be a k-regular graph with girth at least 7 and $k \geq 3$. It can be easily seen that $m(\mu(G)) = 2k + 1$. Hence it is enough to produce a b-coloring using $2k + 1$ colors. Let $\{0, 1, 2, \ldots, 2k\}$ be the set of $2k + 1$ colors. Let $v \in V$, $N_1(v) = \{v_1, v_2, \ldots, v_k\}$, for $1 \leq i \leq k$, $M(v_i) = \{v_{i,1}, v_{i,2}, \ldots, v_{i,k-1}\}$ be the set of neighbors of v_i other than v in G and for $1 \leq i \leq k$, and $1 \leq j \leq k - 1$, let $M(v_{i,j}) = \{v_{i,j,1}, v_{i,j,2}, \ldots, v_{i,j,k-1}\}$ be the neighbors of $v_{i,j}$ other than v_i in G.

Let us first partially color the graph to get c.d.vs. for each of the color classes. This is done by defining a coloring 'c' for $\mu(G)$ as follows.

(i) $c(u) = k$, $c(v) = 0$, $c(v') = 2k$, $c(v_{1,1}) = 2k$

(ii) For $1 \leq i \leq k$
$$c(v_i) = i$$
$$c(v_i') = k + i$$

(iii) For $2 \leq i \leq k - 1, 1 \leq j \leq k - 1$
$$c(v_{i,j}) = \begin{cases} j & \text{for } i > j \\ j + 1 & \text{for } i \leq j \end{cases}$$
$$c(v_{i,j}') = k + j$$
$$c(v_{k,j}) = k + j$$
$$c(v_{k,j}') = j$$

This partial coloring makes $v, v_2, v_3, \ldots, v_k$ as c.d.vs. for the colors $0, 2, 3, \ldots, k$ respectively. We have to extend this partial coloring in such a way that we get c.d.vs. for the remaining colors, namely $1, k + 1, k + 2, \ldots, 2k$. Let us do this by making $v_{1,1}, v_{2,1}, v_{k,1}, v_{k,2}, \ldots, v_{k,k-1}$ as c.d.vs. for the colors $2k, 1, k + 1, k + 2, \ldots, 2k - 1$ respectively. For the case when $k = 3$, let $c(v_{1,1,1}) = 5$ and $c(v_{1,1,2}) = 3$. Let us assign the colors $\{6, 3\}$ to the vertices of $M(v_{2,1})$ and the colors $\{5, 2\}$ to the vertices of $M(v_{3,1})$. By (ii) of Observation 1, $M(v_{1,1})$ can only be adjacent to at most one vertex in $M(v_{3,1})$ and one vertex in $M(v_{2,1})$ and hence we can permute the colors to get a proper partial coloring. Next, let us assign the colors $\{4, 1\}$, $\{0, 2\}$, $\{0, 4\}$, $\{0, 1\}$ and $\{0, 2\}$ to the vertices of $M(v_{3,2})$, $M(v_{1,1})'$, $M(v_{2,1})'$, $M(v_{3,1})'$ and $M(v_{3,2})'$ respectively. Again for the same reason, we can permute the colors to get a proper coloring. In this case, it can be seen that $v_{1,1}, v_{2,1}, v_{3,1}, v_{3,2}$ are c.d.vs. for the colors $6, 1, 4, 5$ respectively.

When $k \geq 4$, a similar strategy but with a little more involvement will yield the desired extensions by using the facts given in Observation 1. Finally, the remaining vertices can be colored by using greedy coloring technique. □

Let us recall the concept of System of Distinct Representatives (SDR) for a family of subsets of a given finite set. Let $\mathcal{F} = \{A_\alpha : \alpha \in J\}$ be a family of sets. An SDR for the family $\mathcal{F}$ is a set of elements $\{x_\alpha : \alpha \in J\}$ such that $x_\alpha \in A_\alpha$ for every $\alpha \in J$ and $x_\alpha \neq x_\beta$ whenever $\alpha \neq \beta$. Theorem 2 gives a necessary and sufficient condition for the existence of a SDR for a given family of finite sets.

Theorem 2 [5]. *Let $\mathcal{F} = \{A_i : 1 \leq i \leq r\}$ be a family of finite sets. Then $\mathcal{F}$ has an SDR if and only if the union of any k members of $\mathcal{F}$, $1 \leq i \leq r$, contains at least k elements.*

Let us next consider k-regular graphs with girth 6 and a subclass of k-regular graphs with girth 5.

Theorem 3. *If G is a k-regular graph with girth 6, then $k + \lfloor \frac{k+1}{2} \rfloor \leq b(\mu(G)) \leq 2k + 1$.*

Theorem 4. *For $k \geq 3$, if G is a k-regular graph with girth 5, diameter at least 5 and containing no cycles of length 6, then $b(\mu(G)) = 2k + 1$.*

Theorems 3 and 4 can be proved by adopting similar techniques used in Theorem 1 and by using the concept of SDR and its necessary and sufficient condition given in Theorem 2.

In [1], it has been proved that every k-regular graph with girth at least 6 having no cycles of length 7 is b-continuous. By using this together with the strategy used in Theorem 1, we have Theorem 5.

Theorem 5. *If G is a k-regular graph with girth at least 8, then $\mu(G)$ is b-continuous.*

3   Exact Value of $b(\mu(G))$ for Some Families of Graphs

In Sect. 3, we shall find the exact values of $b(\mu(G))$ for some families of graphs. In [9], J. Kratochvíl et.al. have determined the complete characterization of a graph with $b(G) = 2$. With the help of this characterization, we have proved Theorem 6.

Theorem 6. *If G is a graph with $b(G) = 2$, then $b(\mu(G)) = 3$.*

Next let us consider the Mycielskian of split graphs.

Theorem 7. *For a split graph G, $b(\mu(G)) = b(G) + 1 = \omega(G) + 1$.*

Proof. Let G be a split graph. Then the vertex set $V(G)$ can be partitioned into two sets, one inducing a clique and the other inducing an independent set. Let $V(G) = A \cup B$ where A induces a maximum clique and B is an independent set. Clearly $|A| = \omega(G)$.

Suppose $b(\mu(G)) = \ell \geq \omega(G) + 2$, then there exists a b-coloring say 'c' of $\mu(G)$ using ℓ colors. Let $\{1, 2, \ldots, \ell\}$ be the set of colors. Without loss of generality, let $1, 2, \ldots, \omega(G)$ be the colors assigned to the vertices of A. Since the degree of any vertex in B is at most $\omega(G) - 1$, none of the vertex in B' can be a c.d.v. in $\mu(G)$.

Case (i) B contains a c.d.v.

Let $v \in B$ be a c.d.v. of the color, say $\omega(G) + 1$. Since A is a maximum clique, there exists at least one vertex $w \in A$ which is not adjacent to v. It can also be observed that $N(v) \subseteq N(w)$, and hence v cannot be adjacent to a vertex whose color is $c(w)$, a contradiction.

Case (ii) B contains no c.d.vs.

This concludes that all the c.d.vs. must be in $A \cup A' \cup \{u\}$. Since $|A| = \omega(G)$ and $\ell \geq \omega(G) + 2$, it can be seen that A' contains at least one c.d.v., say, w_1' of a color $p \geq \omega(G) + 1$. Then $c(u) = c(w_1)$ and hence the c.d.v. of $\omega(G) + 2$ must also be in A', say w_2'. This again forces $c(u) = c(w_2)$, a contradiction. $\square$

It can be observed that, not all k-regular graphs of girth 4 have $b(G) = k+1$. While considering k-regular graphs with girth 4 and $b(G) = k+1$, we shall show that it is not necessary that $b(\mu(G))$ is very close to $2k+1$.

Theorem 8. *If $G = K_{n,n} - PM$ where PM is a perfect matching of the graph, then $b(\mu(G)) = n + \left\lceil \frac{n-1}{2} \right\rceil$, for $n \geq 3$.*

Proof. Let $G = K_{n,n} - PM$ where PM is a perfect matching of the graph. Let $V(G) = X \cup Y$ where $X = \{x_1, x_2, \ldots, x_n\}$ and $Y = \{y_1, y_2, \ldots, y_n\}$ be the bipartition classes of G and $\{x_1y_1, x_2y_2, \ldots, x_ny_n\}$ be the PM.

Let us now show that we can define a b-coloring 'c' using $n + \left\lceil \frac{n-1}{2} \right\rceil$ colors as follows. Let $\{1, 2, \ldots, n + \left\lceil \frac{n-1}{2} \right\rceil\}$ be the set of colors.

(i) $c(u) = 1$

(ii) $c(x_i) = i$ for $1 \leq i \leq n$

(iii) $c(y_i) = \begin{cases} n + i - 1 & \text{for } 2 \leq i \leq \left\lceil \frac{n-1}{2} \right\rceil + 1 \\ i & \text{for } \left\lceil \frac{n-1}{2} \right\rceil + 2 \leq i \leq n \ \& \ i = 1 \end{cases}$

(iv) $c(x_i') = \begin{cases} 2 & \text{for } i = 1 \\ n + i - 1 & \text{for } 2 \leq i \leq \left\lceil \frac{n-1}{2} \right\rceil + 1 \\ i - \left\lfloor \frac{n-1}{2} \right\rfloor & \text{for } \left\lceil \frac{n-1}{2} \right\rceil + 2 \leq i \leq n \end{cases}$

(v) $c(y_i') = \begin{cases} n + 1 & \text{for } i = 1 \\ i & \text{for } 2 \leq i \leq \left\lceil \frac{n-1}{2} \right\rceil + 1 \\ i + \left\lfloor \frac{n-1}{2} \right\rfloor & \text{for } \left\lceil \frac{n-1}{2} \right\rceil + 2 \leq i \leq n \end{cases}$

In a routine way, one can check that the given coloring c is proper and $x_1, x_2, x_3, \ldots, x_n, y_2, y_3, \ldots, y_{\lceil \frac{n-1}{2} \rceil + 1}$ are the c.d.vs. for the colors $1, 2, 3, \ldots, n, n+1, n+2, \ldots, n + \left\lceil \frac{n-1}{2} \right\rceil$ respectively. Thus $b(\mu(G)) \geq n + \left\lceil \frac{n-1}{2} \right\rceil$. Moreover, without much difficulty it can be seen that the given b-coloring c is optimal and that there exist no b-coloring using ℓ colors, for any $\ell > n + \left\lceil \frac{n-1}{2} \right\rceil$. Hence $b(\mu(G)) \leq n + \left\lceil \frac{n-1}{2} \right\rceil$. $\qquad\square$

Acknowledgment. For the first author, this research was supported by SERB DST Project, Government of India, File no: EMR/2016/007339. For the second author, this research was supported by UGC - BSR, Research Fellowship, Government of India.

References

1. Balakrishnan, R., Kavaskar, T.: b-coloring of Kneser graphs. Discret. Appl. Math. **160**(1–2), 9–14 (2012)
2. Balakrishnan, R., Raj, S.F.: Bounds for the b-chromatic number of the Mycielskian of some families of graphs. Ars Comb. **122**, 89–96 (2015)
3. Balakrishnan, R., Raj, S.F., Kavaskar, T.: b-coloring of Cartesian product of odd graphs. Ars Comb. **131**, 285–298 (2017)
4. Francis, P., Raj, S.F.: On b-coloring of powers of hypercubes. Discret. Appl. Math. **225**, 74–86 (2017)
5. Hall, P.: On representatives of subsets. J. London Math. Soc. **1**(1), 26–30 (1935)
6. Irving, R.W., Manlove, D.F.: The b-chromatic number of a graph. Discret. Appl. Math. **91**(1–3), 127–141 (1999)

7. Kouider, M.: b-chromatic number of a graph, subgraphs and degrees. Rapport interne LRI 1392 (2004)
8. Kouider, M., El Sahili, A.: About b-colouring of regular graphs. Rapport de Recherche 1432 (2006)
9. Kratochvíl, J., Tuza, Z., Voigt, M.: On the b-chromatic number of graphs. In: Goos, G., Hartmanis, J., van Leeuwen, J., Kučera, L. (eds.) WG 2002. LNCS, vol. 2573, pp. 310–320. Springer, Heidelberg (2002). https://doi.org/10.1007/3-540-36379-3_27
10. Mycielski, J.: Sur le coloriage des graphs. Colloq. Math. 3(2), 161–162 (1955)
11. West, D.B.: Introduction to Graph Theory. Prentice-Hall of India Private Limited, Delhi (2005)

Drawing Bipartite Graphs in Two Layers with Specified Crossings

Ajit A. Diwan[1($\boxtimes$)], Bodhayan Roy[2], and Subir Kumar Ghosh[3]

[1] Department of Computer Science and Engineering,
Indian Institute of Technology Bombay, Powai, Mumbai 400076, India
aad@cse.iitb.ac.in
[2] Faculty of Informatics, Masaryk University, Brno, Czech Republic
b.roy@fi.muni.cz
[3] Department of Computer Science,
Ramakrishna Mission Vivekananda Educational and Research Institute,
Belur, Howrah 711202, India
ghosh@tifr.res.in

Abstract. We give a polynomial-time algorithm to decide whether a bipartite graph admits a two-layer drawing in the plane such that a specified subset of pairs of edges cross. This is a generalization of the problem of recognizing permutation graphs, and we generalize the characterization of permutation graphs.

Keywords: Abstract topological graph · Two-layer drawing · Bipartite graph · Permutation graph

1 Introduction

An *abstract topological graph* is a graph $G(V, E, F)$, with vertex set V, edge set E, and a collection F of unordered pairs of edges. A *realization* of an abstract topological graph is a drawing of the graph G in the plane satisfying the following properties.

1. Vertices are represented by distinct points in the plane.
2. An edge is represented by the line segment joining the points representing its endvertices and it does not contain any point representing a vertex in its interior.
3. The line segments representing distinct edges $a, b \in E$ intersect in their interior if and only if $\{a, b\} \in F$. In this case, we say the edges a, b cross in the drawing.

Note that if two edges cross in a drawing, they cannot have any common endvertex. We may thus assume that F contains only pairs of non-adjacent edges, otherwise no realization is possible.

S. K. Ghosh—The author's work is funded by SERB, Government of India through a grant under MATRICS.

S. P. Pal and A. Vijayakumar (Eds.): CALDAM 2019, LNCS 11394, pp. 97–108, 2019.
https://doi.org/10.1007/978-3-030-11509-8_9

A *weak realization* of an abstract topological graph is a drawing in which pairs of edges in F cross, but pairs of edges not in F may or may not cross.

In general, deciding whether an abstract topological graph is realizable, or weakly realizable, is NP-hard [6]. In fact, even when G is a complete graph, the realizability problem is $\exists\mathbb{R}$-complete [7]. We consider only bipartite graphs, with a fixed bipartition of the vertices, and a special kind of drawing called a *two-layer drawing*. In such a drawing, all vertices in a part are represented by distinct points contained in a horizontal line, with distinct horizontal lines used for the two parts. Thus, whether two edges cross or not is determined completely by the ordering of the vertices along the two lines, and the actual coordinates do not matter. Such drawings have been extensively studied [2]. An abstract topological bipartite graph G has a two-layer realization if there exists a two-layer drawing of G that realizes it.

Consider two horizontal lines and a collection of line segments with distinct endpoints, such that each segment has an endpoint in both the lines. The intersection graph of such a collection of segments is called a permutation graph [4]. The problem of recognizing and characterizing permutation graphs has been well-studied, and several efficient algorithms and characterizations are known [4,8,9]. This problem is a special case of deciding whether a given abstract topological bipartite graph has a two-layer realization.

Suppose an abstract topological bipartite graph $G(V, E, F)$ is a matching. Consider the graph $L(G)$ whose vertices are edges of G, and a is adjacent to b if and only if $\{a, b\} \in F$. Then the abstract topological graph $G(V, E, F)$ has a two-layer realization if and only if $L(G)$ is a permutation graph. Any representation of $L(G)$ as the intersection graph of line segments with endpoints in two horizontal lines gives a two-layer drawing of G, and vice-versa.

For arbitrary bipartite graphs G, this condition is necessary but not sufficient, as shown by the example in Fig. 1. This happens because in a representation of the permutation graph $L(G)$ as an intersection graph of line segments with endpoints in two horizontal lines, the endpoints of line segments representing edges having a common endvertex in G may not appear consecutively on the

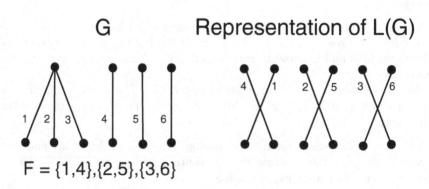

Fig. 1. Example showing insufficiency

horizontal line. We need to add more conditions to ensure this constraint is satisfied. We show that the basic algorithm for recognizing permutation graphs can be extended to achieve this.

In contrast, the problem of deciding whether an abstract topological bipartite graph has a weak two-layer realization is known to be NP-complete [3].

In [1], we had presented some simpler results on this problem. In particular, for a connected abstract topological bipartite graph, there is a much simpler way of deciding two-layer realizability. However, this does not include the case of matchings, which is equivalent to recognizing permutation graphs. In this paper, we consider arbitrary graphs, and thus generalize the known characterization of permutation graphs.

2   Characterization

In this section, we give a characterization of abstract topological bipartite graphs that have a two-layer realization. We assume that there are no isolated vertices in the graph, as these do not affect realizability. As mentioned in the introduction, a two-layer drawing of a bipartite graph may be described combinatorially by a linear order $<$ defined on the vertices in the two parts (say X and Y) of $V(G)$. A drawing defined by a linear order $<$ is a two-layer realization of an abstract topological bipartite graph $G(X, Y, E, F)$ iff it satisfies the following property.

Criterion 1. *For any two non-adjacent edges* $a = x_1 y_1$ *and* $b = x_2 y_2$ *in* G, $\{a, b\} \in F$ *iff either* $x_1 < x_2$ *and* $y_2 < y_1$ *or* $x_2 < x_1$ *and* $y_1 < y_2$.

Thus the problem of finding a two-layer realization reduces to that of finding a linear order $<$ on X and Y, satisfying Criterion 1. A linear ordering on the vertices of G also induces a linear ordering on the edges of G, by lexicographic ordering. For two distinct edges $x_1 y_1, x_2 y_2$, define $x_1 y_1 < x_2 y_2$ iff $x_1 < x_2$ or $x_1 = x_2$ and $y_1 < y_2$. Informally, for edges a, b in G, $a < b$ if the endvertex of a in X is to the left of the endvertex of b in X in the drawing, or, if they have the same endvertex in X, then the endvertex of a in Y is to the left of the endvertex of b in Y. The required property (Criterion 1) of the ordering of vertices, leads to some constraints on the ordering of edges. Also, some constraints arise due to the fact that the ordering on edges is induced by an ordering on the vertices. Our approach is to find a linear ordering on the edges of G satisfying these constraints. We show that this can be done in a way similar to that for recognizing permutation graphs.

Let $G(X, Y, E, F)$ be an abstract topological bipartite graph. The *colored graph representation* of G is the complete graph $C(G)$ with vertex set E whose edges are assigned colors as follows:

1. If a, b are non-adjacent edges in E, then the edge ab in $C(G)$ is colored black if $\{a, b\} \in F$, otherwise it is colored white.
2. If a, b have a common endvertex in X, then the edge ab in $C(G)$ is colored red, otherwise if they have a common endvertex in Y, then it is colored green.

Note that the colored graph $C(G)$ uniquely represents the abstract topological bipartite graph $G(X, Y, E, F)$, and every edge in $C(G)$ is assigned a color. Also, the edges of red color form a spanning subgraph of $C(G)$ in which each component is a complete graph corresponding to a vertex in X. Similarly, the green edges form a subgraph with each component being complete, corresponding to each vertex in Y. We will henceforth work with this representation of the abstract topological bipartite graph.

The main intuition behind the solution can be described as follows. Suppose a, b are non-adjacent edges in G that are required not to cross. Then in any realization, either a is completely to the left of b or vice-versa. Now suppose c is another edge not adjacent to b that is required not to cross b, but either is required to cross a or has an endvertex in common with a. Then in any realization, a and c must be on the same side of b. This forces certain constraints on the ordering of the edges in any realization. These are described by the "forcing" relation $\Leftrightarrow$ defined on ordered pairs of vertices in $C(G)$. Intuitively, if a pair (c, d) is forced by a pair (a, b), then in any ordering of the edges that realizes the graph, $a < b$ if and only if $c < d$. The forcing rules are described below, and these can be obtained by simple geometric arguments, as shown later. Note that "forcing" is a symmetric relation, although we describe the rule only one way, with the first edge being black or white.

Let a, b, c be 3 distinct vertices in $C(G)$.

Rule 1. If edges ab and bc have the same color in $C(G)$ but ac has a different color then $(a, b) \Leftrightarrow (c, b)$, and also $(b, a) \Leftrightarrow (b, c)$.

Rule 2. If edge ab is black or white, bc is green, and ac has a color different from that of ab then $(a, b) \Leftrightarrow (c, b)$ and $(b, a) \Leftrightarrow (b, c)$.

Rule 3. If edge ab is white, bc is red, and ac is not white then $(a, b) \Leftrightarrow (c, b)$, and also $(b, a) \Leftrightarrow (b, c)$.

Rule 4. If edge ab is black, bc is red, and ac is not black, then $(a, b) \Leftrightarrow (b, c)$, and $(b, a) \Leftrightarrow (c, b)$.

Let $\equiv$ denote the reflexive, symmetric and transitive closure of the forcing relation. Thus $\equiv$ is an equivalence relation defined on ordered pairs of distinct vertices of $C(G)$. An *orientation* of $C(G)$ is an assignment of directions to the edges of $C(G)$. An orientation is said to be *acyclic* if the resulting directed graph has no directed cycles. Note that since $C(G)$ is a complete graph, an orientation of $C(G)$ is acyclic iff it is transitive, that is, if there are directed edges from a to b and from b to c, there is also a directed edge from a to c. An orientation of $C(G)$ is said to be *consistent* if for all ordered pairs of distinct vertices $(a, b), (c, d)$ such that $(a, b) \equiv (c, d)$, the edge ab in $C(G)$ is directed from a to b iff the edge cd is directed from c to d.

We can now state the main result.

Theorem 1. *Let $G(X, Y, E, F)$ be an abstract topological bipartite graph and $C(G)$ its colored graph representation. G has a two-layer realization iff $C(G)$ has a consistent acyclic orientation.*

Note that if G is a matching, the graph $C(G)$ has only black and white edges, and only Rule 1 is applicable. In this case, G is realizable iff the spanning subgraph of $C(G)$ formed by black edges is a permutation graph. It is well-known that a graph H is a permutation graph if and only if both H and $\overline{H}$, the complement of H, are comparability graphs [4,8], that is, have a transitive orientation. This is equivalent to Theorem 1 when only black and white edges are present. Thus Theorem 1 may be considered to be a generalization. The proof follows the same arguments as the characterization of permutation graphs.

Proof (Theorem 1). Suppose $G(X, Y, E, F)$ has a two-layer realization and consider any fixed two-layer drawing that realizes $G(X, Y, E, F)$. Define a linear order $<$ on E by $a < b$ iff either the endvertex of a in X is to the left of the endvertex of b in X, or they have the same endvertex in X but the endvertex of a in Y is to the left of the endvertex of b in Y. Assign direction to the edge ab in $C(G)$ from a to b if $a < b$. Clearly, this is an acyclic orientation of $C(G)$. We show that it is also consistent. It is enough to show that if $(a, b) \Leftrightarrow (c, b)$ or $(a, b) \Leftrightarrow (b, c)$, then the assigned directions are consistent.

Case 1   Suppose ab and bc have the same color but ac has a different color. Since the edges of red or green color form a subgraph of $C(G)$ consisting of disjoint cliques, the common color of ab and bc must be either black or white. Suppose it is black, which implies the edges a, b and b, c cross in the drawing. If we had either $a < b < c$ or $c < b < a$, then a and c must also cross in the drawing, contradicting the fact that the edge ac is not black. Therefore $a < b$ if and only if $c < b$. The same argument holds if both ab and bc are white, in which case, $a < b < c$ or $c < b < a$ would imply the 3 edges are pairwise non-crossing (Fig. 2(a)).

Case 2   Suppose ab is black or white, bc is green, and ac has a color different from that of ab. Suppose ab is black, which implies edges a and b cross in the drawing. Since c has the same endvertex as b in Y, if $a < b < c$ or $c < b < a$, then a, c are non-adjacent edges that cross in the drawing, contradicting the fact that the edge ac is not black. A similar argument holds if ab is white, in which case, $a < b < c$ or $c < b < a$ would imply that a and c are non-adjacent edges that do not cross (Fig. 2(b)).

Case 3   Suppose ab is white and bc is red, and ac is not white. This implies b, c have the same endvertex in X. Again, we see that if $a < b < c$ or $c < b < a$, then the edges a and c are non-adjacent and do not cross, a contradiction (Fig. 2(c)).

Case 4   Suppose $(a, b) \Leftrightarrow (b, c)$. This can happen only due to Rule 4 in the definition of forcing. In this case, ab must be black, bc is red, and ac is not black. If $a < b$ and $c < b$, or $b < a$ and $b < c$, we observe that the edge ac must be black, a contradiction (Fig. 2(d)). Therefore $a < b$ iff $b < c$ and the orientation is consistent.

This completes the proof of the fact that the orientation defined by the $<$ ordering of the edges is consistent.

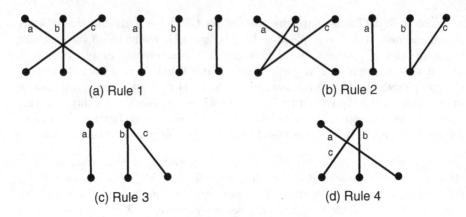

(a) Rule 1   (b) Rule 2

(c) Rule 3   (d) Rule 4

Fig. 2. Forcing rules

To prove the converse, suppose there is a consistent acyclic orientation of $C(G)$. Let D denote the resulting directed graph. We will construct a two-layer realization of the abstract topological bipartite graph. Define a linear order $<$ on E, by $a < b$ if the edge ab is directed from a to b in D.

Claim. Let $a < b < c$ be 3 edges in E such that ac is red in $C(G)$. Then both ab and bc must be red in $C(G)$.

Proof. If not, then at least one of ab, bc must be black or white, since the green and red edges form disjoint cliques. If both are black, or both white, then $(a, b) \Leftrightarrow (c, b)$ (Rule 1), which contradicts the consistency of the orientation. If exactly one of them is black, then either $(b, a) \Leftrightarrow (a, c)$ (if ab is black), or $(b, c) \Leftrightarrow (c, a)$ (Rule 4). Again, this contradicts consistency. The only other possibility is that one of them is white and the other green. In this case, either $(a, b) \Leftrightarrow (c, b)$ (if ab is white and bc is green), or $(b, c) \Leftrightarrow (b, a)$ (Rule 2), contradicting the consistency of the orientation.   □

This implies that all edges in G incident with a common vertex in X, which induce a clique of red edges in $C(G)$, form an interval in the linear order. Thus this defines a linear order on the vertices in X, where $x_1 < x_2$ if $x_1y_1 < x_2y_2$ for some edges x_1y_1, x_2y_2. Since all edges incident with a vertex in X occur consecutively in the $<$ order, the ordering of X is well-defined, independent of the choice of the edges x_1y_1, x_2y_2.

A similar method is used to construct the ordering of the vertices in Y. Construct a new orientation of $C(G)$ from D, say D', by reversing the directions of all black edges in $C(G)$. We claim that D' is also acyclic. If not, it must contain a directed triangle. At least one but not all of the edges in the triangle must be black, otherwise D itself would not be acyclic. Suppose a, b, c is a directed triangle in D'. If the edges ab and bc are black, then D contains edges (c, b) and (b, a). But in this case, $(a, b) \Leftrightarrow (c, b)$ (Rule 1), contradicting the consistency of D. Suppose ab is the only black edge. Then $b < c < a$. If at least one of bc, ac is white, then either $(c,b) \Leftrightarrow (c,a)$ (if bc is white), or $(a,c) \Leftrightarrow (b,c)$ (Rules 1, 2, 3), contradicting consistency. Otherwise, at least one of the edges bc, ac must be red. Then either $(a,b) \Leftrightarrow (b,c)$ (if bc is red) or $(b,a) \Leftrightarrow (a,c)$ (Rule 4), again a contradiction. Thus after reversing the directions of the black edges, we get a new linear order $\prec$ on E.

Claim. Let $a \prec b \prec c$ be 3 edges in E such that ac is colored green in $C(G)$. Then both ab and bc must be colored green.

Proof. Since ac is green and $a \prec c$, we have $a < c$. If both ab and bc are black, we must have $b < a$ and $c < b$, contradicting $a < c$. If both are white, we have $a < b < c$, which contradicts the fact that $(a,b) \Leftrightarrow (c,b)$ (Rule 1). If they have different colors, and one of them is black, say ab, then we have $b < a$, contradicting the fact that $(b,a) \Leftrightarrow (c,a)$ (Rule 2). If bc is black, we have $c < b$ again contradicting the fact that $(b,c) \Leftrightarrow (a,c)$ (Rule 2). The only other possibility is that one of them is white and the other red. In this case, we have $a < b < c$, but either $(a,b) \Leftrightarrow (c,b)$ (if ab is white) or the other way (Rule 3), again giving a contradiction.   □

Again, this implies that edges that have a common endvertex in Y form an interval in the $\prec$ linear order. We can now define an order on the vertices in Y as $y_1 \prec y_2$ iff $x_1 y_1 \prec x_2 y_2$ for some edges $x_1 y_1, x_2 y_2$.

We now construct a realization of the abstract topological graph, by placing vertices in X on the line $y = 1$ in left to right order according to the $<$ relation. Similarly, place vertices in Y on the line $y = 0$ in left to right order according to the $\prec$ ordering. It remains to show that this ordering satisfies Criterion 1.

Let a, b be any two non-adjacent edges in E and suppose $\{a, b\} \in F$. The edge ab is black in $C(G)$ and hence $a < b$ iff $b \prec a$. This implies that the endvertex of a in X is to the left of the endvertex of b in X iff the endvertex of a in Y is to the right of the endvertex of b in Y. Thus the edges a, b must cross in the constructed drawing. A similar argument holds if $\{a, b\} \notin F$. In this case, the edge ab is white, and we have $a < b$ iff $a \prec b$.

This completes the proof of Theorem 1.   □

3   Algorithm

In this section, we give a simple algorithm to check in polynomial-time the characterization given by Theorem 1. Again, the algorithm is an extension of the standard algorithm to test whether a graph has a transitive orientation, that is, to recognize comparability graphs [4]. The algorithm follows easily from the following theorem.

Theorem 2. *Let $C(G)$ be the colored graph representation of an abstract topological bipartite graph G. Then $C(G)$ has a consistent acyclic orientation iff for all ordered pairs (a, b) of distinct vertices in $C(G)$, $(a, b) \not\equiv (b, a)$.*

Before proving Theorem 2, we need a few definitions and lemmas. We say an equivalence class of $\equiv$ is consistent if for all ordered pairs (a, b) of distinct vertices of $C(G)$, at most one of the pairs $(a, b), (b, a)$ belongs to the class.

Let a, b, c be 3 distinct vertices in $C(G)$. We say the triangle a, b, c is *good* if either $(a, b) \equiv (a, c)$, or $(b, a) \equiv (b, c)$ or $(c, a) \equiv (c, b)$. A triangle that is not good is said to be *bad*. Informally, a good triangle cannot form a directed cycle in a consistent orientation of $C(G)$.

Lemma 1. *Any triangle containing two edges with different colors is good.*

Proof. Any triangle containing two edges of different colors must contain a black or white edge. If it contains 2 black, or 2 white edges, without loss of generality ab and bc, then $(b, a) \Leftrightarrow (b, c)$ (Rule 1) and hence the triangle is good. Otherwise it must contain a red or green edge. If the triangle has a white edge, say ab, and bc is the red or green edge then $(b, a) \equiv (b, c)$ (Rules 2,3). If there is no white edge, there must be a black edge, say ab, and a green edge bc. Again $(b, a) \equiv (b, c)$ (Rule 2), and the triangle is good. $\square$

Lemma 2. *Let a, b, c be a bad triangle with all edges red, or all edges green. Then for all ordered pairs (x, y) with $x \in \{a, b, c\}$, $y \in \{a, b, c\} \setminus \{x\}$, the equivalence class of $\equiv$ containing (x, y) does not contain any other ordered pair.*

Proof. Suppose without loss of generality that the equivalence class containing (a, b) also contains some other pair. Then there is an edge ad such that $(a, d) \Leftrightarrow (a, b)$ or $(d, a) \Leftrightarrow (a, b)$, or there is an edge bd such that $(d, b) \Leftrightarrow (a, b)$ or $(b, d) \Leftrightarrow (a, b)$. Suppose there is such an edge ad. Then ad must be black or white and bd has a color different from that of ad. Suppose $(a, d) \Leftrightarrow (a, b)$. If the edge cd has a color different from that of ad, then $(a, d) \Leftrightarrow (a, c)$, hence $(a, b) \equiv (a, c)$, contradicting the assumption that the triangle is bad. If edge cd has the same color as ad, then $(a, d) \Leftrightarrow (c, d)$, and $(c, d) \Leftrightarrow (c, b)$. Thus $(a, b) \equiv (c, b)$ and hence $(b, a) \equiv (b, c)$, a contradiction. If $(d, a) \Leftrightarrow (a, b)$, then ad is black, ab is red and bd is not black. If cd is not black, $(d, a) \Leftrightarrow (a, c)$, hence $(a, b) \equiv (a, c)$. If cd is also black, then $(d, a) \Leftrightarrow (d, c)$, and $(d, c) \Leftrightarrow (c, b)$. Again, we get $(b, a) \equiv (b, c)$. A symmetrical argument holds if there is an edge bd such that either $(d, b) \Leftrightarrow (a, b)$ or $(b, d) \Leftrightarrow (a, b)$. $\square$

Lemma 3. *Let a, b, c be a bad triangle with all edges black or white. Let x, y, z be any permutation of a, b, c. If $(x, y) \equiv (p, q)$ for some ordered pair of distinct vertices p, q, then $(z, x) \equiv (z, p)$, $(z, y) \equiv (z, q)$, and z, p, q is a monochromatic triangle with edges of the same color as in the triangle a, b, c. In particular, $z \notin \{p, q\}$.*

Proof. Since $(x, y) \equiv (p, q)$, there exists a sequence of ordered pairs of distinct vertices $(a_1, b_1), (a_2, b_2), \ldots, (a_k, b_k)$ such that $(a_1, b_1) = (x, y)$, $(a_k, b_k) = (p, q)$, and for all $1 \leq i < k$, either $(a_i, b_i) \Leftrightarrow (a_{i+1}, b_{i+1})$ or $(a_{i+1}, b_{i+1}) \Leftrightarrow (a_i, b_i)$. We will prove the statement by induction on the length of the sequence. It is true by assumption for $k = 1$. Assume it is true for sequences of length $k - 1$.

Then $(z, x) \equiv (z, a_{k-1})$, $(z, y) \equiv (z, b_{k-1})$, and z, a_{k-1}, b_{k-1} is a monochromatic triangle with edges of the same color as in the triangle a, b, c.

Since $a_{k-1}b_{k-1}$ is either a black or white edge, we may assume $(a_{k-1}, b_{k-1}) \Leftrightarrow (a_k, b_k)$. Suppose $b_{k-1} = b_k$. If $z = a_k$, then $(x, y) \equiv (a_{k-1}, b_{k-1}) \equiv (z, b_{k-1}) \equiv (z, y)$, contradicting the fact that a, b, c is a bad triangle.

Suppose za_k has the same color as za_{k-1}. Then $(z, a_{k-1}) \Leftrightarrow (z, a_k)$, and hence $(z, x) \equiv (z, a_k)$. If $a_k b_k$ has the same color as za_k, then the required property holds for (a_k, b_k). Suppose $a_k b_k$ has a different color. Then $(z, b_{k-1}) \Leftrightarrow (z, a_k)$. Hence $(z, x) \equiv (z, a_k) \equiv (z, b_{k-1}) \equiv (z, y)$, contradicting the assumption that x, y, z is a bad triangle.

Suppose za_k has a different color than za_{k-1}. Then $(z, b_{k-1}) \Leftrightarrow (a_k, b_{k-1})$ and hence $(x, y) \equiv (a_k, b_{k-1}) \equiv (z, b_{k-1}) \equiv (z, y)$, implying x, y, z is a good triangle, a contradiction.

Suppose $b_{k-1} = a_k$. This can happen only when $a_{k-1}b_{k-1}$ is black and $a_k b_k$ is red. This implies $b_k \neq z$. If zb_k is also black then $(z, b_{k-1}) \Leftrightarrow (z, b_k)$ and also $(z, a_{k-1}) \Leftrightarrow (z, b_k)$. This implies $(z, x) \equiv (z, a_{k-1}) \equiv (z, b_{k-1}) \equiv (z, y)$, a contradiction. If zb_k is not black, then $(z, b_{k-1}) \Leftrightarrow (a_k, b_k)$, hence $(x, y) \equiv (a_k, b_k) \equiv (z, b_{k-1}) \equiv (z, y)$, a contradiction.

A symmetrical argument holds if $a_{k-1} = a_k$ or $a_{k-1} = b_k$.   □

Lemma 4. *If an equivalence class of* $\equiv$ *is consistent, then it is transitive, that is* $(a, b) \equiv (b, c)$ *implies* $(a, b) \equiv (a, c)$.

Proof. Suppose $(a, b) \equiv (b, c)$ and the triangle a, b, c is good. If either $(a, b) \equiv (a, c)$ or $(c, b) \equiv (c, a)$, then we get the required property. The only other possibility is $(b, a) \equiv (b, c)$. However this implies $(c, b) \equiv (a, b) \equiv (b, c)$, contradicting the consistency of the equivalence class. If the triangle a, b, c is bad, then by Lemma 1, it must be monochromatic. If all edges in the triangle are red or green, then by Lemma 2, the equivalence class of (a, b) cannot contain any other pair, contradicting $(a, b) \equiv (b, c)$. If all edges in the triangle are black or white, then $(a, b) \equiv (b, c)$ contradicts Lemma 3, with $x = a, y = b, z = c, p = b, q = c$. Thus the triangle must be good and $(a, b) \equiv (a, c)$.   □

Proof (Theorem 2). The necessity of the condition is clear. If $(a, b) \equiv (b, a)$ for some pair of vertices a, b in $C(G)$, then there is no way of assigning a direction to the edge ab to satisfy the consistency requirement. We may assume that each equivalence class of $\equiv$ is consistent. We now show the sufficiency by giving a simple algorithm to construct a consistent acyclic orientation of $C(G)$.

Algorithm for consistent acyclic orientation of C(G)

```
1. While there exists an undirected edge ab in C(G)
2.      For all pairs (p,q) equivalent to (a,b)
        assign direction from p to q to edge pq.
3.      While there exists an undirected edge xy such that
        there are directed edges from x to z and from z to y
4.          For all pairs (p,q) equivalent to (x,y)
            assign direction from p to q to edge pq.
```

We claim that at the start of every execution of the while loop in line 1 of the algorithm, the directed graph formed by edges that have been assigned a direction satisfies the following properties.

1. It is acyclic and transitive, that is, if (a, b) and (b, c) are directed edges, then so is (a, c).
2. It is consistent, that is, if an edge ab has been assigned a direction from a to b, then every edge pq such that $(a, b) \equiv (p, q)$ is assigned a direction from p to q.

This is clearly true when no edges have been assigned a direction. Now we show that every execution of the while loop preserves this property. It is easy to see that consistency is maintained at each step. Whenever an undirected edge is assigned a direction, in lines 2 or 4, all edges in its equivalence class are assigned the appropriate direction to maintain consistency. Since the initial assignment was consistent, all these edges must have been initially undirected, so no edge is assigned conflicting directions.

We show that no directed cycles are created in any step. Let D be the directed graph at the beginning of the while loop in line 1 and let E be the set of edges assigned directions in line 2. Thus E is the equivalence class of some pair (a, b). D is acyclic by assumption and E is acyclic as each equivalence class is transitive, by Lemma 4. Suppose $D \cup E$ contains a cycle and consider the shortest possible cycle. Since both D and E are transitive, the cycle cannot contain two consecutive edges in D or E, otherwise we get a shorter cycle in $D \cup E$. Therefore there must exist vertices p, q, r, s in the cycle such that (p, q) and (r, s) are edges in E and (q, r) is an edge in D. Suppose the triangle p, q, r is good. If $(p, q) \equiv (p, r)$, then (p, r) is also an edge in E, giving a shorter cycle in $D \cup E$. If $(q, r) \equiv (q, p)$, the consistency of D implies that $(q, p) \in D$, contradicting the fact that edges in E were not assigned directions in D. If $(r, q) \equiv (r, p)$, then the consistency of D implies (p, r) is an edge in D, and we again get a shorter cycle. Therefore the triangle p, q, r must be bad. Then the fact that $(p, q) \equiv (r, s)$ contradicts either Lemma 2 or Lemma 3, depending on the color of the edges in the triangle p, q, r. Therefore after execution of line 2, the resulting directed graph D_1 is acyclic and consistent.

We next show that at the start of every execution of the while loop in line 3, the directed graph D_1 formed by edges that have been assigned directions is acyclic and consistent. By the previous argument, this is true for the first execution. The consistency follows by the same argument, since in line 4, all edges in an equivalence class are assigned directions, and none of them could have had a direction assigned earlier. We show that no cycles are created. Let E be the set of edges that are assigned directions in line 4. E is the equivalence class of a pair (x, y) such that (x, z) and (z, y) are edges in D_1.

Let (p, q) be any edge in E, that is $(x, y) \equiv (p, q)$. We claim that (p, z) and (z, q) are edges in D_1. Suppose the triangle x, y, z is good. If $(x, y) \equiv (x, z)$ the consistency of D_1 implies that (x, y) is an edge in D_1, contradicting the assumption that it was not assigned a direction. The same argument holds if $(y, z) \equiv (y, x)$. On the other hand, if $(x, z) \equiv (y, z)$, it contradicts the consistency of D_1 since $(x, z), (z, y)$ are edges in D_1. Therefore x, y, z must be a bad triangle. If all edges in the triangle x, y, z are red or green, then (x, y) is the only edge in E by Lemma 2, and the claim holds. On the other hand, if all edges in the triangle x, y, z are black or white, Lemma 3 implies $(z, p) \equiv (z, x)$ and $(z, q) \equiv (z, y)$. The consistency of D_1 then implies that (p, z) and (z, q) are edges in D_1.

Now if $D_1 \cup E$ contains a cycle, we can replace every edge (p, q) of the cycle that is in E, by the path $(p, z), (z, q)$. This gives a closed walk in D_1 containing at least one edge, implying that D_1 itself contains a cycle, a contradiction.

Thus after every execution of the while loop in line 3, the resulting directed graph is acyclic and consistent. When this loop terminates, there is no edge xy that is not assigned a direction but there exist edges (x, z) and (z, y) in the directed graph. This implies that after one execution of the while loop in line 1, the resulting directed graph is acyclic, consistent and transitive. The algorithm thus terminates when all edges of $C(G)$ have been directed, and we get a consistent acyclic orientation of $C(G)$.

This completes the proof of Theorem 2.   □

4   Conclusion

There are several ways in which this work can be extended. Instead of two-layer drawings of bipartite graphs, we could consider circular drawings of arbitrary graphs. In this case, the vertices are represented by points on a circle. If the graph is a matching, this reduces to recognizing circle graphs. It would be interesting to see if the characterizations and algorithms for circle graphs can be extended to this case. Similarly, instead of two layers, we could consider graphs with vertices partitioned into layers, with all edges joining vertices in consecutive layers.

Permutation graphs have a nice forbidden subgraph characterization, which follows from the characterization of comparability graphs. Let a 'cycle' in a graph be a sequence of vertices, not necessarily distinct, $v_1, v_2, \ldots, v_n$ such that $v_i v_{i+1}$ and $v_1 v_n$ are edges in the graph, $(v_i, v_{i+1}) \neq (v_j, v_{j+1})$, $(v_i, v_{i+1}) \neq (v_n, v_1)$, for $1 \leq i < j < n$. A triangular chord of the cycle is an edge of the form $v_i v_{i+2}$ or $v_2 v_n$ or $v_1 v_{n-1}$. A graph is a comparability graph if and only if every odd 'cycle' has a triangular chord [5]. In terms of colored graphs with only black and white edges, this means in every black or white odd 'cycle' there is a triangular chord of the same color. However, this is not sufficient when there are edges of red and/or green color, as shown by the example in Fig. 1.

We have not made any attempt to improve the efficiency of our algorithm. There are faster algorithms known for recognizing permutation graphs [9], and those techniques may also be applicable here. Also, as shown in [1], there is a much simpler algorithm for connected graphs, and any connected graph has essentially a unique realization, if it has one. It may be possible to use this to combine the realizations of individual components to get a realization of the whole graph, in time depending on the number of components rather than the total number of edges.

References

1. Diwan, A.A., Roy, B., Ghosh, S.K.: Two-layer drawings of bipartite graphs. Electron. Notes Discret. Math. **61**, 351–357 (2017). https://doi.org/10.1016/j.endm.2017.06. 059
2. Eades, P., Whitesides, S.: Drawing graphs in two layers. Theor. Comput. Sci. **131**(2), 361–374 (1994). https://doi.org/10.1016/0304-3975(94)90179-1
3. Finocchi, I.: Crossing-constrained hierarchical drawings. J. Discret. Algorithms **4**, 299–312 (2006). https://doi.org/10.1016/j.jda.2005.06.001
4. Golumbic, M.C.: Algorithmic Graph Theory and Perfect Graphs. Annals of Discrete Mathematics, vol. 57, 2nd edn. Elsevier, New York (2004). https://doi.org/10.1016/ S0167-5060(04)80051-7
5. Gilmore, P.C., Hoffman, A.J.: A characterization of comparability graphs and of interval graphs. Can. J. Math. **16**, 539–548 (1964)
6. Kratochvíl, J.: String graphs II. Recognizing string graphs is NP-Hard. J. Combin. Theory Ser. B **52**, 67–78 (1991). https://doi.org/10.1016/0095-8956(91)90091-W
7. Kynčl, J.: Simple realizability of complete abstract topological graphs in P. Discret. Comput. Geom. **45**, 383–399 (2011). https://doi.org/10.1007/s00454-010-9320-x
8. Pnueli, A., Lempel, A., Even, S.: Transitive orientation of graphs and identification of permutation graphs. Can. J. Math. **23**, 160–175 (1971). https://doi.org/10.4153/ CJM-1971-016-5
9. Spinrad, J.: On comparability and permutation graphs. SIAM J. Comput. **14**(3), 658–670 (1985). https://doi.org/10.1137/0214048

Minimal-Perimeter Polyominoes: Chains, Roots, and Algorithms

Gill Barequet and Gil Ben-Shachar[✉]

Department of Computer Science, Technion—IIT, Haifa, Israel
{barequet,gilbe}@cs.technion.ac.il

Abstract. A polyomino is a set of edge-connected squares on the square lattice. We investigate the combinatorial and geometric properties of minimal-perimeter polyominoes. We explore the behavior of minimal-perimeter polyominoes when they are "inflated," i.e., expanded by all empty cells neighboring them, and show that inflating all minimal-perimeter polyominoes of a given area create the set of all minimal-perimeter polyominoes of some larger area. We characterize the roots of the infinite chains of sets of minimal-perimeter polyominoes which are created by inflating polyominoes of another set of minimal-perimeter polyominoes, and show that inflating any polyomino for a sufficient amount of times results in a minimal-perimeter polyomino. In addition, we devise two efficient algorithms for counting the number of minimal-perimeter polyominoes of a given area, compare the algorithms and analyze their running times, and provide the counts of polyominoes which they produce.

1 Introduction

A polyomino is a set of edge-connected squares on the square lattice. The study of polyominoes began in the 1950s, where it was studied in parallel as topics in combinatorics [7] and in statistical physics [5]. Typically, polyominoes are enumerated by their area, i.e., the number of lattice cells they contain. The number of polyominoes with area n is usually denoted in the literature by $A(n)$. No formula for $A(n)$ is known as of today, but there exist a few algorithms for computing values of $A(n)$, such as those of Redelmeier [13] and Jensen [8]. Using the latter algorithm, values of $A(n)$ were computed up to $n = 56$. The existence of the *growth constant* of $A(n)$, namely, $\lambda := \lim_{n \to \infty} \sqrt[n]{A(n)}$, was shown by Klarner [9]. More than 30 years later, Madras [11] showed that $\lim_{n \to \infty} A(n+1)/A(n)$ exists and, thus, is equal to λ. The best known lower and upper bounds on λ are 4.0025 [4] and 4.6496 [10], resp. The currently best *estimate* of λ is 4.0625696 ± 0.0000005 [8].

Some works studied polyominoes by their *perimeter*. The perimeter of a polyomino is the set of empty (unoccupied) cells adjacent to the polyomino. (Sometimes, when the meaning is clear from the context, we used the term "perimeter"

Work on this paper by both authors has been supported in part by ISF Grant 575/15 and by BSF (joint with NSF) Grant 2017684.

S. P. Pal and A. Vijayakumar (Eds.): CALDAM 2019, LNCS 11394, pp. 109–123, 2019.
https://doi.org/10.1007/978-3-030-11509-8_10

to denote the *number* of perimeter cells.) This definition originated in statistical physics, where the numbers of polyominoes with certain areas and perimeters are used for modeling physical phenomena, such as percolation processes. Asinowski et al. [2] recently provided formulae for the number of polyominoes whose perimeter is close to the maximum possible. Here, we complement their work and study *minimal-perimeter polyominoes*, that is, polyominoes with the minimum possible perimeter size for their area.

Several works discussed the minimum perimeter of a polyomino with area n. Wang and Wang [17] were the first to address the subject, showing a sequence of polyominoes with increasing areas, whose perimeter is minimal. Altshuler et at. [1] later characterized all polyominoes with *maximal* area for a given perimeter. (As we explain below, there is a close relation between minimal-perimeter (for a fixed area) and maximal-area (for a fixed perimeter) polyominoes.) Minimal-perimeter polyominoes also appear in certain game-theoretic problems, where formulae for minimum perimeter of polyominoes of area n [14], or, similarly, for polyiamonds and polyhexes (shapes on the planar triangular and hexagonal lattices, resp.) [6], were developed.

In a recent work [3], we addressed the question of how many minimal-perimeter polyominoes of a given area exist. In that work, we defined the inflated version of a polyomino to be the union of the polyomino and the set of its perimeter cells. We showed that this operation induces a bijection between the set of minimal-perimeter polyominoes of a given area, and the set of the minimal-perimeter polyominoes of the area of the inflated polyominoes. By inflating repeatedly polyominoes in such a set, we obtain an infinite chain of sets of minimal-perimeter polyominoes, all of which have the same cardinality. We call these sequences "inflation chains." A natural interesting question is "For which areas n, the set of minimal-perimeter polyominoes of area n cannot be generated by inflating a set of polyominoes of a smaller area $n' < n$?" We call such areas n "inflation-chains roots" and give a full characterization of these values. Inspired by inflation chains, we also show that any polyomino becomes a minimal-perimeter polyomino after a finite amount of inflation steps.

We also investigate the question of how many minimal-perimeter polyominoes of area n exist for a general value of n (not necessarily an inflation-chain root). The inflation chains provide a partial answer to this question. If n is not a root, we reduce the problem to computing the number of minimal-perimeter polyominoes of area n', for some $n' < n$. However, the question remains open for inflation-chain roots. Hence, we present two algorithms for counting minimal-perimeter polyominoes of area n. Both algorithms are based on some geometric properties of minimal-perimeter polyominoes.

2   Preliminaries

As mentioned above, this work answers a few questions raised in an earlier work on minimal-perimeter polyominoes [3]. Thus, we provide here the needed definitions and results, and refer the reader to the previous paper for more details.

Let Q be a polyomino. The perimeter of Q, denoted by $\mathcal{P}(Q)$, is the set of all empty cells adjacent to Q. The border of Q, denoted by $\mathcal{B}(Q)$, is the set of all cells of Q which have an adjacent empty cell. The inflation of Q is defined as $I(Q) = Q \cup \mathcal{P}(Q)$, namely, inflating Q is expanding it by adding to it its perimeter. Figure 1 illustrates these concepts.

A minimal-perimeter polyomino has the minimum perimeter size (number of perimeter cells) out of all polyominoes of the same area. Let M_n denote the set of all minimal-perimeter polyominoes with area n, and $\epsilon(n)$ denote the minimum perimeter of a polyomino with area n. Sieben [14] showed that $\epsilon(n) = \lceil 2 + \sqrt{8n - 4} \rceil$.

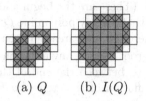

(a) Q (b) $I(Q)$

Fig. 1. An example of a polyomino Q and its inflated polyomino $I(Q)$.

Following our early terminology [3], a cell of Type (d) is a polyomino cell with exactly two adjacent empty cells in two opposite sides (▨▨▨), and a cell of Type (z) is an empty cell with exactly two adjacent occupied cells in two opposite sides (▨▨▨).

Here are a few previous results which we will need later in the current paper.

Lemma 1 [3, Corollary 1]. *For any n, we have $\epsilon(n + \epsilon(n)) = \epsilon(n) + 4$.*

Theorem 1 [3, Theorem 2]. *For a polyomino Q with no holes and no patterns of Types (d) and (z), we have $|\mathcal{P}(Q)| = |\mathcal{B}(Q)| + 4$.*

Lemma 1 and Theorem 1 are given merely for using them in proofs of some claims. However, the next theorem is interesting by its own.

Theorem 2 [3, Theorem 4]. *For all $n \geq 3$, we have $\{I(Q) \mid Q \in M_n\} = M_{n+\epsilon(n)}$.*

This theorem shows the existence of inflation chains which we discuss in the next section.

3   Inflation Chains: Roots and Convergence

In this section we address the question of what numbers are the roots of the inflation chains. In addition, we show that inflating repeatedly any polyomino for a sufficient amount of times results in a minimal-perimeter polyomino.

3.1   Roots of Inflation Chains

Recall that $\epsilon(n)$ is defined as the minimum perimeter of polyominoes of area n. This function is not one-to-one, thus, it does not have an inverse function. To overcome this, we define a pseudo-inverse function $\epsilon^{-1}(p) = \min_{n \in \mathbb{N}} \{n \mid \epsilon(n) = p\}$, that is, the minimum area of minimal-perimeter polyominoes with perimeter p.

Theorem 3. *An integer n is an inflation-chain root if and only if $n = \epsilon^{-1}(p)$ for some p.*

Proof. Notice that the function $\epsilon(n)$ behaves like a step function (see Fig. 3), thus, for any perimeter p, let us define the values $n_b^p = \epsilon^{-1}(p)$ and $n_e^p = \epsilon^{-1}(p+1)-1$. (These are the begin and end of the 'step' correspond to the perimeter p). Note that for any polyomino Q, such that $|Q| \in [n_b^p, n_e^p]$, the area of $I(Q)$ is $|Q| + p$. Therefore, any area between $n_b^p + p$ and $n_e^p + p$ (inclusive) is not an inflation-chain root. Hence, any inflation-chain root must be between $n_e^p + p$ and $n_b^{p+1} + p + 1$ (that is, between the end of the inflation of the step of p and the beginning of the inflation of the step of $p + 1$). By the definitions of n_e^p and n_b^{p+1}, we have that $n_b^{p+1} + p + 1 = n_e^p + p + 2$, thus, the area $n_e^p + p + 1$ is included neither in the sequence ending at $n_e^p + p$, nor by the sequence beginning at $n_b^{p+1} + p + 1$. Therefore, the area $n_e^p + p + 1$ is an inflation-chain root for all p.

From Lemma 1, we know that $\epsilon(n_e^p + p) = p + 4$ and $\epsilon(n_e^p + p + 2) = p + 5$, thus, $n_e^p + p + 1$ equals either n_e^{p+4} or n_b^{p+5}. Altshuler et al. [1] characterized the set of minimal-perimeter polyominoes with *maximal* area and divided it into four classes. These

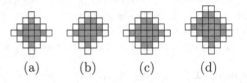

(a)   (b)   (c)   (d)

Fig. 2. Representative minimal-perimeter polyominoes with maximal area.

classes essentially consist of square-like polyominoes, aligned with the diagonals, and differ by the 'corners' which contain one or two cells (see Fig. 2). This set is closed under inflation, thus, the inflation-chain root cannot be n_e^{p+4}, and so we are left with n_b^{p+5} as the root. This means that for any p, the value n_b^p is an inflation-chain root.

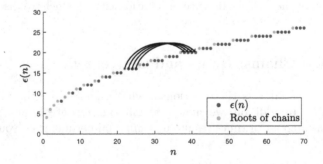

Fig. 3. Values of $\epsilon(n)$. Inflation-chains roots are colored in red. The arrows show the effect of the inflation operation on minimal-perimeter polyominoes. (Color figure online)

Theorem 3 is illustrated in Fig. 3. Altshuler et al. [1] also provided a formula for the values of n_e^p. However, their formula is a function of the area: Given an area k, the function provides the largest $n \leq k$, such that $n = n_e^p$ for some p. We provide an equivalent formula, however, as a function of p.

Theorem 4. $\epsilon^{-1}(p) = \left\lfloor \frac{(p-3)^2}{8} + \frac{3}{2} \right\rfloor$.

Proof. $\epsilon^{-1}(p) = \min_{n \in \mathbb{N}} \{n \mid \epsilon(n) = p\} = \min_{n \in \mathbb{N}} \{n \mid \lceil 2 + \sqrt{8n-4} \rceil \geq p\} = \min_{n \in \mathbb{N}} \{n \mid \sqrt{8n-4} > p - 3\} = \min_{n \in \mathbb{N}} \{n \mid n > \frac{(p-3)^2}{8} + \frac{1}{2}\} = \left\lfloor \frac{(p-3)^2}{8} + \frac{3}{2} \right\rfloor$.

3.2 Convergence of Inflation

We now discuss the structure of an inflated polyomino, and show that after a sufficient number of inflations, it will obey the formula $|\mathcal{P}(Q)| = |\mathcal{B}(Q)| + 4$. Using this, together with Theorem 3, we show that inflating any polyomino (or any set of disconnected lattice cells for that matter), converges to a minimal-perimeter polyomino after a finite amount of inflation steps.

We begin with defining the operator $I^k(Q)$ $(k > 0)$ to be Q inflated k times:

$$I^k(Q) = Q \cup \{c \mid \mathrm{Dist}(c, Q) \leq k\},$$

where $\mathrm{Dist}(c, Q)$ is the *Manhattan distance* from a cell c to a polyomino Q. We will use the notation $Q^k = I^k(Q)$ for brevity. Let $R(Q)$ denote the *diameter* of Q, i.e., the maximal horizontal or vertical distance (L^∞) between two cells of Q. The following lemma shows that some geometric features of a polyomino disappear after inflating it enough times. The proof is omitted here due to space limitations, and is given in the full version of the paper.

Lemma 2. *For any $k > R(Q)$, the polyomino Q^k does not contains (i) holes; (ii) cells of Type (d); and (iii) patterns of Type (z).*

Corollary 1. *After $k = R(Q)$ inflations, we will have $|\mathcal{P}(Q^k)| = |\mathcal{B}(Q^k)| + 4$.*

Proof. This follows at once from Lemma 2 and Theorem 1.

Consider a polyomino Q, and define the function $\phi(Q) = \epsilon^{-1}(|\mathcal{P}(Q)|) - |Q|$. When $\phi(Q) \geq 0$, it counts the cells that should be added to Q, with no change to its perimeter, in order to make it a min.-perimeter polyomino. In particular, if $\phi(Q) = 0$, then Q is a min.-perimeter polyomino. If $\phi(Q) < 0$, then Q is also a min.-perimeter polyomino, and $-\phi(Q)$ cells can be removed from Q while keeping the result a minimal-perimeter polyomino (without changing the perimeter).

Lemma 3. *For any value of p, we have $\epsilon^{-1}(p + 4) - \epsilon^{-1}(p) = p - 1$.*

Proof. Let Q be a min.-perimeter polyomino with area $n_b^p = \epsilon^{-1}(p)$. The area of $I(Q)$ is $n_b^p + p$, thus, by Lemma 1, $\mathcal{P}(I(Q)) = p + 4$. The area n_b^{p+4} is an inflation-chain root, thus, the area of $I(Q)$ cannot be n_b^{p+4}. Except n_b^{p+4}, polyominoes of all other areas between n_b^{p+4} and n_e^{p+4} are created by inflating min.-perimeter polyominoes with perimeter p. The polyomino Q is of area n_b^p, i.e., the area of $I(Q)$ must be the min. area from $\left[n_b^{p+4}, n_e^{p+4} \right]$ which is not an inflation-chain root. Hence, the area of $I(Q)$ is $n_b^{p+4} + 1$. We now equate the two expressions for the area of $I(Q)$: $n_b^p + p = n_b^{p+4} + 1$. I.e., $n_b^{p+4} - n_b^p = p - 1$. The claim follows.

Lemma 4. *If* $|\mathcal{P}(I(Q))| = |\mathcal{P}(Q)| + 4$, *then* $\phi(I(Q)) = \phi(Q) - 1$.

Proof. $\phi(I(Q)) = \epsilon^{-1}(|\mathcal{P}(I(Q))|) - |I(Q)| = \epsilon^{-1}(|\mathcal{P}(Q)| + 4) - (|Q| + |\mathcal{P}(Q)|)$
$= \epsilon^{-1}(|\mathcal{P}(Q)|) + |\mathcal{P}(Q)| - 1 - |Q| - |\mathcal{P}(Q)| = \epsilon^{-1}(|\mathcal{P}(Q)|) - |Q| - 1 = \phi(Q) - 1$.

Corollary 2. (Convergence of inflation). *After a finite amount of inflation steps, any polyomino becomes a minimal-perimeter polyomino.*

Proof. Consider a polyomino Q. By Corollary 1, after $R(Q)$ inflation steps, we will have that $\left| \mathcal{P}(Q^{R(Q)}) \right| = \left| \mathcal{B}(Q^{R(Q)}) \right| + 4$. After each step, the new border is the previous perimeter. Thus, after $R(Q)$ steps, the perimeter will increase by 4 with each additional inflation, and Lemma 4 will hold, i.e., any additional step will decrease the value of $\phi(Q^{R(Q)})$ by 1. After $R(Q)$ steps, $\phi(Q^{R(Q)})$ will have some finite value, φ, thus, after φ more steps, we will necessarily have $\phi(Q^{R(Q)+\varphi}) = 0$. Since a polyomino is a minimal-perimeter polyomino if $\phi(Q) \le 0$, this implies that Q will become a minimal-perimeter polyomino after $R(Q) + \phi(Q^{R(Q)})$ steps.

4   Counting Minimal-Perimeter Polyominoes

The second question which we address is what is the number of minimal-perimeter polyominoes of area n, i.e., what is $|M_n|$. Notice that by using Theorem 2, this question can be reduced to counting the number of minimal-perimeter polyominoes only for inflation-chains roots. An area n is not an inflation-chain root if there exists some n', such that $n = n' + \epsilon(n')$, i.e., if there exist minimal-perimeter polyominoes of area n', that, after inflation, become polyominoes of area n. Lemma 1 tells us that if such n' exists, then $\epsilon(n) = \epsilon(n') + 4$, thus, we can easily check if such n' exists and if so, reduce the problem to area n'.

We were not able to provide a closed-form formula for $|M_n|$. However, we present here two algorithms which compute $|M_n|$ by using the geometric structure of min.-perimeter polyominoes. The running times of both algorithms are polynomial with n. We discuss the properties that guide the algorithms, present, analyze, and compare the algorithms, and provide the produced polyomino counts.

4.1   First Algorithm: Diagonal Counting

The first algorithm builds the polyominoes under consideration diagonal by diagonal.

A *diagonally-convex* polyomino is a polyomino whose cells along every diagonal (both descending [top-left to bottom-right] and ascending [bottom-left to top-right]) are consecutive without gaps. The first algorithm is based solely on the following theorem. The proof of this theorem is omitted here and is given in the full version of the paper.

Theorem 5. *Any minimal-perimeter polyomino is diagonally-convex.*

Based on Theorem 5, we devise an algorithm which counts minimal-perimeter polyominoes of area n. The algorithm builds the polyomino one diagonal at a time while maintaining the current perimeter. Here are the main details of the algorithm.

Building Polyominoes. The algorithm builds polyominoes along *descending* diagonals, one diagonal at a time. In the first iteration, we iterate through all possible sizes of the most extreme diagonal. In each of the following iterations, we iterate through both all possible sizes of the current diagonal and all possible positions of the current diagonal with respect to the previous one. The algorithm stops if the area of the polyomino reaches a target value n, or if the perimeter size exceeds $\epsilon(n)$. A generated polyomino is counted if the area of the polyomino is n, and its perimeter is $\epsilon(n)$.

Maintaining the Perimeter. Throughout the execution of the algorithm, we maintain the current polyomino perimeter. When a new diagonal is introduced, we add the number of newly-created perimeter cells. The potential number of perimeter cells created by a diagonal of length ℓ is $2(\ell+1)$. From this amount, we subtract the number of cells already occupied (by either polyomino or perimeter cells), as well as the number of perimeter cells eliminated by the new diagonal. All these computations are performed in constant time, given the locations of the first and last cells of the new diagonal and of the last two diagonals.

Keeping the Polyomino Connected. While building the polyomino, we keep track of the disconnected sets of cells. These cells must be connected by the next diagonal, thus, we make sure that this indeed happens. This is illustrated in Fig. 4.

Fig. 4. A partial polyomino. The next diagonal must occupy the red cells in order to keep the polyomino connected. (Color figure online)

Narrowing the Range of Diagonals. Using the fact that Theorem 5 holds for diagonals in both directions, we can narrow the range of possible diagonals to only those that do not create a nonconvex *ascending* diagonal. This does not change the asymptotic complexity of the algorithm, but in practice it speeds up the algorithm substantially.

Memoization. Like many other recursive algorithms, this algorithm can be accelerated using memoization. In each partial state of the polyomino, the number of possible completions of the polyomino to a minimal-perimeter polyomino of area n depends on the current area n', the current perimeter p', and the locations of the first and last cells of the last two diagonals (s_0, t_0, s_1, t_1). The last two diagonals can be represented in a canonical form, for example, with the last diagonal starting at 0, and, thus, each partial polyomino will be represented by the tuple $(n', p', t_0 - s_0, s_1 - s_0, t_1 - s_0)$. Using this representation, one can store the number of possible completions of a partial polyomino, and look it up when a partial polyomino with the same signature is built.

Complexity Analysis. Consider the graph of states of the algorithm, in which each node stores a state s represented by its memoizing tuple, and there is a directed edge $s_1 \to s_2$ if the algorithms performs a recursive call from state s_1 to state s_2. The time spent in a single node s for the first time is $O(d_{out}(s))$, i.e., the number of edges outgoing from s, and $O(1)$ if s was already memoized. Since we use memoization, any outgoing edge needs to be traversed only once, thus, the amount of times a state, s, is called is at most $d_{in}(s)$. Therefore, the total time complexity is $\sum_s O(1 + d_{out} + d_{in}) = O(V + E)$, where V, and E are the number of edges and vertices in the graph respectively. This means that the total complexity of the computation is the same as the complexity of the computation graph.

Since we look for minimal-perimeter polyominoes, the formula for $\epsilon(n)$ dictates that $p = \Theta(\sqrt{n})$ in the root call to the algorithm. Recall that each state is represented by a 5-tuple, in which the first entry is the area of the polyomino (which is $O(n)$), the second entry is its perimeter (which is $O(\sqrt{n})$), and the last three entries represent the first and last cells along diagonals. Since any added diagonal increase the perimeter (or preserve it), the length of each diagonal is bounded from above by the perimeter we seek to achieve, otherwise, the perimeter of the whole polyomino will be greater than the target perimeter. Therefore, the last three entries are also $O(\sqrt{n})$. We conclude that the total number of nodes is $O(n^3)$.

In each node, the algorithm considers all possible sizes of the next diagonal (which length is at most the size of the perimeter, i.e., it is $O(\sqrt{n})$, and all starting points of the next diagonal with respect to the current diagonal. Since each diagonal must be connected to the previous one, the starting point parameter is also $O(\sqrt{n})$. In total, we have $O(n)$ outgoing edges from each node, thus the total number of edges is $O(n^4)$. The total complexity of the graph, and hence, the total time complexity of the algorithm, is $O(n^4)$. (Note that this is a pseudo-polynomial complexity, since the only input to the algorithm is the *number* n, which can be specified by using $\log n$ bits.)

4.2   Second Algorithm: Bulk Counting

Our second approach for counting minimal-perimeter polyominoes is based on the fact that every such polyomino can be divided into an "octagonal" center-piece and "caps." We refer to the main eight directions by the terms North, West, South, East, and their combinations.

Settings. Let Q be a minimal-perimeter polyomino. In addition, let c_x, c_y denote, resp., the x- and y-coordinates of a lattice cell c. Finally, let D_{ne} denote the sum of the x- and y-coordinates of the most extreme cells of Q in the north-east direction, that is, $D_{ne} = \max_{c \in Q} \{c_x + c_y\}$. Similarly, $D_{nw} = \min_{c \in Q} \{c_x - c_y\}, D_{sw} = \min_{c \in Q} \{c_x + c_y\}$, and $D_{se} = \max_{c \in Q} \{c_x - c_y\}$. Using a construction similar to that of Altshuler et al. [1, Theorem 7], we define $S(Q)$, the squaring of Q, as the set of cells found in the 45°-oriented bounding box of Q, that is,

$$S(Q) = \{c \mid (D_{sw} \leq c_x + c_y \leq D_{ne}) \wedge (D_{nw} \leq c_x - c_y \leq D_{se})\}.$$

This construction is illustrated in Fig. 5.

Notice that each perimeter cell of $S(Q)$ can be mapped uniquely to a perimeter cell of Q, therefore, $|\mathcal{P}(S(Q))| \leq |\mathcal{P}(Q)|$. In addition to this, since, obviously, $Q \subset S(Q)$, the monotonicity of $\epsilon(n)$ implies that if Q is a minimal-perimeter polyomino, then $|\mathcal{P}(S(Q))| = |\mathcal{P}(Q)|$.

It is easy to verify that if an entire diagonal is removed from a squared polyomino, its perimeter is reduced by 1. Let d be the length of the smallest diagonal of $S(Q)$.

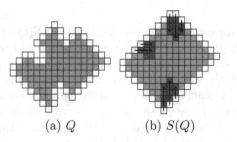

(a) Q (b) $S(Q)$

Fig. 5. The squaring construction. Arrows indicate correspondence between newly-created perimeter cells of $S(Q)$ and destroyed perimeter cells of Q.

Then, the minimum perimeter of a polyomino of area $|S(Q)| - d$ is at most $|\mathcal{P}(S(Q))| - 1$. Thus, the area of a minimal-perimeter polyomino Q is at least $|S(Q)| - d + 1$. In order to get a feeling of the decomposition, notice that $d = O(\sqrt{|S(Q)|})$, thus, we may say that the shape of a minimal-perimeter polyomino is always similar to that of a squared polyomino.

In the sequel, consider Q as a minimal-perimeter polyomino. We now describe constructively the decomposition of Q into a "centerpiece" and "caps" (see Fig. 6). Consider the highest row of Q that spans the entire $S(Q)$ (the highest row that contains grey cells in Fig. 6(a)). Cells above this row will be marked as 'caps.' The same process is applied to the lowest row, the rightmost column, and the leftmost column of Q. Without the caps in all directions, we are left with a truncated version of $S(Q)$, which in general looks like an octagon, as seen in Fig. 6(a). (Obviously, the octagon can be degenerate in some directions. For example, Fig. 6(b) shows a triangular "octagon.") Note that since the area of Q is at least $|S(Q)| - d + 1$ (recall that d is the length of the smallest diagonal of Q), at most d rows or columns can be marked as caps (as seen in Fig. 6(b)), and so we are sure to be left with some cells marked as centerpiece. Taking a closer look at the centerpiece, we see that the only place where it differs from $S(Q)$ is in the truncated parts. It is easy to see that filling the truncated parts back to form $S(Q)$ would not change the perimeter size. Instead of a linear sequence of

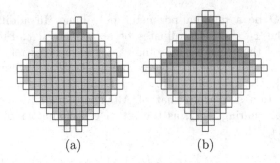

(a)   (b)

Fig. 6. An example of the decomposition of two minimal-perimeter polyomino to the centerpiece (gray), and caps (red). (Color figure online)

perimeter cells, there is a triangle in $S(Q)$, but the size of the perimeter of the triangle is exactly the same as the size of the perimeter of the centerpiece, thus the size of the perimeter of the centerpiece is equal to the size of the perimeter of $S(Q)$, which we already know to be identical to the size of the perimeter of Q.

In conclusion, any minimal perimeter polymino Q can be decomposed into a centerpiece (with the same perimeter as Q) and caps which do not change the perimeter. Looking at this process backwards, we can construct all the minimal-perimeter polyominoes of area n by iterating over all possible center-pieces with perimeter $\epsilon(n)$, and counting all possible combinations of 'caps' that make, together with the centerpieces, polyominoes of area n.

The Algorithm. Given an integer, n we want to compute the value of $|M_n|$. Based on the previous section, any minimal-perimeter polyomino of area n, can be decomposed to a centerpiece with perimeter $\epsilon(n)$ and caps, thus, by iterating over all the possible centerpieces with perimeter $\epsilon(n)$, and counting all the possible completions of these centerpieces to a polyomino of area n we can compute the value of $|M_n|$. Due to space limitations, a full discussion of the representation of the centerpieces and of how they are traversed is omitted here. Formally, let B_p be the set of all centerpieces with perimeter p, and given a certain center-piece b, let $C(b)$ be the number of ways to add to b caps with total area of $n - |b|$. Then, we have that

$$|M_n| = \sum_{b \in B_{\epsilon(n)}} C(b). \tag{1}$$

However, we need to be cautious with the definition of $C(b)$. If we allow any cap to be counted in $C(b)$, we may count some polyominoes more than once. Therefore, we define the notion of "proper caps." In a *proper cap*, the first layer does not span the entire length of the centerpiece edge it is attached to (see Fig. 7 for an illustration). By counting only proper caps, polyominoes are counted only once since they are counted only through the completions of the maximal centerpiece (in terms of containment).

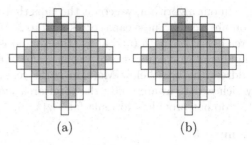

Fig. 7. Caps: (a) proper; (b) nonproper.

Counting Caps. Caps of a minimal-perimeter polyomino consist of cells added on top of the flat edges of the centerpiece without altering the perimeter of the centerpiece. By Theorem 5, we deduce that each layer of the cap must be properly "contained" within the previous level (i.e., as in ▪▪▪▪ and not as in ▪▪▪▪). For example, consider a cell c in the north cap. The cells below c to its right and to its left must also belong to the polyomino, otherwise, the diagonal connecting c to the centerpiece will have a gap, in contradiction with Theorem 5.

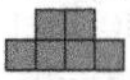

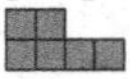

These caps can be represented by Motzkin paths. A Motzkin path of length n is a grid path starting at $(0,0)$ and ending at $(0,n)$, where each step (represented by a vector) belongs to the set $\{(1,-1),(1,0),(1,1)\}$, that is, right-down, right, or right-up, and the path never goes below the x-axis. The number of Motzkin paths of length n is known as the nth Motzkin number. We can map every cell in the envelop of the cap to a step in a Motzkin path. Consider a cap of the north edge. An occupied cell with an empty cell to its left (i.e., a cell which starts a new sequence of steps) corresponds to a right-up step; an occupied cell with an occupied cell to its left corresponds to a right step; and an empty cell with an occupied cell to its left corresponds to a right-down step. This correspondence is shown in Fig. 8. Notice that a cap of length k is mapped to a Motzkin path of length $k+1$.

Denote by $\mathrm{Mz}(k)$ the kth Motzkin number. It is well known that

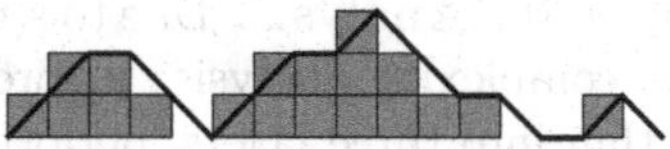

Fig. 8. An example of a cap and the corresponding Motzkin path.

$$\mathrm{Mz}(k) = \mathrm{Mz}(k-1) + \sum_{i=0}^{k-2} \mathrm{Mz}(i)\mathrm{Mz}(k-i-2).$$

The first element of the recursion corresponds to a path starting with a right step; The rest of the path is itself a Motzkin path of length $(n-1)$. The other element of the recursion (the summation) corresponds to Motzkin paths that start with a right-up step. Such a path must return to the x-axis at some point (either in the middle or at the end of the path), and the recursion sums over the possible locations where the path returns to the x-axis for the first time.

In our algorithm, we count the caps that fit on an edge of length k, and that contain r cells. These caps correspond to Motzkin paths of length $k+1$, while the area below the path (and above the x-axis) is equal to r. Denote by $\mathrm{Mz}(k,r)$ the number of Motzkin paths of length k and of area of r below them. Several works address the area below Motzkin paths (e.g., [12,15,16]), however, all works of which we are aware study the total area under *all* Motzkin paths of length k, but do not provide a formula for $\mathrm{Mz}(k,r)$.

Lemma 5.

$$\mathrm{Mz}(k,r) = \mathrm{Mz}(k-1,r) + \sum_{i=0}^{k-2} \sum_{j=0}^{r-i-1} \mathrm{Mz}(i,j)\mathrm{Mz}(k-2-i,r-i-1-j).$$

The proof of Lemma 5 is given in the full version of the paper.

Counting only proper caps is simple when we think of it in terms of Motzkin paths. We want to count only "proper paths," i.e., to avoid paths that start with a right-up step and do not touch the x-axis until the last step. This kind of paths is counted in the last element of the recursive formula for Motzkin paths. Thus, the formula for proper Motzkin paths is

$$\widetilde{\mathrm{Mz}}(k,r) = \mathrm{Mz}(k-1,r) + \sum_{i=0}^{k-3} \sum_{j=0}^{r-i-1} \mathrm{Mz}(i,j)\mathrm{Mz}(k-2-i,r-i-1-j),$$

where $\widetilde{\mathrm{Mz}}(k,r)$ is the number of proper Motzkin path with length k and area r.

With this formula at hand, we can devise a formula for $C(b)$. For a specific centerpiece b, our aim is to complete b to a polyomino of area n, thus, we need to add to it $n - |b|$ cells. Let d_i (for $1 \le i \le 4$) be the lengths of the four "flat" edges of b. Then, $C(b) = \displaystyle\sum_{\substack{n_1,n_2,n_3,n_4 \\ n_1+n_2+n_3+n_4=n-|b|}} \prod_{i=1}^{4} \widetilde{\mathrm{Mz}}(d_i - 1, n_i)$. That is, we sum over all partitions of the $n - |b|$ additional cells into the four caps.

Complexity Analysis. Due to space limitations, we provide here only a sketch of the complexity analysis. In order to facilitate the analysis, we divide the algorithm into three layers, perform each one of them separately, and store the results in a re-usable memory. First, we compute all relevant values of $\widetilde{\mathrm{Mz}}(n,k)$. Second, we compute all relevant values of $C(b)$. Finally, we use the stored data and compute $|M_n|$ according to Eq. 1. In fact, we use this layered analysis only in order to bound from above the time complexity of the algorithm. In practice, we perform a lazy evaluation, that is, we compute values of each function on demand, and memoize the results.

Computing the values of $\widetilde{\mathrm{Mz}}(n,k)$ takes $O(n^3)$ time. The computation of $C(b)$ takes $O(n^4)$ time. Finally, the computation of $|M_n|$ takes $O(n^{2.5})$ time. The three steps are independent, thus, the total time complexity is $O(n^4)$. The full details are given in the full version of the paper.

4.3   Counts of Minimum-Perimeter Polyominoes

The "Bulk Counting" algorithm computes the first 1,000 values of $|M_n|$ in about one hour on a 2.8 GHz, single processor, 16 GB RAM machine. The counts are shown in Table 1 and are plotted in Fig. 9. A detailed discussion of these counts is provided in the full version of the paper.

Table 1. Values in the four columns of $|M_n|$. Each value in the table is $\left|M_{\epsilon^{-1}(p)}\right|$ (an inflation-chain root), where p is the sum of the number in the left column and the number in the header of the column.

Perimeter	+0	+1	+2	+3
4	1	0	$4(2)^{\dagger}$	4
8	9	4	22	28
12	52	28	106	124
16	206	124	392	456
20	719	456	1,276	1,464
24	2,206	1,464	3,738	4,252
28	6,226	4,252	10,180	11,468
32	16,368	11,468	25,984	29,112
36	40,751	29,112	63,116	70,268
40	96,691	70,268	146,714	162,584
44	220,574	162,584	328,702	362,672
48	485,846	362,672	712,736	783,508
52	1,038,209	783,508	1,502,024	1,645,536
56	2,159,068	1,645,536	3,084,988	3,369,832
60	4,382,886	3,369,832	6,192,586	6,745,780
64	8,703,908	6,745,780	12,173,336	13,228,364
68	16,945,143	13,228,364	23,480,084	25,457,184
72	32,393,530	25,457,184	44,504,628	48,152,268
76	60,899,182	48,152,268	83,011,376	89,642,052
80	112,731,886	89,642,052	152,546,840	164,439,468
84	205,712,776	164,439,468	276,483,464	297,542,816
88	370,405,262	297,542,816	494,690,890	531,547,196
92	658,692,094	531,547,196	874,505,946	938,293,884
96	1,157,749,216	938,293,884	1,528,537,056	1,637,794,712

† The inflation-chain root of perimeter 6 is area 2, and $|M_2| = 2$. However, when M_2 is inflated, the next element in the chain is M_8, but $|M_8| = 4$. This exception is caused by the two shapes having two diagonally-placed cells ▪ and ▪) which have a minimal perimeter but do not count as polyominoes. Due to this anomaly, Theorem 2 applies only for $n \geq 3$.

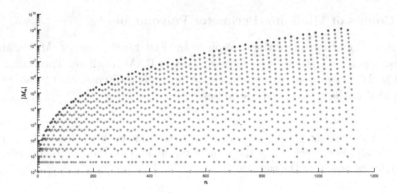

Fig. 9. Values of $|M_n|$. Inflation-chain roots are colored in red. (Color figure online)

5   Conclusion

In this paper, we have shown some results concerning minimal-perimeter polyominoes. First, we characterized the roots of the inflation chains of polyominoes. Second, we showed that any polyomino, when inflated enough times, becomes a minimal-perimeter polyomino. This opens the opportunity to tie between minimal-perimeter polyominoes and all polyominoes by dividing polyominoes into classes, where all members in a class eventually inflate into the same minimal-perimeter polyomino. Characterizing this class may, in turn, shed some light on the total number of polyominoes. Third, we provided two algorithms for computing $|M_n|$. Finding a closed formula for $|M_n|$ remains an open problem.

An interesting open problem is the behavior of $|M_n|$ when one considers only inflation-chain roots. Figure 9 shows that $|M_n|$ increases monotonically with a decreasing slope. Does $|M_n|$ have an asymptotic slope? And if so, what is it?

References

1. Altshuler, Y., Yanovsky, V., Vainsencher, D., Wagner, I.A., Bruckstein, A.M.: On minimal perimeter polyminoes. In: Kuba, A., Nyúl, L.G., Palágyi, K. (eds.) DGCI 2006. LNCS, vol. 4245, pp. 17–28. Springer, Heidelberg (2006). https://doi.org/10.1007/11907350_2
2. Asinowski, A., Barequet, G., Zheng, Y.: Enumerating polyominoes with fixed perimeter defect. In: Proceedings of 9th European Conference on Combinatorics, Graph Theory, and Applications, vol. 61, pp. 61–67. Elsevier, Vienna, August 2017
3. Barequet, G., Ben-Shachar, G.: Properties of minimal-perimeter polyominoes. In: Wang, L., Zhu, D. (eds.) COCOON 2018. LNCS, vol. 10976, pp. 120–129. Springer, Cham (2018). https://doi.org/10.1007/978-3-319-94776-1_11
4. Barequet, G., Rote, G., Shalah, M.: $\lambda > 4$: an improved lower bound on the growth constant of polyominoes. Commun. ACM **59**(7), 88–95 (2016)
5. Broadbent, S., Hammersley, J.: Percolation processes: I. Crystals and Mazes. In: Mathematical Proceedings of the Cambridge Philosophical Society, vol. 53, pp. 629–641. Cambridge University Press, Cambridge (1957)

6. Fülep, G., Sieben, N.: Polyiamonds and polyhexes with minimum site-perimeter and achievement games. Electron. J. Comb. **17**(1), 65 (2010)
7. Golomb, S.: Checker boards and polyominoes. Am. Math. Mon. **61**(10), 675–682 (1954)
8. Jensen, I.: Counting polyominoes: a parallel implementation for cluster computing. In: Sloot, P.M.A., Abramson, D., Bogdanov, A.V., Gorbachev, Y.E., Dongarra, J.J., Zomaya, A.Y. (eds.) ICCS 2003. LNCS, vol. 2659, pp. 203–212. Springer, Heidelberg (2003). https://doi.org/10.1007/3-540-44863-2_21
9. Klarner, D.: Cell growth problems. Canad. J. Math. **19**, 851–863 (1967)
10. Klarner, D., Rivest, R.: A procedure for improving the upper bound for the number of n-ominoes. Canad. J. Math. **25**(3), 585–602 (1973)
11. Madras, N.: A pattern theorem for lattice clusters. Ann. Combin. **3**(2), 357–384 (1999)
12. Pergola, E., Pinzani, R., Rinaldi, S., Sulanke, R.: A bijective approach to the area of generalized Motzkin paths. Adv. Appl. Math. **28**(3–4), 580–591 (2002)
13. Redelmeier, D.H.: Counting polyominoes: yet another attack. Discret. Math. **36**(2), 191–203 (1981)
14. Sieben, N.: Polyominoes with minimum site-perimeter and full set achievement games. Eur. J. Comb. **29**(1), 108–117 (2008)
15. Sulanke, R.A.: Moments of generalized Motzkin paths. J. Integer Seq. **3**(001), 1 (2000)
16. Sulanke, R.A.: Bijective recurrences for Motzkin paths. Adv. Appl. Math. **27**(2–3), 627–640 (2001)
17. Wang, D.L., Wang, P.: Discrete isoperimetric problems. SIAM J. Appl. Math. **32**(4), 860–870 (1977)

On Rectangle Intersection Graphs
with Stab Number at Most Two

Dibyayan Chakraborty[1(✉)], Sandip Das[1], Mathew C. Francis[2], and Sagnik Sen[3]

[1] Indian Statistical Institute, Kolkata, India
dibyayancg@gmail.com
[2] Indian Statistical Institute, Chennai, India
[3] Ramakrishna Mission Vivekananda Educational and Research Institute,
Howrah, India

Abstract. Rectangle intersection graphs are the intersection graphs of axis-parallel rectangles in the plane. A graph G is said to be a *k-stabbable rectangle intersection graph*, or *k-SRIG* for short, if it has a rectangle intersection representation in which k horizontal lines can be placed such that each rectangle intersects at least one of them. The stab number of a graph G, denoted by $stab(G)$, is the minimum integer k such that G is a k-SRIG. In this paper, we introduce some natural subclasses of 2-SRIG and study the containment relationships among them. We also give a linear time recognition algorithm for one of those classes. In this paper, we prove that the CHROMATIC NUMBER problem is NP-complete even for 2-SRIGs. This strengthens a result by Imai and Asano [13]. We also show that triangle-free 2-SRIGs are three colorable.

1 Introduction

Study of *interval graphs* (intersection graphs of intervals on the real line) dates back to 1950's when Benzer established a direct relation between interval graphs and arrangements of genes in the chromosome [2]. Due to several elegant characterisations, there are efficient algorithms for RECOGNITION PROBLEM, CHROMATIC NUMBER, MAXIMUM INDEPENDENT SET etc. on interval graphs [8,14,16]. However, the scenario is different in two dimensional counter part i.e. *rectangle intersection graphs*.

A *rectangle intersection representation* of a graph is a collection of axis-parallel closed rectangles in the plane such that each rectangle in the collection represents a vertex of the graph and two rectangles intersect if and only if the vertices they represent are adjacent in the graph. The graphs that have rectangle intersection representations are called *rectangle intersection graphs*. The RECOGNITION PROBLEM, CHROMATIC NUMBER, MAXIMUM INDEPENDENT SET are known to be NP-Hard for rectangle intersection graphs [13,15].

Recently, rectangle intersection graphs are being studied in a new approach using a parameter called *"stab number"* [3,4,6]. A *k-stabbed rectangle intersection representation* is a rectangle intersection representation along with a collection of k horizontal lines called *stab lines* such that every rectangle intersects

© Springer Nature Switzerland AG 2019
S. P. Pal and A. Vijayakumar (Eds.): CALDAM 2019, LNCS 11394, pp. 124–137, 2019.
https://doi.org/10.1007/978-3-030-11509-8_11

at least one of the stab lines. A graph G is a k-*stabbable rectangle intersection graph* (k-*SRIG*), if there exists a k-stabbed rectangle intersection representation of G. The *stab number*, $stab(G)$, of a rectangle intersection graph, is the minimum integer k such that G is a k-SRIG. A k-*exactly stabbed rectangle intersection representation* is a k-stabbed rectangle intersection representation in which every rectangle intersects precisely one of the stab lines. A graph G is a k-*exactly stabbable rectangle intersection* graph, or k-*ESRIG* for short, if there exists a k-exactly stabbed rectangle intersection representation of G. The *exact stab number*, $estab(G)$, of a rectangle intersection graph, is the minimum integer k such that G is k-ESRIG. We use the notation k-SRIG and k-ESRIG also to denote the class of all k-SRIGs and the class of all k-ESRIGs respectively. In this paper, we limit our study on 2-SRIG and its subclasses.

For a 2-stabbed rectangle intersection representation $\mathcal{R}$ of a graph G, let $y = a_1$ and $y = a_2$ with $a_1 < a_2$ be the stab lines. Let $\mathcal{R}_t$ and $\mathcal{R}_b$ be the sets of intervals obtained by projecting the rectangles that intersect $y = a_2$ and $y = a_1$ respectively on the x-axis. By putting restrictions on $\mathcal{R}_t$ and $\mathcal{R}_b$, we have several subclasses of 2-SRIG as defined below. A set of intervals is *proper* if no interval in the set is a subset of another. Let $\mathcal{I}$ denote the set of all sets of intervals, $\mathcal{P}$ the set of all proper sets of intervals and $\mathcal{U}$ the set of all sets of unit length intervals. For $\mathcal{X}, \mathcal{Y} \subseteq \mathcal{I}$, let $\mathcal{R}$ be a 2-stabbed rectangle intersection representation of G. If $\mathcal{R}_t \in \mathcal{X}$ and $\mathcal{R}_b \in \mathcal{Y}$ then $\mathcal{R}$ is a $(\mathcal{X}, \mathcal{Y})$-representation of G. Graphs having a $(\mathcal{X}, \mathcal{Y})$-representation are $(\mathcal{X}, \mathcal{Y})$-graphs. Here we consider $\mathcal{X}, \mathcal{Y}$ to be $\mathcal{I}, \mathcal{P}$ or $\mathcal{U}$. Note that the class of $(\mathcal{I}, \mathcal{I})$-graphs is the same as 2-SRIG, while the classes $(\mathcal{I}, \mathcal{P})$, $(\mathcal{I}, \mathcal{U})$, $(\mathcal{P}, \mathcal{P})$, $(\mathcal{P}, \mathcal{U})$ and $(\mathcal{U}, \mathcal{U})$-graphs are all subclasses of 2-SRIG. A graph G is a *2-stabbable unit square intersection graph* or 2-SUIG, if G has a 2-stabbed rectangle intersection representation $\mathcal{R}$ in which all rectangles are unit squares. Our first result establishes containment relationships among the above classes.

Theorem 1. *2-SUIG* $= (\mathcal{U}, \mathcal{U})$-*graphs* $\subset (\mathcal{P}, \mathcal{U})$-*graphs* $= (\mathcal{P}, \mathcal{P})$-*graphs* $\subset (\mathcal{I}, \mathcal{U})$-*graphs* $= (\mathcal{I}, \mathcal{P})$-*graphs* $\subset$ *2-ESRIG* $=$ *2-SRIG*.

Given a triangle-free graph it is possible to figure out if it is a $(\mathcal{P}, \mathcal{P})$-graph or not in linear time.

Theorem 2. *Let G be a triangle-free graph. There is an $O(|V(G)|)$ time algorithm to decide if G is a $(\mathcal{P}, \mathcal{P})$-graph.*

To prove Theorem 2, we actually show that triangle-free $(\mathcal{P}, \mathcal{P})$-graphs are subclasses of *outerplanar graphs*. This implies that the *chromatic number* of triangle-free $(\mathcal{P}, \mathcal{P})$-graphs is at most three. In fact, we manage to extend this observation to the class of triangle-free 2-SRIG.

Theorem 3. *Triangle-free 2-SRIGs are 3-colorable.*

Indeed for a fixed c and any 2-SRIG G, it is possible to decide if G is c-colorable. To prove this, observe that when a 2-SRIG graph G is c-colorable, the treewidth [10] of G is at most $2c$. Now using a standard dynamic programming on the tree decomposition [10] of G, it is possible to decide if G is c-colorable with running time that is polynomial in the number of vertices. However, the problem of finding the chromatic number of 2-SRIG in general remains NP-Hard.

Theorem 4. *The problem of finding the chormatic number is NP-complete for 2-SRIGs.*

In Sect. 2, we give some definitions and notation that will be used throughout the paper. In Sects. 3 and 4, we prove Theorems 1 and 2, respectively. In Sect. 5, we prove Theorem 3 and give only a proof sketch for Theorem 4. In Sect. 6, conclusions are drawn.

2   Preliminaries

For standard notations and terminologies we refer to the book of West [18].

Let G be a rectangle intersection graph with rectangle intersection representation $\mathcal{R}$. A rectangle in $\mathcal{R}$ corresponding to the vertex v is denoted as $r_v = [x_v^-, x_v^+] \times [y_v^-, y_v^+]$. Thus we have standardized a notation for the boundary points of the intervals corresponding to the projection of r_v on X-axis and Y-axis, respectively, by x_v^-, x_v^+ and y_v^-, y_v^+.

Let G be a 2-SRIG with a 2-stabbed rectangle intersection representation $\mathcal{R}$ in which the stab lines are $y = a_1$ and $y = a_2$ where $a_1 < a_2$. The *top* (resp. *bottom*) stab line of $\mathcal{R}$ is the stab line $y = a_2$ (resp. $y = a_1$). A vertex $u \in V(G)$ is "on" a stab line if r_u intersects that stab line.

A *c-coloring* of G is a mapping $\phi: V(G) \to \{1, 2, \ldots, c\}$ such that $\phi(u) \neq \phi(v)$ when $uv \in E(G)$. A graph is *c-colorable* if it has a c-coloring. The *chromatic number* $\chi(G)$ of G is the minimum c such that G is c-colorable. The CHROMATIC NUMBER problem is to decide, onss given a graph G and an integer c as input, whether G is c-colorable.

3   Proof of Theorem 1

The proof Theorem 1 is contained in several observations, propositions and lemmas presented in this section. The equality 2-SRIG = 2-ESRIG is easy to prove.

Observe that as 2-SRIG = 2-ESRIG, $(\mathcal{I}, \mathcal{P})$-graphs $\subseteq$ 2-ESRIG. To show that the class of $(\mathcal{I}, \mathcal{P})$-graphs is a proper subclass of 2-ESRIG, we show that the (3×4)-grid depicted in Fig. 2(a) is a 2-ESRIG but not a $(\mathcal{I}, \mathcal{P})$-graph. A 2-exactly stabbed representation of the (3×4)-grid given in Fig. 2(b) proves the former. For proving the later we need to work a bit more. First of all we need to define some notations and terminologies.

Let G be a 2-SRIG with a 2-stabbed rectangle intersection representation $\mathcal{R}$. The *span* of a vertex u is $span(u) = [x_u^-, x_u^+]$. The span of $S \subseteq V(G)$ is $span(S) = \cup_{u \in S} span(u)$. Observe that when $G[S]$ is connected, $span(S)$ is an interval. The span of an edge $uv \in E(G)$ is $span(uv) = span(u) \cap span(v)$. We write $I_1 = [a_1, b_1] < I_2 = [a_2, b_2]$ if $b_1 < a_2$. A set of vertices of G have a *common stab* if all of them are on a particular stab line. A *bridge edge* $uv \in E(G)$ in $\mathcal{R}$ is an edge such that there are two consecutive stab lines having u on one of them and v on the other. Whenever the 2-stabbed rectangle intersection representation of a graph G under consideration is clear from the context, the terms r_u, x_u^-, x_u^+, y_u^-, y_u^+, for every vertex $u \in V(G)$ and usages such as "on a stab line", "have a common stab", "span" etc. are considered to be defined with respect to this representation.

Observation A. *Let $\mathcal{R}$ be a 2-exactly stabbed representation of a graph G. Let uv be a bridge edge in $\mathcal{R}$ and let $S = \{w \in V(G): span(w) \cap span(uv) \neq \emptyset\}$. Let $a, b \in V(G)$ such that $span(a) < span(uv) < span(b)$. Then a and b are in different connected components of $G - S$.*

Proof. Suppose for the sake of contradiction that a and b are in the same connected component C of $G - S$. As $span(V(C))$ is an interval that contains both $span(a)$ and $span(b)$, it is clear that $span(V(C))$ also contains $span(uv)$. But this means that C contains some vertex w such that $span(w) \cap span(uv) \neq \emptyset$, which is a contradiction. □

Note that in the above observation, if $a \notin N[u] \cup N[v]$ and $b \notin N[u] \cup N[v]$, then because $S \subseteq N[u] \cup N[v]$, we can conclude that a and b are in different connected components of $G - (N[u] \cup N[v])$. We shall use this form of Observation A in several places.

Observation B. *Let $\mathcal{R}$ be a 2-exactly stabbed rectangle intersection representation of a triangle-free graph G. Let $e_1, e_2 \in E(G)$ be two bridge edges in $\mathcal{R}$. Then, $span(e_1) \cap span(e_2) = \emptyset$.*

Proof. Suppose for the sake of contradiction that $I = span(e_1) \cap span(e_2) \neq \emptyset$. Let $e_1 = uv$, $e_2 = ab$ and u, a being distinct vertices. Then, we have $I \subseteq [x_u^-, x_u^+]$, $I \subseteq [x_v^-, x_v^+]$, $I \subseteq [x_a^-, x_a^+]$, and $I \subseteq [x_b^-, x_b^+]$. Let us assume without loss of generality that u and a are on the bottom stab line and that v and b are on the top stab line. Now observe that if $y_u^+ \geq y_a^+$, then u, a, b form a triangle in G ($(I \times [y_b^-, y_a^+]) \subseteq r_u \cap r_a \cap r_b$) and that if $y_u^+ < y_a^+$, then a, u, v form a triangle in G ($(I \times [y_v^-, y_u^+]) \subseteq r_a \cap r_u \cap r_v$). □

Proposition 1. *In any 2-exactly stabbed rectangle intersection representation of a cycle of order greater than 3, there are exactly two bridge edges.*

Proof. Let G be a cycle of order greater than 3. Clearly, all the vertices of G cannot have a common stab as G is not an interval graph. This implies that in any 2-exactly stabbed rectangle intersection representation of G, there are at least two bridge edges. Suppose for the sake of contradiction assume that is a 2-exactly stabbed rectangle intersection representation $\mathcal{R}$ of G that have more than two bridge edges. As G is triangle-free, we can use Observation B to conclude that the bridge edges in $\mathcal{R}$ can be ordered as $e_1, e_2, \ldots, e_k$, where $k \geq 3$, such that $span(e_1) < span(e_2) < \cdots < span(e_k)$. Let $e_i = u_i v_i$ for all i. As $span(e_1) < span(e_2) < span(e_k)$, it is clear from the definition of $span(e_i)$

that there exists a vertex $w_1 \in \{u_1, v_1\}$ and a vertex $w_2 \in \{u_k, v_k\}$ such that $span(w_1) < span(e_2) < span(w_k)$. We can now apply Observation A to conclude that w_1 and w_k are in different connected components of $G - (N[u_2] \cup N[v_2])$. But this is a contradiction as in any cycle of order greater than 3, it is not possible to remove the closed neighbourhoods of two consecutive vertices to obtain a disconnected non-empty graph. □

Fig. 1. A graph belonging to the $W_{6,2}$ family and v is the central vertex.

The graph family $W_{n+1,d}$ with $n \geq 4, d \geq 2, n \geq d$ consists of triangle free graphs that are isomorphic to a cycle of order n with d vertices adjacent to a new *central vertex*. Figure 1 shows a graph belonging to the $W_{6,2}$ family.

Proposition 2. *Let $n \geq 4, d \geq 2$ be two integers and $\mathcal{R}$ be any 2-exactly stabbed rectangle intersection representation of a graph $G \in W_{n+1,d}$ with central vertex v. Then the number of bridge edges incident on v is $|N(v)| - 1$. Moreover, if $d \geq 3$ and uv be an edge such that u, v have a common stab, then $span(v) \subset span(u)$.*

Proof. Consider an arbitrary 2-exactly stabbed rectangle intersection representation $\mathcal{R}$ of G. Let C be the cycle obtained by removing the central vertex v from G and let $u_1, u_2, \ldots, u_d$ be the neighbours of v on C in the cyclic order. Let P_i denote the subpath of C from u_i to u_{i+1} (where $u_{d+1} = u_1$) that does not contain any neighbour of v as an internal vertex. Notice that each of C, $C_1 = G[V(P_1) \cup \{v\}]$, $C_2 = G[V(P_2) \cup \{v\}]$, $\ldots$, $C_d = G[V(P_d) \cup \{v\}]$ are induced cycles of G that are of order greater than 3. Applying Proposition 1 to each of $C, C_1, C_2, \ldots, C_d$, we can conclude that each of them contain exactly two bridge edges. Let us first consider C_i, for some $i \in \{1, 2, \ldots, d\}$. Suppose that the two bridge edges of C_i are also in P_i. Then these two edges are exactly the two bridge edges of C, implying that none of the paths in $P_1, P_2, \ldots, P_d$ other than P_i contain any bridge edges. This means that the two bridge edges of C_{i+1} (where again, $C_{d+1} = C_1$) are vu_i and vu_{i+1}. But then vu_i is a third bridge edge in C_i other than the two bridge edges on P_i, which is a contradiction. So we can assume that there is at most one bridge edge in P_i, for each i. As C has exactly two bridge edges, it follows that there are exactly two values in $\{1, 2, \ldots, d\}$, say t and t', such that P_t and $P_{t'}$ contain a bridge edge each. This tells us that for $i \in \{1, 2, \ldots, d\} \setminus \{t, t'\}$, the edges vu_i and vu_{i+1} are the two bridge edges in C_i. Now if t and t' are not consecutive (i.e., $t - 1 \neq t' \neq t + 1$), then by our previous observation, both vu_t and vu_{t+1} are bridge edges, which is a contradiction as we would then have three bridge edges in C_t. Therefore, we can conclude that t and t' are consecutive. Let us assume without loss of generality that $t' = t + 1$. By our earlier observation, we know that every edge in vu_i, where $i \in \{1, 2, \ldots, d\} \setminus \{t+1\}$ is a bridge edge, as it belongs to some C_j, where $j \in \{1, 2, \ldots, d\} \setminus \{t, t'\}$. Also, we can see that vu_{t+1} is not a bridge edge as otherwise, the cycles C_t and $C_{t'}$ will have more than two bridge edges. Therefore, the set of bridge edges incident on v is $\{u_1, u_2, \ldots, u_d\} \setminus \{u_{t+1}\}$.

We will now show that $span(v) \subset span(u_{t+1})$. For ease of notation, let $a = u_t$, $b = u_{t+1}$ and $c = u_{t+2}$. Let us assume without loss of generality that v is on the bottom stab line. Then, we know that a and c are both on the top stab line as va and vc are bridge edges. Since a and c are nonadjacent, it follows that $span(a) \cap span(c) = \emptyset$. Let us assume by symmetry that $span(a) < span(c)$. Since v is a neighbour of both a and c, it follows that $span(v)$ intersects both $span(a)$ and $span(c)$, or in other words, $[x_a^+, x_c^-] \subseteq span(v)$. Notice that there are no bridge edges in the path $P = P_{t+2} \cup P_{t+3} \cup \cdots \cup P_d \cup P_1 \cup P_2 \cup \cdots P_{t-1}$. Therefore, all the vertices in P are have a common stab line, in particular, have a common stab line with a and c. As $a, c \in P$, we have that $span(V(P))$ contains both $span(a)$ and $span(c)$, which implies that $[x_a^+, x_c^-] \subseteq span(V(P))$. Let ww' and zz' be the bridge edges on P_t and P_{t+1} respectively, where w, z, a, c have a common stab line and w', z', b, v have a common stab line. Let P_t' be the path $P_t - \{a, b\}$ and P_{t+1}' the path $P_{t+1} - \{b, c\}$. As no vertex of P_t' is adjacent to any vertex of P or to v, we can conclude that $span(V(P_t')) \cap [x_a^+, x_c^-] = \emptyset$. As there is a neighbour of a on P_t', $span(V(P_t'))$ intersects $span(a)$, leading us to the conclusion that $span(V(P_t')) < [x_a^+, x_c^-]$. Since at least one of w, w' is on P_t', this means that $span(ww') < [x_a^+, x_c^-]$. With the same kind of arguments, we can also deduce that $[x_a^+, x_c^-] < span(V(P_{t+1}'))$ and that $[x_a^+, x_c^-] < span(zz')$. By Observation B, we know that the spans of any two bridge edges of G are disjoint. Since it is clear that $span(va) \cap [x_a^+, x_c^-] \neq \emptyset$ and $span(vc) \cap [x_a^+, x_c^-] \neq \emptyset$, we now have $span(ww') < span(va) < span(vc) < span(zz')$ (recall that $span(a) < span(c)$). As $span(ww') < span(va)$, there exists a vertex $w'' \in \{w, w'\}$ such that $span(w'') < span(va)$. Let $S = \{u \in V(G) : span(u) \cap span(va) \neq \emptyset\}$. It is easy to see that $S \subseteq N[v] \cup N[a]$. Now, by Observation A, w'' and c are in two connected components of $G - S$ (note that $c \notin S$). This implies that $b \in S$, or in other words, $span(b) \cap span(va) \neq \emptyset$. Using the same kind of reasoning for the bridge edges vc and zz', we can conclude that $span(b) \cap span(vc) \neq \emptyset$. Together, we get $[x_a^+, x_c^-] \subseteq span(b)$. Recall that $[x_a^+, x_c^-] \subseteq span(v)$. As b and v have a common stab line and because $a, c \in N(v) \setminus N(b)$, we can conclude that $y_b^+ < y_v^+$. Now suppose that $x_v^- \leq x_b^-$. Let b' be the neighbour of b on P_t'. As $span(V(P_t')) < [x_a^+, x_c^-]$, we have $span(b') < [x_a^+, x_c^-]$. But then, $r_{b'}$ cannot intersect r_b without intersecting r_v. This contradiction lets us conclude that $x_b^- < x_v^-$. Arguing symmetrically, we can also derive $x_b^+ > x_v^+$. This shows that $span(v) \subset span(b)$. $\qquad\square$

The (h, w)-*grid* is the undirected graph G with $V(G) = \{(x, y) : x, y \in \mathbb{Z}, 1 \leq x \leq h, 1 \leq y \leq w\}$ and $E(G) = \{(u, v)(x, y) : |u - x| + |v - y| = 1\}$.

Observation C. *Consider the $(3, 4)$-grid graph H as shown in Fig. 2(a). In any 2-exactly stabbed rectangle intersection representation of H, the edge $v_1 v_2$ is a bridge edge.*

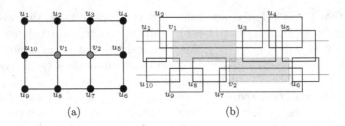

Fig. 2. A 2-exactly stabbed rectangle intersection representation of $(3,4)$-grid graph.

Proof. Suppose for the sake of contradiction that there is a 2-exactly stabbed rectangle intersection representation $\mathcal{R}$ of H such that v_1 and v_2 have a common stab. In H, both the subsets $\{v_1, u_1, u_2, u_3, v_2, u_7, u_8, u_9, u_{10}\}$ and $\{v_2, u_2, u_3, u_4, u_5, u_6, u_7, u_8, v_1\}$ induce subgraphs belonging to the $W_{9,4}$ family. Hence, by Proposition 2, for each $i \in \{1,2\}$, there is exactly one vertex $w_i \in N(v_i)$ that have a common stab line as v_i and $span(v_i) \subset span(w_i)$ in $\mathcal{R}$. Then by definition of w_1 and w_2, we have $w_1 = v_2$ and $w_2 = v_1$. Now we can use our earlier observation to infer that $span(v_1) \subset span(v_2) \subset span(v_1)$, which is a contradiction. $\square$

Let $\mathcal{R}$ be a 2-exactly stabbed rectangle intersection representation of a graph G along with the stab lines $y = a_1$ and $y = a_2$ where $a_1 < a_2$. Recall that $\mathcal{R}_t$ and $\mathcal{R}_b$ are the sets of intervals obtained by projecting the rectangles that intersect $y = a_2$ and $y = a_1$ respectively on the x-axis.

Lemma 1. *The family of $(\mathcal{I}, \mathcal{P})$-graphs is a proper subset of 2-ESRIG.*

Proof. By definition, an $(\mathcal{I}, \mathcal{P})$-graph is a 2-ESRIG. Let H be the $(3,4)$-grid as shown in Fig. 2(a). Clearly, there is a 2-exactly stabbed rectangle intersection representation of H (Fig. 2(b)). In H, both the subsets $\{v_1, u_1, u_2, u_3, v_2, u_7, u_8, u_9, u_{10}\}$ and $\{v_2, u_2, u_3, u_4, u_5, u_6, u_7, u_8, v_1\}$ induce subgraphs belonging to the $W_{9,4}$ family. Hence, by Proposition 2, in any 2-exactly stabbed rectangle intersection representation of H, for each $i \in \{1,2\}$, there is exactly one vertex $w_i \in N(v_i)$ that have a common stab line as v_i and $span(v_i) \subset span(w_i)$. Moreover, by Observation C, in any 2-exactly stabbed rectangle intersection representation of H, the edge v_1v_2 is a bridge edge. This implies that, in any 2-exactly stabbed rectangle intersection representation $\mathcal{R}$ of H, none of the sets $\mathcal{R}_t$ and $\mathcal{R}_b$ is a proper set of interval. Hence, H is not an $(\mathcal{I}, \mathcal{P})$-graph. $\square$

Lemma 2. *The family of $(\mathcal{I}, \mathcal{P})$-graphs is equivalent to the family of $(\mathcal{I}, \mathcal{U})$-graphs.*

Proof. By definition, an $(\mathcal{I}, \mathcal{U})$-graph is a $(\mathcal{I}, \mathcal{P})$-graph. Let G be an $(\mathcal{I}, \mathcal{P})$-graph and $\mathcal{R}$ be a $(\mathcal{I}, \mathcal{P})$-representation of G. We shall assume without loss of generality that $\mathcal{R}_b$ is a proper set of intervals and for any two vertices $u, v \in$

$V(G)$ we have $\{x_v^-, x_v^+\} \cap \{x_u^-, x_u^+\} = \emptyset$. We shall use V_1 and V_2 to denote the sets of vertices that are on the top and bottom stab lines respectively. Note that $\mathcal{R}_b$ is a proper interval representation of $G[V_2]$. Let $p_1, p_2, \ldots, p_{2|V_2|}$ be the endpoints of the intervals in $\mathcal{R}_b$ written in ascending order. We now use the fact that every graph that has a proper interval representation also has a unit interval representation in which the endpoints of the intervals are in the same order [5]. Let $\mathcal{U}$ be the unit interval representation corresponding to $\mathcal{R}_b$ and let $p_1', p_2', \ldots, p_{2|V_2|}'$ be the endpoints of the intervals in $\mathcal{U}$ written in ascending order. We now construct a 2-exactly stabbed rectangle intersection representation $\mathcal{R}' = \{r_u' = [x_u'^-, x_u'^+] \times [y_u'^-, y_u'^+]\}_{u \in V(G)}$ of G such that at least one of $\mathcal{R}_t'$ and $\mathcal{R}_b'$ is a set of unit intervals as follows. In the representation $\mathcal{R}'$, the rectangle corresponding to a particular vertex intersect the same stab line as it intersects in $\mathcal{R}$. We define $[y_u'^-, y_u'^+] = [y_u^-, y_u^+]$ for every vertex $u \in V(G)$. For each vertex $u \in V_2$, we let $x_u'^+ = p_i'$ and $x_u'^- = p_j'$, where $x_u^+ = p_i$ and $x_u^- = p_j$. Define $f : \bigcup_{u \in V_1} \{x_u^+, x_u^-\} \to \mathbb{R}$ as follows:

$$f(p) = p_i' + \frac{j}{t+1}(p_{i+1}' - p_i')$$

where $p_i < p < p_{i+1}$, and $q_1, q_2, \ldots, q_{k-1}, (q_k = p), q_{k+1}, \ldots, q_t$ are the points in $\{x_u^+ : u \in V_1, p_i < x_u^+ < p_j\} \cup \{x_u^- : u \in V_1, p_i < x_u^- < p_j\}$, in ascending order. For each vertex $u \in V_1$, we let $x_u'^+ = f(x_u^+)$ and $x_u'^- = f(x_u^-)$. It is not difficult to verify that $\mathcal{R}'$ is a 2-exactly stabbed rectangle intersection representation of G such that $\mathcal{R}_b'$ is a set of unit intervals. $\qquad \square$

From the proof of Lemma 2, it is clear that the left and right edges of the rectangles of $\mathcal{R}'$ are in the same order as they are in $\mathcal{R}$. Let G be a $(\mathcal{P}, \mathcal{P})$-graph and $\mathcal{R}$ be a $(\mathcal{P}, \mathcal{P})$-representation of G. The construction procedure described in Lemma 2 when applied on $\mathcal{R}$ gives us a 2-exactly stabbed rectangle intersection representation $\mathcal{R}'$ of G such that one of $\mathcal{R}_t'$ and $\mathcal{R}_b'$ is a set of unit intervals and the other is a proper set of intervals. This gives us the following lemma.

Lemma 3. *The family of $(\mathcal{P}, \mathcal{P})$-graphs is equivalent to the family of $(\mathcal{P}, \mathcal{U})$-graphs.*

Lemma 4. *The family of $(\mathcal{P}, \mathcal{P})$-graphs is a proper subset of the family of $(\mathcal{I}, \mathcal{P})$-graphs.*

Proof. By definition, a $(\mathcal{P}, \mathcal{P})$-graph is an $(\mathcal{I}, \mathcal{P})$-graph. Consider the labelled $(3,3)$-grid graph H shown in Fig. 3(a). Clearly, there is a $(\mathcal{I}, \mathcal{P})$-representation of H (Fig. 3(b)). Note that H is a graph belonging to the $W_{9,4}$ family. By Proposition 2, in any 2-exactly stabbed rectangle intersection representation of H, there is a vertex $w \in N(v)$ such that w, v have a common stab and $span(v) \subset span(w)$. Hence, H is not a $(\mathcal{P}, \mathcal{P})$-graph. $\qquad \square$

Lemma 5 states that the family of $(\mathcal{U}, \mathcal{U})$-graphs is a proper subset of the family of $(\mathcal{P}, \mathcal{U})$-graph. Essentially, we can show that the graph shown in Fig. 4(a) is a $(\mathcal{P}, \mathcal{U})$-graph (representation given in Fig. 4(b)) but not a $(\mathcal{U}, \mathcal{U})$-graph.

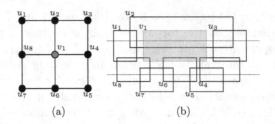

Fig. 3. 2-exactly stabbed rectangle intersection representation $\mathcal{R}$ of $(3,3)$-grid graph such that $\mathcal{R}_b$ is a proper set of intervals.

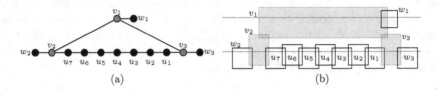

Fig. 4. A $(\mathcal{P},\mathcal{U})$-representation (b) of graph shown in (a).

Lemma 5. *The family of $(\mathcal{U},\mathcal{U})$-graphs is a proper subset of the family of $(\mathcal{P},\mathcal{U})$-graph.*

We will be done by proving that 2-SUIG is equivalent to the class of $(\mathcal{U},\mathcal{U})$-graphs whose proof follows easily from the definitions. For sake of completion, we have given the proof below.

Lemma 6. *The family of 2-SUIG is equivalent to the family of $(\mathcal{U},\mathcal{U})$-graphs.*

Proof. By definition, a 2-SUIG is a $(\mathcal{U},\mathcal{U})$-graph. Let G be a $(\mathcal{U},\mathcal{U})$-graph and $\mathcal{R}$ be a $(\mathcal{U},\mathcal{U})$-representation of G. Let I be the interval graph induced by the set of intervals $\{[y_v^-, y_v^+]\}_{v \in V(G)}$. Notice that, cardinality of the maximum independent set of I is at most two. Therefore, I must be an unit interval graph and $\mathcal{I} = \{y_v'^-, y_v'^+\}_{v \in V(H)}$ be an unit interval representation of I. Consider the set of rectangles $\mathcal{R}' = \{[x_v^+, x_v^-] \times [y_v'^-, y_v'^+]\}_{v \in V(G)}$. Clearly, $\mathcal{R}'$ is a 2-exactly stabbed rectangle intersection representation of G and all rectangles are unit squares. Hence, G is a 2-SUIG.

4   Proof of Theorem 2

A planar graph G is said to have an *LL-drawing* if G has a straight line planar embedding such that the point corresponding to a vertex of G lies on one of two given horizontal lines. A planar graph G is an *LL-graph* if G has an *LL-drawing*. The following observation is easy to see.

Observation D. *A graph G is an LL-graph if and only if there exists a partition of $V(G)$ into two ordered sets $A = \{a_1, a_2, \ldots, a_k\}$ and $B = \{b_1, b_2, \ldots, b_t\}$ such that:*

(a) there does not exist $1 \leq i < j \leq k$ and $1 \leq i' < j' \leq t$ with the property that $a_i b_{j'}, a_j b_{i'} \in E(G)$, and
(b) for any a_i, $N(a_i) \cap A \subseteq \{a_{i-1}, a_{i+1}\}$ and for any b_i, $N(b_i) \cap B \subseteq \{b_{i-1}, b_{i+1}\}$.

Indeed, the two ordered sets A and B referred to in the above observation consist of the vertices lying on each of the two horizontal lines in an *LL*-drawing of the graph, sorted according to their increasing x-coordinate. Lemma 7 states that all *LL*-graphs are $(\mathcal{P}, \mathcal{P})$-graphs and Lemma 8 states that all triangle-free $(\mathcal{P}, \mathcal{P})$-graphs are *LL*-graphs. See Fig. 5 for an example.

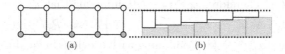

(a)   (b)

Fig. 5. (a) An example *LL*-graph. (b) a $(\mathcal{P}, \mathcal{P})$-representation of (a).

Lemma 7. *If a graph G is an LL-graph then G is also a $(\mathcal{P}, \mathcal{P})$-graph.*

Proof. Let $(A = \{a_1, a_2, \ldots, a_k\}, B = \{b_1, b_2, \ldots, b_t\})$ be the partition of $V(G)$ as given by Observation D. For $a \in A$, define $l(a) = \min\{i : b_i \in N(a)\}$ and $r(a) = \max\{i : b_i \in N(a)\}$ if $N(a) \cap B \neq \emptyset$ and $l(a) = r(a) = 0$ otherwise. It then follows from Observation D that for any a_i, a_j having $N(a_i) \cap B \neq \emptyset$ and $N(a_j) \cap B \neq \emptyset$, where $i < j$, $r(a_i) \leq l(a_j)$ and $a_i a_j$ or $b_i b_j$ is an edge only if $|i - j| = 1$. Let us define a $(\mathcal{P}, \mathcal{P})$-representation $\mathcal{R} = \{[x_u^-, x_u^+] \times [y_u^-, y_u^+]\}_{u \in V(G)}$ having stab lines $y = 0$ and $y = t + 1$ as follows.

For each b_i, we set

$$x_{b_i}^- = i, \quad x_{b_i}^+ = \begin{cases} i + \frac{1}{2} & \text{if } b_i b_{i+1} \notin E(G) \\ i + 1 & \text{otherwise} \end{cases}$$
$$y_{b_i}^- = 0, \quad y_{b_i}^+ = \begin{cases} i & \text{if } N(b_i) \cap A \neq \emptyset \\ 0 & \text{otherwise} \end{cases}$$

Let $\epsilon = \frac{1}{|V(G)|}$ and

$$x_{a_1}^- = 1, \qquad\qquad x_{a_1}^+ = \begin{cases} r(a_1) & \text{if } N(a_1) \cap B \neq \emptyset \\ 1 + \epsilon & \text{otherwise} \end{cases}$$
$$y_{a_1}^- = \begin{cases} t + 1 - l(a_1) & \text{if } N(a_1) \cap B \neq \emptyset \\ t + 1 & \text{otherwise} \end{cases}, \quad y_{a_1}^+ = t + 1$$

For each a_i, $i > 1$, we inductively define

$$x_{a_i}^- = \begin{cases} x_{a_{i-1}}^+ & \text{if } a_{i-1} a_i \in E(G) \\ x_{a_{i-1}}^+ + \epsilon & \text{otherwise} \end{cases}, \quad x_{a_i}^+ = \begin{cases} r(a_i) & \text{if } N(a_i) \cap B \neq \emptyset \\ x_{a_i}^- + \epsilon & \text{otherwise} \end{cases}$$
$$y_{a_i}^- = \begin{cases} t + 1 - l(a_i) & \text{if } N(a_i) \cap B \neq \emptyset \\ t + 1 & \text{otherwise} \end{cases}, \quad y_{a_i}^+ = t + 1$$

It is straightforward to verify that the $(\mathcal{P}, \mathcal{P})$-representation defined as above is a valid representation of the graph G.

Lemma 8. *If a triangle-free graph G is a $(\mathcal{P}, \mathcal{P})$-graph then G is an LL-graph.*

Proof. Let $\mathcal{R} = \{r_u = [x_u^-, x_u^+] \times [y_u^-, y_u^+]\}_{u \in V(G)}$ be a $(\mathcal{P}, \mathcal{P})$-representation of G having stab lines $y = 0$ and $y = 1$. Let $A = \{u : r_u \text{intersects the stab line} y = 1\}$ and $B = V(G) \setminus A$. Clearly, the rectangles corresponding to each vertex in B intersects the stab line $y = 0$. Let $k = |A|$ and $t = |B|$. Also let $a_1, a_2, \ldots, a_k$ be the vertices of A and let $b_1, b_2, \ldots, b_t$ be the vertices of B, such that for $1 \leq i < j \leq k$, $x_{a_i}^- \leq x_{a_j}^-$ and for $1 \leq i < j \leq t$, $x_{b_i}^- \leq x_{b_j}^-$. Suppose that the sets A and B with these orderings on their vertices violate condition (a) of Observation D. Then there exist i, j, i', j' such that $1 \leq i < j \leq k$ and $1 \leq i' < j' \leq t$ and $a_i b_{j'}, a_j b_{i'} \in E(G)$. Therefore, $r_{a_i} \cap r_{b_{j'}} \neq \emptyset$, implying that $x_{a_i}^+ \geq x_{b_{j'}}^-$. Similarly, we have $x_{b_{i'}}^+ \geq x_{a_j}^-$. As $x_{a_i}^- \leq x_{a_j}^-$, $x_{b_{i'}}^- \leq x_{b_{j'}}^-$, and $\mathcal{R}$ is a $(\mathcal{P}, \mathcal{P})$-representation, we have $x_{a_i}^+ \leq x_{a_j}^+$ and $x_{b_{i'}}^+ \leq x_{b_{j'}}^+$. Combining with previous inequalities, we now get $x_{a_i}^+ \geq x_{b_{j'}}^- \geq x_{b_{i'}}^-$ and $x_{a_j}^+ \geq x_{a_j}^- \geq x_{a_i}^-$. This implies that the intervals $[x_{a_i}^-, x_{a_i}^+]$ and $[x_{b_{i'}}^-, x_{b_{i'}}^+]$ intersect. Similarly, we get $x_{a_j}^+ \geq x_{a_i}^+ \geq x_{b_{j'}}^-$ and $x_{b_{j'}}^+ \geq x_{b_{i'}}^+ \geq x_{a_j}^-$ which implies that the intervals $[x_{a_j}^-, x_{a_j}^+]$ and $[x_{b_{j'}}^-, x_{b_{j'}}^+]$ intersect. Now let us consider the case when $y_{b_{i'}}^+ \leq y_{b_{j'}}^+$ (the case when $y_{b_{i'}}^+ > y_{b_{j'}}^+$ is symmetric and will not be discussed). As $r_{b_{i'}} \cap r_{a_j} \neq \emptyset$, we have $y_{b_{i'}}^+ \geq y_{a_j}^-$ and therefore $y_{b_{j'}}^+ \geq y_{a_j}^-$. Combined with our previous observation that $[x_{a_j}^-, x_{a_j}^+] \cap [x_{b_{j'}}^-, x_{b_{j'}}^+] \neq \emptyset$, this implies that $r_{a_j} \cap r_{b_{j'}} \neq \emptyset$ (recall that $0 \in [y_{b_{j'}}^-, y_{b_{j'}}^+]$ and $1 \in [y_{a_j}^-, y_{a_j}^+]$). Since this means that $a_j b_{j'} \in E(G)$, we have $a_i a_j, b_{i'} b_{j'} \notin E(G)$ (otherwise, either $a_i a_j b_{j'}$ or $b_{i'} b_{j'} a_j$ would be a triangle in G). Therefore, we can conclude that $x_{a_i}^+ < x_{a_j}^-$ and $x_{b_{i'}}^+ < x_{b_{j'}}^-$. Combining these with our earlier observation that $x_{a_i}^+ \geq x_{b_{j'}}^-$, we get $x_{a_j}^- > x_{b_{i'}}^+$, which contradicts the fact that $x_{b_{i'}}^+ \geq x_{a_j}^-$. This shows that the sets A and B do not violate condition (a) of Observation D.

Now suppose that $a_i a_j \in E(G)$, where $i < j$, then $x_{a_i}^+ \geq x_{a_j}^-$. Since $x_{a_i}^- \leq x_{a_{i+1}}^- \leq \cdots \leq x_{a_j}^-$, the rectangle r_{a_i} will intersect each of the rectangles $r_{a_{i+1}}, r_{a_{i+2}}, \ldots, r_{a_j}$, implying that $\{a_i, a_{j-1}, \ldots, a_j\}$ induces a triangle in G, which is a contradiction. We can conclude that $j = i + 1$. It can be similarly proven that if $b_i b_j \in E(G)$, where $i < j$, we have $j = i + 1$. Therefore, the sets A and B also satisfy condition (b) of Observation D, proving that G is an LL-graph. □

Due to Lemmas 7 and 8, we can infer that a triangle-free graph G is a $(\mathcal{P}, \mathcal{P})$-graph if and only if G is an LL-graph. Cornelsen et al. [9] proved that given a graph G, there is an $O(|V(G)|)$ time algorithm to decide if G is an LL-graph. Combining this fact with Lemmas 7 and 8, we have the proof of Theorem 2.

5 Proof of Theorem 3 and 4

Let H be a triangle-free 2-SRIG. First we prove the following lemma.

Lemma 9. *Let H be a triangle-free 2-SRIG. Then H is a planar graph.*

Proof. Observe that it is possible to find 2-stabbed rectangle intersection representation $\mathcal{R}$ of H such that given any two vertices $u, v \in V(H)$, $r_v \setminus r_u$ is nonempty and connected. Thus H is a planar graph due to Perepelitsa (Theorem 7 [17]). □

Proof of Theorem 3 follows from Grötzsch's Theorem, which says every triangle-free planar graph (hence any triangle-free 2-SRIG) is 3-colorable. Moreover, we can get such a coloring in $O(|V(H)|)$ time due to Dvořák et al. [11].

Proof sketch of Theorem 4: Given a 2-stabbed rectangle intersection representation $\mathcal{R}$ of a 2-SRIG G and an integer k, the 2-SRIG COLORING problem is to decide whether the chromatic number of G is at most k. *Circular arc graphs* are intersection graphs or arcs of a circle. Given a circular arc representation $\mathcal{C}$ of a circular arc graph G and an integer k, the CIRCULAR ARC COLORING problem is to decide whether the chromatic number of G is at most k. We shall reduce the NP-complete CIRCULAR ARC COLORING [12] problem to the 2-SRIG COLORING problem. An overview of the reduction procedure is given below. The detailed proof of Theorem 4 is given in the extended version.

Overview of the reduction: Given a circular arc representation $\mathcal{C}$ of a circular arc graph G, we shall cut the circle at two points to obtain two collections of intervals that can be thought of as the interval representations of two interval graphs G_1 and G_2 (see Fig. 6(b)). While doing this, each arc in $\mathcal{C}$ is divided into at most three parts, implying that $|V(G_1) \cup V(G_2)| \leq 3 \cdot |V(G)|$. Then we shall introduce new disjoint complete graphs (whose number of vertices would depend on k) and "join" them with "certain" subgraphs of G_1 and G_2. We shall show that the graph H constructed in this way is k-colorable if and only if G is k-colorable (see Fig. 6(c)). Moreover, H is a 2-SRIG and the number of vertices in H is $O(nk)$ where $n = |V(G)|$.

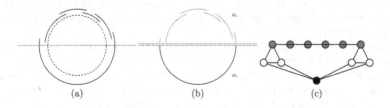

Fig. 6. Overview of the reduction procedure. (a) Circular arc representation of a graph G. (b) Each arcs in (a) are divided into at most three parts to get two interval graphs G_1 and G_2. (c) The constructed graph for $k = 3$.

Remark 1. Asinowski et al. [1] proved that finding CHROMATIC NUMBER of B_0-VPG graphs (intersection graphs of vertical and horizontal line segments on the plane) is NP-Complete. Given a circular arc graph G, our reduction procedure gives a graph which is a B_0-VPG graph having a representation where the horizontal line segments lie on one of two horizontal lines. So, we can also strengthen the result of Asinowski et al. [1] using the same reduction procedure. This also proves a claim made by Chaplick et al. [7].

Remark 2. For any integer $k \geq 2$, consider the graph G_k on $3k - 1$ vertices arranged in a circular manner and each vertex is adjacent to $k - 1$ many vertices that come after it. It is easy to see that for $k \geq 2$, G_k is a circular arc graph with $\omega(G) = k$ and $\chi(G) = \lceil \frac{3k-1}{2} \rceil$. Our construction implies that there are 2-SRIG graphs H such that $\chi(H) = \lceil \frac{3\omega(H)-1}{2} \rceil$.

6   Conclusions

In this paper, we focus our study on graphs that have stab number at most 2 and its subclasses. Since $(\mathcal{P},\mathcal{P})$-graphs are proper subsets of 2-SRIG, a direction of further research could be to investigate the class of $(\mathcal{P},\mathcal{P})$-graphs and try to characterize this class of graphs.

Question 1. Develop a forbidden structure characterization and/or a polynomial-time recognition algorithm for $(\mathcal{P},\mathcal{P})$-graphs.

For a 2-SRIG H, notice that $\chi(H) \leq 2\omega(H)$. Moreover, there are 2-SRIG graphs H with $\chi(H) = \lceil \frac{3\omega(H)-1}{2} \rceil$. Therefore, the following is a natural question in this direction.

Question 2. Is it true that for any 2-SRIG H, $\chi(H) \leq \lceil \frac{3\omega(H)}{2} \rceil$?

References

1. Asinowski, A., Cohen, E., Golumbic, M.C., Limouzy, V., Lipshteyn, M., Stern, M.: Vertex intersection graphs of paths on a grid. J. Graph Algorithms Appl. **16**(2), 129–150 (2012)
2. Benzer, S.: On the topology of the genetic fine structure. Proc. Natl. Acad. Sci. **45**(11), 1607–1620 (1959)
3. Bhore, S.K., Chakraborty, D., Das, S., Sen, S.: On a special class of boxicity 2 graphs. In: Ganguly, S., Krishnamurti, R. (eds.) CALDAM 2015. LNCS, vol. 8959, pp. 157–168. Springer, Cham (2015). https://doi.org/10.1007/978-3-319-14974-5_16
4. Bhore, S., Chakraborty, D., Das, S., Sen, S.: On local structures of cubicity 2 graphs. In: Chan, T.-H.H., Li, M., Wang, L. (eds.) COCOA 2016. LNCS, vol. 10043, pp. 254–269. Springer, Cham (2016). https://doi.org/10.1007/978-3-319-48749-6_19

5. Bogart, K.P., West, D.B.: A short proof that 'proper = unit'. Discret. Math. **201** (1–3), 21–23 (1999)
6. Chakraborty, D., Francis, M.C., On the stab number of rectangle intersection graphs. CoRR, abs/1804.06571 (2018)
7. Chaplick, S., Cohen, E., Stacho, J.: Recognizing some subclasses of vertex intersection graphs of 0-bend paths in a grid. In: Kolman, P., Kratochvíl, J. (eds.) WG 2011. LNCS, vol. 6986, pp. 319–330. Springer, Heidelberg (2011). https://doi.org/10.1007/978-3-642-25870-1_29
8. Corneil, D.G., Olariu, S., Stewart, L.: The LBFS structure and recognition of interval graphs. SIAM J. Discret. Math. **23**(4), 1905–1953 (2009)
9. Cornelsen, S., Schank, T., Wagner, D.: Drawing graphs on two and three lines. In: Goodrich, M.T., Kobourov, S.G. (eds.) GD 2002. LNCS, vol. 2528, pp. 31–41. Springer, Heidelberg (2002). https://doi.org/10.1007/3-540-36151-0_4
10. Cygan, M., et al.: Parameterized Algorithms, vol. 3. Springer, Cham (2015). https://doi.org/10.1007/978-3-319-21275-3
11. Dvořák, Z., Kawarabayashi, K., Thomas, R.: Three-coloring triangle-free planar graphs in linear time. ACM Trans. Algorithms **7**(4), 41:1–41:14 (2011)
12. Garey, M.R., Johnson, D.S., Miller, G.L., Papadimitriou, C.H.: The complexity of coloring circular arcs and chords. SIAM J. Algebr. Discret. Methods **1**(2), 216–227 (1980)
13. Imai, H., Asano, T.: Finding the connected components and a maximum clique of an intersection graph of rectangles in the plane. J. Algorithms **4**(4), 310–323 (1983)
14. Kleinberg, J., Tardos, E.: Algorithm Design. Pearson Education India, Bangalore (2006)
15. Kratochvíl, J.: A special planar satisfiability problem and a consequence of its NP-completeness. Discret. Appl. Math. **52**(3), 233–252 (1994)
16. Lekkerkerker, C., Boland, J.: Representation of a finite graph by a set of intervals on the real line. Fundamenta Mathematicae **51**(1), 45–64 (1962)
17. Perepelitsa, I.G.: Bounds on the chromatic number of intersection graphs of sets in the plane. Discret. Math. **262**(1–3), 221–227 (2003)
18. West, D.B.: Introduction to Graph Theory, 2nd edn. Prentice Hall, Upper Saddle River (2000)

Dominating Induced Matching in Some Subclasses of Bipartite Graphs

B. S. Panda[✉] and Juhi Chaudhary

Computer Science and Application Group, Department of Mathematics,
Indian Institute of Technology Delhi, Hauz Khas, New Delhi 110016, India
bspanda@maths.iitd.ac.in, chaudhary.juhi5@gmail.com

Abstract. Given a graph $G = (V, E)$, a set $M \subseteq E$ is called a *matching* in G if no two edges in M share a common vertex. A matching M in G is called an *induced matching* if $G[M]$, the subgraph of G induced by M, is same as $G[S]$, the subgraph of G induced by $S = \{v \in V \mid v$ is incident on an edge of $M\}$. An induced matching M in a graph G is dominating if every edge not in M shares exactly one of its endpoints with a matched edge. The *dominating induced matching (DIM)* problem (also known as *Efficient Edge Domination*) is a decision problem that asks whether a graph G contains a dominating induced matching or not. This problem is NP-complete for general graphs as well as for bipartite graphs. In this paper, we show that the DIM problem is NP-complete for perfect elimination bipartite graphs and propose polynomial time algorithms for star-convex, triad-convex and circular-convex bipartite graphs which are subclasses of bipartite graphs.

Keywords: Matching · Dominating induced matching ·
Graph algorithms · NP-completeness · Polynomial time algorithms

1 Introduction

Given a graph $G = (V, E)$, a set of edges $M \subseteq E$ is called a *matching* in G if no two edges of M share a common vertex. Vertices that are incident on the edges of a matching M are called *saturated* vertices. An edge is said to *dominate* itself and all the edges that share a common vertex with it. An *induced matching* in G is a subset of edges such that each edge of G is dominated by at most one edge of the subset. In this paper, we study an important variant of matching called *dominating induced matching*. A matching M in G is called a *dominating induced matching* if every edge of the graph G is saturated by M exactly once. The dominating induced matching (DIM) problem asks whether a graph G admits a dominating induced matching or not. Figure 1 contains a graph admitting a DIM and a graph that does not admit a DIM. DIM problem is NP-complete for general

J. Chaudhary wants to thank the Department of Science and Technology (INSPIRE) for their support.

S. P. Pal and A. Vijayakumar (Eds.): CALDAM 2019, LNCS 11394, pp. 138–149, 2019.
https://doi.org/10.1007/978-3-030-11509-8_12

Fig. 1. Graph in (a) admits a DIM consisting of dotted edges whereas graph in (b) does not admit a DIM

graphs [5] as well as for bipartite graphs [11]. DIM problem is polynomial time solvable for P_k-free graphs for $k \leq 8$, hole-free graphs [1], chordal graphs [10], generalized series parallel graphs [10], weakly chordal graphs [2] etc. The DIM problem has its applications in resource allocation problems, parallel processing systems, encoding theory and network routing problems [5,9]. The alternative definition of the DIM problem asks to determine if the vertex set of a graph G admits a partition into two subsets W and B such that W is an independent set and B induces a 1-regular graph. Here, vertices of sets W and B are called white and black respectively. A graph G has a DIM if and only if G admits a W-B coloring. We call a coloring a *valid partial coloring* if only a part of the vertices of V are assigned with white and black colors such that no two white vertices are adjacent and no black vertex has more than one black neighbor. If all the vertices in a graph can be assigned black and white colors such that no two white vertices are adjacent and every black vertex has exactly one black neighbor then it is called a *valid total coloring*. The class of perfect elimination bipartite graphs is an intermediate graph class between bipartite graphs and chordal bipartite graphs and was introduced by Golumbic and Goss [4]. Since the DIM problem is NP-complete for bipartite graphs and polynomial time solvable for chordal bipartite graphs, it is interesting to know its complexity status for perfect elimination bipartite graphs. The class of circular-convex bipartite graphs was introduced by Liang and Blum [7] and has been studied recently by researchers. In this paper, we will also consider tree-convex bipartite graphs which generalizes both chordal bipartite graphs and convex bipartite graphs. Tree-convex bipartite graphs are recognizable in linear time, and the associated tree T can also be constructed in linear-time [12]. Since star-convex bipartite graphs and triad-convex bipartite graphs are proper subclasses of tree-convex bipartite graphs and triad-convex bipartite graphs are also sandwiched between convex bipartite graphs and tree-convex bipartite graphs, the following interesting question arises naturally that where is the boundary between tractability and intractability of graph problems in these classes. The triad-convex bipartite graphs and star-convex bipartite graphs are studied in [3,8,13]. The relationship between different graph classes is shown in Fig. 2.

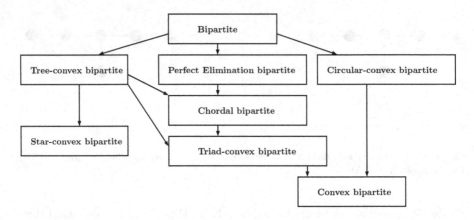

Fig. 2. The hierarchical relationship between subclasses of bipartite graphs.

2 Preliminaries

In a graph $G = (V, E)$, the open and closed neighborhood of a vertex $v \in V$ are denoted by $N(v)$ and $N[v]$ respectively, where $N(v) = \{w | wv \in E\}$ and $N[v] = N(v) \cup \{v\}$. For a graph $G = (V, E)$, the subgraph of G induced by $S \subseteq V$ is denoted by $G[S]$, where $G[S] = (S, E_S)$ and $E_S = \{xy \in E | x, y \in S\}$. For a graph $G = (V, E)$, the subgraph of G induced by $E' \subseteq E$ is denoted by $G[E']$, where $G[E'] = (V_{E'}, E')$ and $V_{E'} = \{x \in V | x$ is incident on an edge in $E'\}$. A graph G is said to be *bipartite* if its vertex set V can be partitioned into two independent sets X and Y such that every edge of G joins a vertex in X to a vertex in Y. An edge $e = xy$ is called a *bisimplicial edge* if $N(x) \cup N(y)$ induces a complete bipartite subgraph. Let $\sigma = x_1 y_1, x_2 y_2, \ldots, x_k y_k$ be a sequence of pairwise nonadjacent edges of G. Denote $S_j = \{x_1, x_2, \ldots x_j\} \cup \{y_1, y_2, \ldots, y_j\}$ and let $S_0 = \emptyset$. Then σ is said to be a *perfect edge elimination scheme* for G if each edge $x_{j+1} y_{j+1}$ is bisimplicial in $G = [(X \cup Y) \setminus S_j]$ for $j = 0, 1, \ldots, k-1$ and $G = [(X \cup Y) \setminus S_k]$ has no edge. A graph for which there exists a perfect edge elimination scheme is a *perfect elimination bipartite graph*. A bipartite graph $G = (X, Y, E)$ is said to be *convex* if there exists a linear ordering of vertices in X (or Y), such that for every vertex in Y (or X), the $N(y)$ (or $N(x)$) is consecutive in the linear ordering of X (or Y). A bipartite graph $G = (X, Y, E)$ is said to be *circular-convex* if there exists a circular ordering of vertices in X (or Y), such that for every vertex in Y (or X), the $N(y)$ (or $N(x)$) are consecutive in the circular ordering of X (or Y). It can be easily seen that convex graphs are circular-convex but the converse need not be true. A *star* is a complete bipartite graph of the form $K_{1,n}$. A *triad* is a tree with 3 leaves that are connected to a single vertex of degree three with paths of length i, j, k, respectively, where $i, j, k \geq 0$. A bipartite graph $G = (X, Y, E)$ is said to be *tree convex*, if a tree $T = (X, E_X)$ can be defined, such that for every vertex y in Y, the $N(y)$ induces a subtree of T. If T is a star, then G is called a *star-convex bipartite graph* and if T is a triad, then G is called a *triad-convex bipartite graph*.

3   Dominating Induced Matching in Perfect Elimination Bipartite Graphs

In this section, we will study the hardness result of DIM problem in perfect elimination bipartite graphs.

Theorem 1. *The Dominating Induced Matching (DIM) problem is NP-complete for perfect elimination bipartite graphs.*

Proof. Clearly, the DIM problem is in NP for perfect elimination bipartite graphs. To show the NP-completeness, we give a polynomial reduction from the DIM problem for bipartite graphs [11], which is already known to be NP-complete.

Construction: Given a bipartite graph $G = (X, Y, E)$, where $X = \{x_1, x_2, \ldots, x_p\}$ and $Y = \{y_1, y_2, \ldots, y_l\}$, construct a bipartite graph $G' = (X', Y', E')$ in the following way: For each $y_i \in Y$, add a path $P_i = y_i, a_i, b_i, c_i$ of length 3. Formally, $X' = X \cup \{a_i, c_i | 1 \leq i \leq l\}$, $Y' = Y \cup \{b_i | 1 \leq i \leq l\}$, and $E' = E \cup \{y_i a_i, a_i b_i, b_i c_i | 1 \leq i \leq l\}$. G' is a perfect elimination bipartite graph as $(c_1 b_1, c_2 b_2, \ldots, c_l b_l, a_1 y_1, a_2 y_2, \ldots, a_l y_l)$ is a perfect edge elimination ordering of G'. Figure 3 illustrates the construction of G' from G.

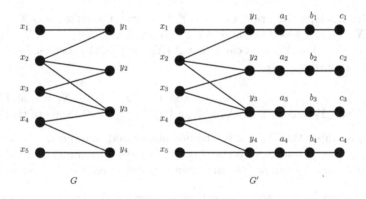

Fig. 3. An illustration to the construction of G' from G.

Claim. G has a dominating induced matching if and only if G' has a dominating induced matching.

Proof of Claim. Let M be a DIM in G. For every $y_i \in Y$, either y_i is saturated by M or not. If y_i is saturated, add $b_i c_i$ to M, otherwise add $a_i b_i$ to M and call it M'. Clearly, M' is a DIM in G' as all edges are dominated exactly once by M'. Conversely, if M' is a DIM in G', it can be seen that either $a_i b_i$ or $b_i c_i$ should be part of M' and none of $y_i a_i$ should be part of M'. Consider the matching $M = M' \cap E$, it can be seen that M is a DIM in G.   ◇

□

4   Dominating Induced Matching in Star-Convex Bipartite Graphs

In this section, we will study about the DIM problem in star-convex bipartite graphs.

Lemma 1. *A bipartite graph $G = (X, Y, E)$ is a star-convex bipartite graph if and only if there exists a vertex x_0 in X such that every vertex y in Y is either a pendant vertex or is adjacent to x_0.*

Proof. Let $G = (X, Y, E)$ be a star-convex bipartite graph. So, there exists a star $T = (X, E_X)$ such that $T[N_G(y)]$ is a subtree of T for each $y \in Y$. Let x_0 be the center of the star T. If $|X| \geq 3$ then for any $y \in Y$ such that $d(y) \geq 2$ $N_G(y)$ contains x_0 as $T[N_G(y)]$ is a subtree of T. Hence, y is either pendant or is adjacent to x_0. Conversely, let there exists a vertex $x_0 \in X$ such that for every $y \in Y$, y is either pendant or is adjacent to x_0. Then, define a star T by taking this x_0 as the center of the star and other vertices as pendant vertices of the star T. In this way for every $y \in Y$, $T[N_G(y)]$ will be a subtree of T. $\square$

Theorem 2. *The Dominating Induced Matching problem is polynomial time solvable for star-convex bipartite graphs.*

Proof. Given a bipartite graph $G = (X, Y, E)$, divide the vertex sets X and Y as follows: $X_1 = \{x|\ x$ has a neighbor y s.t. y is pendant$\}$, $X_2 = X \setminus X_1$, $Y_1 = \{y|\ d(y) = 1\}$, $Y_2 = Y \setminus Y_1$. It is clear that $N(X_2) = Y' \subseteq Y_2$. Let us consider the following exhaustive cases:

1. $X_1 = \phi$
2. $X_2 = \phi$
3. $Y_1 = \phi$
4. $Y_2 = \phi$
5. $X_1 = \phi$ and $Y_1 = \phi$
6. $X_1 = \phi$ and $Y_2 = \phi$
7. $X_2 = \phi$ and $Y_1 = \phi$
8. $X_2 = \phi$ and $Y_2 = \phi$
9. None of them is ϕ.

When X_1 is empty then Y_1 can not be non empty and vice versa, so cases 1 and 3 reduces to case 5 and cases 6 and 7 does not have a meaning. Similarly when Y_2 is empty then X_2 can not be non empty, so case 4 reduces to case 8.

Claim 1. When $X_1 = \phi$ and $Y_1 = \phi$ then G have a DIM if and only if for every $y \in Y_2$, there exists at least one neighbor $x \in X_2$ such that $d(x) = 1$.

Proof of Claim 1. Let G have a DIM and let $y \in Y_2$ be a vertex such that it does not have any neighbor of degree 1 in X_2. Since $x_0 \in X_2$ and $d(y) > 1$, so let $x_0 y$ and $x'y$ are edges in G and since $d(x') > 1$ (as y does not have any degree 1 neighbor) let $x'y'$ is also an edge in G. In this way $x_0 y x' y$ forms an induced C_4. As it is well known that none of the edges of a C_4 can be part of a DIM, the only way in which a C_4 can be dominated by a matching is by saturating the opposite vertices of the cycle as illustrated in Fig. 4. But in both the possibilities that are shown in Fig. 4, when we try to dominate the C_4, we get a contradiction as illustrated in Fig. 5. So, for every $y \in Y_2$, there exists at least one neighbor $x \in X_2$ such that $d(x) = 1$. Conversely, if for every $y \in Y_2$ there exists a neighbor x in X_2 of degree 1, then by taking these pendant edges

"xy" into a matching, we will be able to saturate all $y \in Y$ (as $Y_1 = \phi$). Since only pendant edges are part of the matching, so all edges are dominated exactly once and thus this matching will be a DIM.   ◇

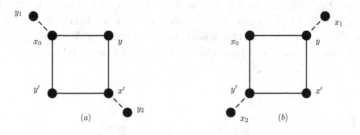

Fig. 4.

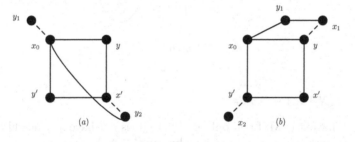

Fig. 5.

Claim 2. When $X_2 = \phi$, then G always admits a DIM.

Proof of Claim 2. When $X_2 = \phi$, then all x in X will have a neighbor y in Y_1 of degree 1. By taking these pendant edges "xy" into a matching, we will be able to saturate all $x \in X$ (as $X_2 = \phi$). Since only pendant edges are part of the matching, so all edges are dominated exactly once and thus this matching will be a DIM.   ◇

When none of X_i or Y_i, $i = 1, 2$ is empty, we divide the problem into the following subcases:

Subcase 1: $|X_2| = 1$ and $x_0 \in X_2$.

Claim 3. G have a DIM if and only if for all $y \in Y_2$, there exists exactly one neighbor $x \in X_1$ (unique for all y) such that $xy \in E$.

Proof of Claim 3. Let G have a DIM. Since $d(y) > 1$ for all $y \in Y_2$ and $|X_2| = 1$, so each $y \in Y_2$ will have at least one neighbor in X_1. Let us assume that there exists a $y \in Y_2$ such that it has two neighbors, namely x_1 and x_2 in X_1. In this case, we get an induced graph as shown in Fig. 6 that can never admits a DIM. In this way, we get a contradiction and thus each $y \in Y_2$ will have exactly one neighbor in X_1. Conversely, let us assume that for every y in Y_2 there exists exactly one neighbor x in X_1. If for every $y \in Y_2$ we take edges of the form "xy" in matching where x is the unique neighbor of y in X_1, then all edges are dominated exactly once and thus this matching will be a DIM.   ◇

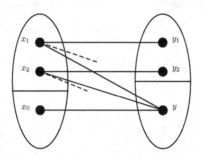

Fig. 6.

Subcase 2: $|X_2| = 1$ and $x_0 \in X_1$.

Claim 4. G have a DIM if and only if $y' \in Y_2$ does not have any neighbor in X_1 except x_0 (where y' is the unique neighbor of $x' \in X_2$).

Proof of Claim 4. Let G have a DIM and let x' be the only member of X_2. Then $d(x') = 1$ because if it is not true then by assuming that there are two neighbors of x', namely y_1 and y_2, a cycle of length 4 ($x'y_1x_0y_2$) will be formed. It is easy to see that there is no way to dominate this cycle with a matching which is a contradiction and thus $d(x') = 1$. Let y' be the unique neighbor of x'. It can be seen that y' can not have a neighbor in X_1 except x_0 otherwise we will get an induced graph that cannot be dominated. Conversely, for every $y \in Y_2$, $d(y) > 1$ and since $d(x') = 1$, every $y \in Y_2$ except y' must have a neighbor in X_1 other than x_0. So, if we take edges of the form "xy" such that $x \in X_1$ and y is the unique degree 1 neighbor of x in Y_1 and x_0y' in a matching, then this matching will be a DIM.   ◇

Subcase 3: $|X_2| > 1$ and $x_0 \in X_2$.

Note: In this case the necessary condition for a graph G to have a DIM is that for every $y \in Y_2$, either it should have a neighbor in X_1 or should have a neighbor of degree one in X_2 but this condition is not sufficient. Moreover, if for every

$y \in Y_2$, either it has a *unique* neighbor in X_1 or has a neighbor of degree one in X_2 then G always have a DIM but this is not a necessary condition. Let $X_2' \subseteq X_2$ be such that it contains vertices of degree greater than 1. Now, the following cases arise:

If $|X_2'| < 1$ (only pendant vertices in X_2) then $|Y_2| = 1$ and G have a DIM if and only if $y \in Y_2$ has a unique neighbor in X_1.

If $|X_2'| > 1$ then G have a DIM if and only if for every $y \in Y_2$ either y have a unique neighbor in X_1 or y have a pendant neighbor in X_2.

If $|X_2'| = 1$ then if two or more $y's \in Y_2$ have same neighbor or any y has more than one neighbor in X_1 then G have a DIM if and only if there exists exactly one pendant vertex in X_2. Otherwise, G will always have a DIM.

The proof of the above cases is straightforward and hence has been omitted.

Subcase 4: $|X_2| > 1$ and $x_0 \in X_1$.

Claim 5. G have a DIM if and only if $|Y'| = 1$ and the unique vertex in Y' does not have any neighbor in X_1 except x_0.

Proof of Claim 5. Let G have a DIM and let us assume that $|Y'| > 1$. Now, there are two possible cases as illustrated in Fig. 7 and it can be easily seen that a DIM cannot exist in both the cases which is a contradiction. Thus, $|Y'| = 1$. Now, this case becomes similar to the subcase 2 so, the rest of the proof is same.   ◇

□

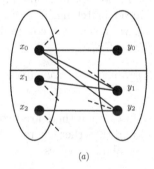
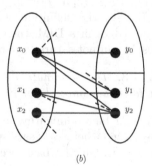

(a)   (b)

Fig. 7.

5   Dominating Induced Matching in Triad-Convex Bipartite Graphs

In this section, we will study about the DIM problem in triad-convex bipartite graphs which is a proper subclass of tree-convex bipartite graphs. The DIM problem for convex graphs which is another proper subclass of tree-convex graphs can be solved in polynomial time [6] and the proof is based on partitioning the vertex set into two colored subsets $W(White)$ and $B(Black)$ such that W is an independent set and B induces a 1-regular graph. We extended the proof in a similar manner for triad-convex bipartite graphs. Let $G = (X, Y, E)$ be a triad convex bipartite graph with a triad $T = (X, E_X)$ defined on X such that for every vertex $y \in Y$, $T[N(y)]$ is a subtree of T. For each $1 \leq i \leq 3$, let $P_i = x_0, x_{i,1}, x_{i,2}, \ldots, x_{i,n_i}$ is a path in $T = (X, E_X)$ where n_i is the length of each path P_i. Let $X = \{x_0\} \cup X_1 \cup X_2 \cup X_3$ be such that for each $i, 1 \leq i \leq 3, X_i = \{x_{i,1}, x_{i,2}, \ldots, x_{i,n_i}\}$ It can be noted that x_0 is a common vertex in all the paths P_1, P_2 and P_3 i.e. $x_{i,0} = x_0$ for all $i, 1 \leq i \leq 3$. It can be easily seen that if $G = (X, Y, E)$ is a triad convex bipartite graph with the triad $T = (X, E_X)$ centered at x_0, then $G' = (X, Y \setminus N(x_0), E \setminus F)$, where $F = \{xy | x \in X, y \in N(x_0)\}$ is a convex bipartite graph.

Lemma 2. *Let M be a DIM in G and let M does not saturate x_0, but it saturates some of the neighbors of x_0. Then M can not saturate more than 3 neighbors of x_0.*

Proof. Let us suppose that M saturates 4 neighbors of x_0, say y_{k1}, y_{k2}, y_{k3} and y_{k4}. Let $\{x_{k1}y_{k1}, x_{k2}y_{k2}, x_{k3}y_{k3}, x_{k4}y_{k4}\} \subseteq M$. Then at least two vertices from the set $\{x_{k1}, x_{k2}, x_{k3}, x_{k4}\}$ will lie in the same path P_i for some $i, 1 \leq i \leq 3$. Without loss of generality, let $x_{k1}, x_{k2} \in P_1$. Also assume that the distance between x_{k1} and x_0 is less than the distance between x_{k2} and x_0 in the path P_1. Since y_{k2} is adjacent to x_0 as well as x_{k2} and also by the definition of triad convex bipartite graph, $T[N(y_{k2})]$ is a subtree of T. So, y_{k2} must be adjacent to x_{k1}. But, by the definition of dominating induced matching, if $x_{k1}y_{k1}$ and $x_{k2}y_{k2}$ are edges in a DIM, then $x_{k1}y_{k2} \notin E$. So, we reach at a contradiction and thus M can not saturate more than 3 neighbors of x_0. $\qquad\square$

Theorem 3. *The Dominating Induced Matching problem is polynomial time solvable for triad-convex bipartite graphs.*

Proof. Let M be a dominating induced matching in G, then there are two possible cases. One where x_0 is saturated by the DIM and the other where x_0 is not saturated by it. On that basis, we will consider the following cases:

Case 1: x_0 is saturated by a DIM.
Let x_0 is saturated by a DIM M that is $x_0y \in M$ for some $y \in N(x_0)$. In this case, we can color x_0 and y with black color and the neighbors of x_0 and y with white color. If the partial coloring is not valid then x_0y is not a part of DIM, otherwise once we color the neighbors of x_0 and the neighbors of neighbors of x_0, the problem can be reduced to coloring of convex bipartite graph by converting the given triad convex graph into convex bipartite graph by removing $N(x_0)$.

Case 2: x_0 is not saturated by a DIM.
Let M be a DIM that does not saturate x_0 but saturates at most 3 neighbors of x_0. The following 3 subcases arises:

Subcase 2.1: Let M saturates exactly one neighbor y_i of x_0 that is $x y_i$ is in M for some neighbor x of y_i except x_0. In this case, color x and y_i with black color and all neighbors of x and y_i with white color. Here, since x_0 will be marked as white, mark all neighbors of x_0 as black, if the partial coloring is valid, try to extend to complete the total coloring of vertices using convex graph algorithm.

Subcase 2.2: Let M saturates exactly two neighbors y_i and y_j of x_0 that is $x_r y_i$ and $x_s y_j$ is in M for some neighbor x_r of y_i and x_s of y_j except x_0. In this case color x_r, x_s, y_i, y_j with black color and all neighbors of x_r, x_s, y_i, y_j with white color, x_0 will be colored with white. Color all neighbors of x_0 as black. If the partial coloring is valid, try to extend to complete the total coloring of vertices using convex graph algorithm.

Subcase 2.3: Let M saturates exactly three neighbors y_i, y_j and y_k of x_0 that is $x_r y_i$, $x_s y_j$ and $x_t y_k$ is in M for some neighbor x_r of y_i, x_s of y_j and x_t of y_k except x_0. In this case color x_r, x_s, x_t, y_i, y_j, y_k with black color and all neighbors of x_r, x_s, x_t, y_i, y_j, y_k with white color, x_0 will be colored with white. Color all neighbors of x_0 as black, if the partial coloring is valid, try to extend to complete the total coloring of vertices using convex graph algorithm.

It can be noted that we can delete from G those colored vertices that have no neighbors among any uncolored ones (as they have no importance for the completion of the procedure of converting valid partial coloring into total coloring). So, if we can color $N(x_0)$ and the neighbors of $N(x_0)$, then we can easily remove $N(x_0)$ from G and convert the triad convex graph into convex and then apply the already known algorithm [6] to solve the problem.   □

6   Dominating Induced Matching in Circular-Convex Bipartite Graphs

In this section, we will study about the DIM problem in circular-convex bipartite graphs which is a superclass of convex bipartite graphs. The DIM problem for convex graphs can be solved in polynomial time [6] and the proof is based on partitioning the vertex set into two colored subsets $W(White)$ and $B(Black)$ such that W is an independent set and B induces a 1-regular graph. We extended the proof in a similar manner for circular-convex bipartite graphs. Let $G = (X, Y, E)$ be a circular-convex bipartite graph with circular ordering $\sigma = (x_0, x_1, x_2, \ldots, x_p, x_0)$ defined on X such that for every $y \in Y$, the vertices in $N(y)$ forms a circular arc in the ordering σ. Let $S = N(x_0) \cap N(x_p)$, then $G' = (X, Y \setminus \{S\}, E \setminus F)$, where $F = \{x_i y_j | x_i \in X, y_j \in S\}$ is a convex bipartite graph.

Lemma 3. *All the vertices in S should be colored with the same color w.r.t the W-B coloring and when $|S| \geq 3$ then all $v \in S$ should be colored with the white color.*

Proof. It is easy to see that all the vertices in $S = N(x_0) \cap N(x_p)$ should take the same color, otherwise, it will be impossible to color x_0 (or x_p) as they will form an induced C_4 with two differently colored vertices. As it is known that any two adjacent vertices in any induced C_4 cannot be colored with the same color [6] for a valid partial W-B coloring, it will be impossible to give x_0 (or x_p) any color. For the second part, let us suppose that $|S| = 3$ and all the vertices in S are colored black. Let $S = \{y_{q1}, y_{q2}, y_{q3}\}$, then for each y_{qi} there should exists a neighbor x_{qi}, such that $y_{qi}x_{qi}$ is in DIM and colored black. It can be easily seen that the neighborhood of one of the y_{qi} (say y') must be contained in the intersection of the neighborhood of other two $y'_{qi}s$. So, we cannot find any neighbor of y' which can be colored black and we arrive at a contradiction. Thus, if $|S| \geq 3$ then all $v \in S$ should be colored white. $\qquad \square$

Theorem 4. *The Dominating Induced Matching problem is polynomial time solvable for circular-convex bipartite graphs.*

Proof. It can be noted that we can delete from G those colored vertices that have no neighbors among any uncolored ones (as they have no importance for the completion of valid partial coloring into total coloring). So, if we can color S and the neighbors of S, then we can easily remove S from G and convert the circular-convex graph into convex bipartite graph and then apply the already known algorithm [6] to solve the DIM problem for convex bipartite graph and then combine the solution with the coloring of S to get a solution of the DIM problem for circular-convex bipartite graph.

If $|S| \geq 3$, then color all $v \in S$ as white and their neighbors as black. In this way, we get a valid partial coloring of G then try to extend this partial coloring into total coloring by removing S from G.

If $S < 3$, then first color all $v \in S$ as white and their neighbors as black and try to extend this valid partial coloring into total coloring. If we are unable to get a total coloring in this case then we will get a valid partial coloring by coloring all $v \in S$ as black. For all $v \in S$ there should exist a neighbor x of v in X which should also be colored black. To find an appropriate black neighbor of $v \in S$ in X, we have to try all the edges of F one by one and get the valid partial coloring by coloring all the neighbors of these two black vertices as white and then try to extend this valid partial coloring to total coloring. $\qquad \square$

7   Conclusion

In this paper, we showed that the DIM problem is NP-complete for perfect elimination bipartite graphs. We also proved that the DIM problem is polynomial time solvable for star-convex bipartite graphs, triad-convex bipartite graphs and circular-convex bipartite graphs which are various subclasses of bipartite graphs. The complexity status of the DIM problem is still unknown for graph classes like tree-convex bipartite graphs, comb-convex bipartite graphs and P_k-free graphs for $k \geq 9$.

References

1. Brandstädt, A., Hundt, C., Nevries, R.: Efficient edge domination on hole-free graphs in polynomial time. In: López-Ortiz, A. (ed.) LATIN 2010. LNCS, vol. 6034, pp. 650–661. Springer, Heidelberg (2010). https://doi.org/10.1007/978-3-642-12200-2_56
2. Brandstädt, A., Leitert, A., Rautenbach, D.: Efficient dominating and edge dominating sets for graphs and hypergraphs. In: Chao, K.-M., Hsu, T., Lee, D.-T. (eds.) ISAAC 2012. LNCS, vol. 7676, pp. 267–277. Springer, Heidelberg (2012). https://doi.org/10.1007/978-3-642-35261-4_30
3. Chen, H., Lei, Z., Liu, T., Tang, Z., Wang, C., Ke, X.: Complexity of domination, hamiltonicity and treewidth for tree convex bipartite graphs. J. Comb. Optim. 32(1), 95–110 (2016)
4. Golumbic, M.C., Goss, C.F.: Perfect elimination and Chordal bipartite graphs. J. Graph Theory 2(2), 155–163 (1978)
5. Grinstead, D.L., Slater, P.J., Sherwani, N.A., Holmes, N.D.: Efficient edge domination problems in graphs. Inf. Process. Lett. 48(5), 221–228 (1993)
6. Korpelainen, N.: A polynomial-time algorithm for the dominating induced matching problem in the class of convex graphs. Electron. Notes Discret. Math. 32, 133–140 (2009)
7. Liang, Y.D., Blum, N.: Circular convex bipartite graphs: maximum matching and Hamiltonian circuits. Inf. Process. Lett. 56(4), 215–219 (1995)
8. Liu, T., Zhao, L., Ke, X.: Tractable connected domination for restricted bipartite graphs. J. Comb. Optim. 29(1), 247–256 (2015)
9. Livingston, M., Stout, Q.F.: Distributing resources in hypercube computers. In: Proceedings of the Third Conference on Hypercube Concurrent Computers and Applications: Architecture, Software, Computer Systems, and General Issues, vol. 1, pp. 222–231. ACM (1988)
10. Lu, C.L., Ko, M.-T., Tang, C.Y.: Perfect edge domination and efficient edge domination in graphs. Discret. Appl. Math. 119(3), 227–250 (2002)
11. Lu, C.L., Tang, C.Y.: Solving the weighted efficient edge domination problem on bipartite permutation graphs. Discret. Appl. Math. 87(1–3), 203–211 (1998)
12. Bao, F.S., Zhang, Y.: A review of tree convex sets test. Comput. Intell. 28(3), 358–372 (2012)
13. Song, Y., Liu, T., Xu, K.: Independent domination on tree convex bipartite graphs. In: Snoeyink, J., Lu, P., Su, K., Wang, L. (eds.) AAIM/FAW 2012. LNCS, vol. 7285, pp. 129–138. Springer, Heidelberg (2012). https://doi.org/10.1007/978-3-642-29700-7_12

Localized Query: Color Spanning Variations

Ankush Acharyya[1]([✉]), Anil Maheshwari[2], and Subhas C. Nandy[1]

[1] Indian Statistical Institute, Kolkata, India
ankushacharyya@gmail.com
[2] Carleton University, Ottawa, Canada

Abstract. Let P be a set of n points and each of the points is colored with one of the k possible colors. We present efficient algorithms to preprocess P such that for a given query point q, we can quickly identify the smallest color spanning object of the desired type containing q. In this paper, we focus on (i) intervals, (ii) axis-parallel square, (iii) axis-parallel rectangle, (iv) equilateral triangle of fixed orientation, as our desired type of objects.

Keywords: Color-spanning object · Multilevel range searching · Localized query

1 Introduction

Facility location is a widely studied problem in the modern theoretical computer science. Here we have different types of facilities $\mathcal{F}$, and each facility has multiple copies, the distance is measured according to a given distance metric δ, and the target is to optimize certain objective function by choosing the *best* facility depending on the problem. In *localized query* problems, the objective is to identify the *best* facility containing a given query point q. In this paper, we consider the following version of the localized facility location problems.

> Given a set $\mathcal{P} = \{p_1, p_2, \ldots, p_n\}$ of n points, each point is colored with one of the k possible colors. The objective is to pre-process the point set $\mathcal{P}$ such that given a query point q, we can quickly identify a color spanning object of the desired type that contains q having optimum size.

The motivation of this problem stems from the *color spanning* variation of the facility location problems [1,9,11,14,17]. Here facilities of type $i \in \{1, 2, \ldots, k\}$ is modeled as points of color i, and the objective is to identify the locality of desired geometric shape containing at least one point of each color such that the desired measure parameter (width, perimeter, area etc.) is optimized. Abellanas et al. [1] mentioned algorithms for finding the smallest color spanning

A. Maheshwari—Research supported by NSERC.

S. P. Pal and A. Vijayakumar (Eds.): CALDAM 2019, LNCS 11394, pp. 150–160, 2019.
https://doi.org/10.1007/978-3-030-11509-8_13

axis parallel rectangle and arbitrary oriented strip in $O(n(n - k) \log^2 k)$ and $O(n^2\alpha(k) \log k + n^2 \log n)$ time, respectively. These results were later improved by Das et al. [9] to run in $O(n(n - k) \log k)$ and $O(n^2 \log n)$ time, respectively. Abellanas et al. also pointed out that the smallest color spanning circle can be computed in $O(nk \log(nk))$ time using the technique of computing the upper envelope of Voronoi surfaces as suggested by Huttenlocher et al. [12]. The smallest color spanning interval can be computed in $O(n)$ time for an ordered point set of size n with k colors on a real line [18]. Color spanning 2-interval[1], axis-parallel equilateral triangle, axis-parallel square can be computed in $O(n^2)$ [14], $O(n \log n)$ [11], $O(n \log^2 n)$ [17] time respectively.

The problem of constructing an efficient data structure with a given set P of n (uncolored) points to find an object of optimum (maximum/minimum) size (depending on the nature of the problem) containing a query point q is also studied recently in the literature. These are referred to as the *localized query problem*. The complexity results for the different variations of localized query problem are given below.

- Augustine et al. [3] considered the largest empty circle and largest empty rectangle localized query problems. Among a set P of n points, the (space, preprocessing time, query time) complexity for (i) largest empty circle problem are $(O(n^2 \log n), O(n^2 \log^2 n), O(\log^2 n))$, and (ii) largest empty rectangle problem are $(O(n^2 \log n), O(n^2 \log n), O(\log n))$.
- Augustine et al. [2] considered P as a simple polygon with n vertices and the (space, preprocessing time, query time) complexity for the localized largest empty circle query problem are $(O(n \log^2 n), O(n \log n), O(\log n))$.
- In [2], they also proposed two improved algorithms for the largest empty circle localized query problem, with complexity results $(O(n^{\frac{3}{2}} \log^2 n),\ O(n^{\frac{3}{2}} \log n),\ O(\log n \log \log n))$ and $(O(n^{\frac{5}{2}} \log n), O(n^{\frac{5}{2}}), O(\log n))$.
- Kaplan et. al [16] improved the complexity results of both the variations of localized query problem. For the maximal empty rectangle problem the results are (i) $(O(n\alpha(n) \log^4 n), O(n\alpha(n) \log^3 n), O(\log^4 n))$, and $(O(n \log n), O(n \log^2 n),\ O(\log^2 n))$ for the maximal empty circle problem.
- Augustine et al. [4] considered another variation of localized query problem, where a given point set P needs to be preprocessed such that given a query line ℓ, the largest circle centered on ℓ can be reported efficiently. The (space, preprocessing time, query time) complexity of their proposed algorithm are $(O(n^3 \log n), O(n^3), O(\log n))$.
- Kaminker et al. [15] studied a special case where the input is $O(n)$ Delaunay disks of a given point set P. Their (space, preprocessing time, query time) complexity for the largest empty disk localized query problem are $(O(n), O(n \log^3 n), O(\log n))$.
- Gester et al. [10] studied the problem of finding the largest empty square inscribed in a rectilinear polygon $\mathcal{R}$ containing a query point. For $\mathcal{R}$ without

---

[1] Given a colored point set on a line, the goal is to find two intervals that cover at least one point of each color such that the maximum length of the intervals is minimized.

holes, their complexity results are $(O(n), O(n), O(\log n))$. For the polygon $\mathcal{R}$ containing holes, the complexity results are $(O(n), O(n \log n), O(\log n))$. Their results can also be extended to find the largest empty square within a point set containing a query point.

Our Contribution: In this paper, we study the localized query variations of different color spanning objects among a set $P = P_1 \cup P_2 \cup \ldots \cup P_k$ of points where P_i is the set of points of color i, $|P_i| = n_i$, $\sum_{i=1}^{k} n_i = n$. The desired (color-spanning) objects are (i) smallest interval (SCSI), (ii) smallest axis-parallel square (SCSS), (iii) smallest axis-parallel rectangle (SCSR), and (iv) smallest equilateral triangle of fixed orientation (SCST). For problem (i), the points are distributed in $\mathbb{R}$, and for all other problems, the points are placed in $\mathbb{R}^2$. The results are summarized in Table 1.

Table 1. Complexity results for variations of color spanning objects containing q.

Object	Pre-processing		Query Time
	Time	*Space*	
SCSI	$O(n \log n)$	$O(n)$	$O(\log n)$
SCSS	$O\left(N(\frac{\log N}{\log \log N})^2\right), N = \Theta(nk)$	$O(N \log^2 N)$	$O\left((\frac{\log N}{\log \log N})^3\right)$
SCSR	$O(N \log N), N = \Theta((n-k)^2)$	$O\left(N(\frac{\log N}{\log \log N})\right)$	$O\left((\frac{\log N}{\log \log N})^2\right)$
SCST	$O(n \log n)$	$O\left(n(\frac{\log n}{\log \log n})\right)$	$O\left((\frac{\log n}{\log \log n})^2\right)$

2   Preliminaries

We use $x(p)$ and $y(p)$ to denote the x- and y-coordinates of a point p. We use $dist(.,.)$ to define the Euclidean distance between (i) two points, (ii) two lines, or (iii) a point and a line depending on the context. We use q_h and q_v to denote the horizontal and vertical lines through q respectively. Without loss of generality, we assume that all the points in P are in positive quadrant.

Range Searching: A range tree $\mathcal{R}$ is a data structure that supports counting and reporting the points of a given set $P \in \mathbb{R}^d$ ($d \geq 2$) that lie in an axis-parallel rectangular query range R. $\mathcal{R}$ can be constructed in $O(n \log^{d-1} n)$ time and space where the counting and reporting for the query range R can be done in $O(\log^{d-1} n)$ and $O(\log^{d-1} n + k)$ time, respectively, where k is the number of reported points [5]. Later, JaJa et al. [13] proposed a dominance query data structure in $\mathbb{R}^d$, that supports $d(\geq 3)$ dimensional query range $[\alpha_1, \infty] \times [\alpha_2, \infty] \times \ldots \times [\alpha_d, \infty]$ with the following result.

Lemma 1 *[13]. Given a set of n points in $\mathbb{R}^d$ ($d \geq 3$), it can be preprocessed in a data structure T of size $O\left(n(\frac{\log n}{\log \log n})^{d-2}\right)$ in $O(n \log^{d-2} n)$ time which supports counting points inside a query rectangle in $O\left((\frac{\log n}{\log \log n})^{d-1}\right)$ time.*

We have computed all possible color-spanning objects of desired type Ψ (e.g., interval, corridor, square, rectangle, triangle) present on the plane, stored those objects in a data structure, and when the query point q appears, we use those objects to compute a smallest object containing q using a data structure similar to [6,7]. The optimum color spanning object of type Ψ containing q may be any one of the following two types:

Type - I: q lies inside the optimum color-spanning object. Here, we search the smallest one among the pre-computed (during preprocessing) color spanning objects containing q.

Type - II: q lies on the boundary of the optimum color-spanning object. This is obtained by expanding some pre-computed object to contain q.

Comparing the best feasible solutions of Type - I and Type - II, the optimum solution is returned.

3 Smallest Color Spanning Intervals (SCSI)

We first consider the basic one dimensional version of the problem:

Given a set of P of n points on a real line L, where each point has one of the k possible colors; for a query point q on the line report the smallest color spanning interval containing q, efficiently.

Without loss of generality assume that L is horizontal. We sort the points in P from left to right, and for each point, we can find the minimal color-spanning interval in total $O(n)$ time [8]. Next, we compute the following three lists for each point $p_i \in P$.

start: the smallest color spanning interval starting at p_i, denoted by $\mathbf{start}(p_i)$.
end: the smallest color spanning interval ending at p_i, denoted by $\mathbf{end}(p_i)$.
span: the smallest color spanning interval spanning p_i (here p_i is not the starting or ending point of the interval), denoted by $\mathbf{span}(p_i)$.

Observe that, none of the minimal color spanning intervals corresponding to the points $p_i \in P$ can be a proper sub-interval of the color spanning interval of some other point [8].

Lemma 2. *For a given point set P, the aforesaid data structure takes $O(n)$ amount of space and can be computed in $O(n \log n)$ time.*

During the query (with a point q), we identify $p_i, p_{i+1} \in P$ such that $x(p_i) \leq x(q) \leq x(p_{i+1})$. Next, we compute the lengths OPT_C and OPT_E of the Type - I SCSI and Type - II SCSI, respectively, as follows. Finally, we compute $OPT = \min\{OPT_C, OPT_E\}$.

Type - I SCSI: OPT_C is min{$\textbf{span}(p_i), \textbf{span}(p_{i+1})$}.
Type - II SCSI: OPT_E is min{$\textbf{end}(p_i) + dist(p_i, q), \textbf{start}(p_{i+1}) + dist(p_{i+1}, q)$}.

Theorem 1. *Given a colored point set P on a real line L, the smallest color spanning interval containing a query point q can be computed in $O(\log n)$ time, using a linear size data structure computed in $O(n \log n)$ time.*

4   Smallest Color Spanning Axis-Parallel Square (SCSS)

In this section, we consider the problem of finding the *minimum width color spanning axis parallel square* (SCSS) among a set P of n colored points. We first compute the set $\mathcal{S}$ of all possible minimal color spanning squares for the given point set P in $O(n \log^2 n)$ time, and store them in $\Theta(nk)$ space [17]. For a square $s_i \in \mathcal{S}$, we maintain a five tuple $(s_i^\ell, s_i^r, s_i^t, s_i^b, \lambda_i)$, where s_i^ℓ and s_i^r are the x-coordinates of the *left* and *right* boundaries of s_i, s_i^t and s_i^b are the y-coordinates its *top* and *bottom* boundaries respectively, and λ_i is its side length.

4.1   Computation of Type - I SCSS

This needs a data structure $\mathcal{T}_{tb\ell r}$ (see Lemma 1) with the point set {$(s_i^t, s_i^b, s_i^\ell, s_i^r)$, $i = 1, 2, \ldots, n$} in $\mathbb{R}^4$ as the preprocessing. Each internal node of the last level of $\mathcal{T}_{tb\ell r}$ contains the minimum length (λ-value) of the squares stored in the sub-tree rooted at that node. Given the query point q, we use this data structure to identify the square of minimum side-length in the set $\mathcal{S}_{contained}$ in $O\left(\left(\frac{\log n}{\log \log n}\right)^3\right)$ time (see Lemma 1), where $\mathcal{S}_{contained}$ is the subset of $\mathcal{S}$ that contains q.

4.2   Computation of Type - II SCSS

For the simplicity in the analysis, we further split the squares in $\mathcal{S} \setminus \mathcal{S}_{contained}$ into two different subcases, namely $\mathcal{S}_{stabbed}$ and $\mathcal{S}_{not\_stabbed}$, where $\mathcal{S}_{stabbed}$ (resp. $\mathcal{S}_{not\_stabbed}$) denote the subset of $\mathcal{S}$ that is stabbed (resp. not stabbed) by the vertical line q_v or the horizontal line q_h (see Fig. 1).

We first consider a subset $\mathcal{S}_{stabbed}^{tbr}$ of the squares in $\mathcal{S}_{stabbed}$, whose members are intersected by q_h, and to the left of the point q. During the preprocessing, we have created a data structure $\mathcal{T}_{tbr}$ with the points {$(s_i^t, s_i^b, s_i^r), i = 1, 2, \ldots, n$} in $\mathbb{R}^3$ (see Lemma 1). Every internal node of the last level of $\mathcal{T}_{tbr}$ contains the maximum value of s_i^ℓ coordinate among the points in the sub-tree rooted at that node. During the query with the point q, the search is executed in $\mathcal{T}_{tbr}$ to find the disjoint subsets of $\mathcal{S}_{stabbed}^{tbr}$ satisfying the above query rooted at the third level of the tree. For each of these subsets, we consider the maximum s_i^ℓ value (x-coordinate of the left boundary), to find a member $s \in \mathcal{S}_{stabbed}^{tbr}$ whose left boundary is closest to q. The square s produces the smallest one among all the squares obtained by extending the members of $\mathcal{S}_{stabbed}^{tbr}$ so that their right boundary passes through q. Similar procedure is followed to get the smallest square among all the squares obtained by extending the members of $\mathcal{S}_{stabbed}^{tb\ell}$

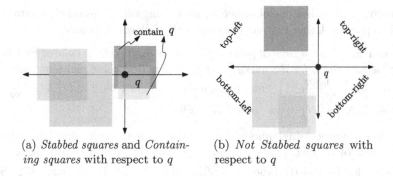

(a) *Stabbed squares* and *Containing squares* with respect to q

(b) *Not Stabbed squares* with respect to q

Fig. 1. Querying for square

(resp. $\mathcal{S}^{\ell r b}_{stabbed}$, $\mathcal{S}^{\ell r t}_{stabbed}$) so that their left (resp. bottom, top) boundary passes through q, where the sets $\mathcal{S}^{tb\ell}_{stabbed}$, $\mathcal{S}^{\ell r b}_{stabbed}$ and $\mathcal{S}^{\ell r t}_{stabbed}$ are defined as the set $\mathcal{S}^{tbr}_{stabbed}$.

Next, we consider the members in $\mathcal{S}_{not\_stabbed}$ that lie in the bottom-left quadrant with respect to the query point q. These are obtained searching the data structure $\mathcal{T}_{tr}$ with points $\{(s_i^t, s_i^r), i = 1, 2, \ldots, n\}$ in $\mathbb{R}^2$, created in the preprocessing phase. Every internal node of $\mathcal{T}_{tr}$ at the second level contains the L_∞ Voronoi diagram (VD_∞) of the bottom-left corners of all the squares in the sub-tree rooted at that node. During the query, it finds the disjoint subsets of $\mathcal{S}^{tr}_{not\_stabbed}$ satisfying the above query rooted at the second level of the tree. Now, in each subset (corresponding to an internal node in the second level of the tree), we locate the closest point of q in VD_∞. The corresponding square produces the smallest one among all the squares in these sets that are obtained by extending the members of $\mathcal{S}^{tr}_{not\_stabbed}$ so that their top/right boundary passes through q. The similar procedure is followed to search for an appropriate element for each of the subsets $\mathcal{S}^{t\ell}_{not\_stabbed}$, $\mathcal{S}^{br}_{not\_stabbed}$ and $\mathcal{S}^{b\ell}_{not\_stabbed}$, defined as in $\mathcal{S}^{tr}_{not\_stabbed}$.

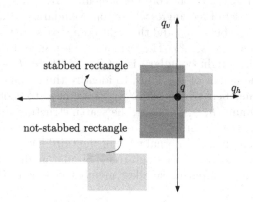

Fig. 2. Different possible positions of Rectangles with respect to q

Finally, the smallest one obtained by processing the aforesaid nine data structures is reported. Using Lemma 1, we have the following theorem:

Theorem 2. *Given a colored point set P, where each point is colored with one of the k possible colors, for the query point q, the smallest color spanning square can be found in $O\left((\frac{\log N}{\log \log N})^3\right)$ time, using a data structure built in $O\left(N(\frac{\log N}{\log \log N})^2\right)$ time and $O(N \log^2 N)$ space, where $N = \Theta(nk)$.*

Proof. The preprocessing time and space complexity results are dominated by the problem of finding Type - I SCSS (the smallest square containing q). The query time for finding (i) Type - I SCSS is $O\left((\frac{\log N}{\log \log N})^3\right)$. (ii) Type - II SCSS in the set $S_{stabbed}$ is $O\left((\frac{\log N}{\log \log N})^2\right)$, and (iii) Type - II SCSS in the set $S_{not\_stabbed}$ is $O(\log^2 N)$ since the search in the 2D range tree[2] is $O(\log N)$, and the search in the L_∞ Voronoi diagram in each of the $O(\log N)$ internal nodes is $O(\log N)$. □

5 Smallest Color Spanning Axis-Parallel Rectangle (SCSR)

Let us first mention that, we consider the perimeter of a rectangle as its size. Among a set of n points with k different colors, the size of the set $\mathcal{R}$ of all possible minimal color spanning axis parallel rectangles is $\Theta((n-k)^2)$, and these can be computed in $O(n(n-k)\log k)$ time [1,9]. As in Sect. 4, (i) we will use $(s_i^\ell, s_i^r, s_i^t, s_i^b)$ to denote the coordinate of left, right, top and bottom side of rectangle $R_i \in \mathcal{R}$, (ii) define the set Type - I and Type - II SCSR (see Fig. 2), and (iii) split the set Type - II SCSR in $\mathcal{R}$ into two subsets $\mathcal{R}_{stabbed}$ and $\mathcal{R}_{not\_stabbed}$ respectively.

The data structures used for finding Type - I SCSR is exactly same as that of Sect. 4; the complexity results are also same. We now discuss the algorithm for computing Type - II SCSR. Let us consider a member $R_i \in \mathcal{R}_{stabbed}^{tbr}$ $(\subseteq \mathcal{R}_{stabbed})$, where $\mathcal{R}_{stabbed}^{tbr}$ is the set of rectangles whose top boundaries are above the line q_h, bottom boundaries are below q_h, and the right boundaries are to the left of q_v. Its bottom-left coordinate is (s_i^ℓ, s_i^b). The size of the rectangle, by extending R_i so that it contains q on its right boundary, is $(x(q)-s_i^\ell)+h_i = (h_i-s_i^\ell)+x(q)$, where $h_i = (s_i^t - s_i^b)$ is the height of R_i. Thus, to identify the minimum size rectangle by extending the members of the set $\mathcal{R}_{stabbed}^{tbr}$, we need to choose a member of $\mathcal{R}_{stabbed}$ with minimum $(h_i - s_i^\ell)$. Thus, the search structure $\mathcal{T}_{stabbed}^{tbr}$ for finding the minimum-sized member in $\mathcal{R}_{stabbed}^{tbr}$ remains same as that of Sect. 4; the only difference is that here in each internal node of the third level of $\mathcal{T}_{stabbed}^{tbr}$, we need to maintain $\min\{(h_i - s_i^\ell)|R_i$ in that sub-tree$\}$. Similar modifications are done in the data structure for computing smallest member in each of the sets $\mathcal{R}_{stabbed}^{tbl}$, $\mathcal{R}_{stabbed}^{\ell rt}$ and $\mathcal{R}_{stabbed}^{\ell rb}$.

---

[2] Here the orthogonal range searching result due to [13] is not applicable since it works for point set in $\mathbb{R}^d$, where $d \geq 3$.

Next, consider a rectangle $R_i \in \mathcal{R}^{tr}_{not\_stabbed}$ ($\subseteq \mathcal{R}_{not\_stabbed}$). The coordinate of its bottom-left corner is (s_i^ℓ, s_i^b). In order to have q on its boundary, we have to extend the rectangle of R_i to the right up to $x(q)$ and to the top up to $y(q)$. Thus, its size becomes $((x(q) - s_i^\ell) + (y(q) - s_i^b)) = ((x(q) + y(q) - (s_i^\ell + s_i^b))$. Hence, instead of maintaining the L_∞ Voronoi diagram at the internal nodes of second level of the data structure $\mathcal{T}^{tr}_{not\_stabbed}$ (see Sect. 4), we maintain $\max\{(s_i^\ell + s_i^b)|R_i$ in the sub-tree rooted at that node$\}$. This leads to the following theorem:

Theorem 3. *Given a set of n points with k colors, the smallest perimeter color spanning rectangle containing the query point q can be reported in $O\left((\frac{\log N}{\log\log N})^2\right)$ time using a data structure built in $O(N\log N)$ time and $O\left(N(\frac{\log N}{\log\log N})\right)$ space, where $N = \Theta((n-k)^2)$.*

6   Smallest Color Spanning Equilateral Triangle (SCST)

Finally, we consider the problem of finding the *minimum width color spanning equilateral triangle of fixed orientation* (SCST). Without loss of generality we consider that the base of the triangles is parallel to the x-axis. For any colored point set P of n points, where each point is colored with one of the k possible colors, the size of the set $\triangle$ of all possible minimal color spanning axis parallel triangles is $O(n)$ and these triangles can be computed in $O(n\log n)$ time [11].

As in earlier sections, we will use s_i^ℓ, s_i^r and s_i^t to denote the *bottom-left, bottom-right* and *top* vertex of a triangle $\triangle_i \in \triangle$. We also define three lines $\triangle_i^l$, $\triangle_i^r$ and $\triangle_i^b$ as the line containing left, right and base arms respectively (see Fig. 3).

We use range query to decide whether the query point q lies inside a triangle $\triangle_i \in \triangle$. For a point $q = (x(q), y(q))$, we consider a six-tuple $(x(q), y(q), x_a(q), y_a(q), x_c(q), y_c(q))$, where $(x(q), y(q))$ is the coordinates of the point q with respect to normal coordinate axes, say ν-axis, $\big(x_a(q) = x(q)\cos 60° - y(q)\sin 60°$, $y_a(q) = x(q)\sin 60° + y(q)\cos 60°\big)$ are the x- and y-coordinates of q with respect to the α-axis, which is the rotation of the axes anticlockwise by an amount of $60°$ centered at the origin, and $\big(x_c(q) = -x(q)\cos 60° - y(q)\sin 60°$, $y_a(q) = -y(q)\sin 60° + x(q)\cos 60°\big)$ are the x- and y-coordinates of q with respect to the β-axis rotated clockwise by an amount of $60°$ centered at the origin.

Thus, we can test whether q is inside a triangle $\triangle abc$ in $O(1)$ time (see Fig. 3) by the above/below test of (i) the point $(x(q), y(q))$ with respect to a horizontal line $\overline{ab}$ along the ν-axis, (ii) the point $(x_a(q), y_a(q))$ with respect to a horizontal line $\overline{ac}$ along the α-axis, and (iii) the point $(x_c(q), y_c(q))$ with respect to a horizontal line $\overline{bc}$ along the β-axis.

Computation for Type - I SCST. Here we obtain the smallest triangle among the members in $\triangle$ that contains q. In other words, we need to identify the smallest triangle among the members in $\triangle$ that are in the same side of q with respect to the lines $\triangle_i^\ell$, $\triangle_i^r$ and $\triangle_i^b$. Similar to Sect. 4, this needs creation of a range searching data structure $\mathcal{T}_{blr}$ with the points $\{(y(s_i^\ell), y_a(s_i^\ell), y_c(s_i^r))), i = 1, 2, \ldots, n\}$

in $\mathbb{R}^3$ as the preprocessing, where s_i^ℓ, s_i^r are two vertices of the triangle $\triangle_i \in \triangle$ as defined above. Each internal node of the third level of this data structure contains the minimum side length (λ-value) of the triangles stored in the sub-tree rooted at that node. Given the query point q, it identifies a subset $\triangle_{contained}$ in the form of sub-trees in $\mathcal{T}_{b\ell r}$, and during this process the smallest triangle is also obtained. The preprocessing time and space are $O(n \log n)$ and $O(n \frac{\log n}{\log \log n})$, and the query time is $O\left((\frac{\log n}{\log \log n})^2\right)$.

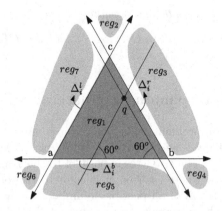

Fig. 3. Different regions with respect to a triangle

Computation for Type - II SCST. For each triangle $\tau_i \in \triangle \setminus \triangle_{contained}$, the lines $\triangle_i^\ell$, $\triangle_i^r$ and $\triangle_i^b$ defines seven regions, namely $\{reg_j, j = 1, 2, \ldots, 7\}$ as shown in Fig. 3, where $q \in \tau_i(reg_i)$ implies $q \in \tau_i$. Thus, in order to consider Type - II SCST, we need to consider those triangles τ_i such that $\tau_i \setminus \tau_i(reg_1)$ regions contain q.

Given the point q, we explain the handling of all triangles satisfying $q \in \tau_i(reg_2)$, and $q \in \tau_i(reg_3)$ separately (see Fig. 4). The cases for reg_4 and reg_6 are handled similar to that of reg_2, and the cases for reg_5 and reg_7 are similar to that of reg_3. The complexity results remain same as for Type - I SCST.

Handling of triangles with q in reg_2. Given a query point q, a triangle τ_i is said to satisfy reg_2-condition if $q \in \tau_i(reg_2)$, or in other words, $y(q) > y(s_i^\ell)$ & $y_a(q) < y_a(s_i^\ell)$ & $y_c(q) > y_c(s_i^r)$ (see Fig. 4(a)). Let $\triangle_2$ be the set of triangles satisfying reg_2-condition with respect to the query point q. Among all equilateral triangles formed by extending a triangle $\tau_i \in \triangle_2$ such that it contains q, the one with top vertex at q will have minimum size. Thus, among all triangles in $\triangle_2 \subset \triangle$, we need to identify the one having maximum $y(s_i^\ell)$ value. If the internal nodes at the third level of $\mathcal{T}_{b\ell r}$ data structure contains maximum among $y(s_i^\ell)$ values of the triangles rooted at that node, then we can identify a triangle in $\triangle_2$ whose extension contains q at its top-vertex and is of minimum size by searching $\mathcal{T}_{b\ell r}$ in $O\left((\frac{\log n}{\log \log n})^2\right)$ time.

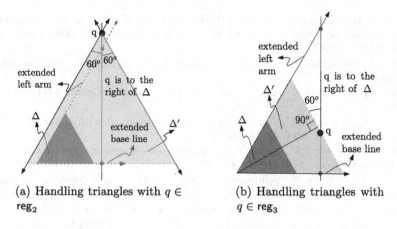

(a) Handling triangles with $q \in$ reg$_2$ (b) Handling triangles with $q \in$ reg$_3$

Fig. 4. Classification of triangles with respect to q

Handling of triangles with q in reg$_3$. Given a query point q, a triangle τ_i is said to satisfy reg$_3$-condition if $q \in \tau_i(\text{reg}_3)$, or in other words, $y(q) > y(s_i^\ell)$ & $y_a(q) > y_a(s_i^\ell)$ & $y_c(q) > y_c(s_i^r)$ (see Fig. 4(b)). Let $\triangle_3$ be the set of triangles satisfying reg$_3$-condition with respect to the query point q. Among all equilateral triangles formed by extending a triangle $\tau_i \in \triangle_3$ such that it contains q, the one whose right-arm passes through q will have the minimum size. Thus, among all triangles in $\triangle_3 \subset \triangle$, we need to identify the one having maximum $y_c(s_i^\ell)$ value. If the internal nodes at the third level of $\mathcal{T}_{b\ell r}$ data structure contains maximum among $y_c(s_i^\ell)$ values of the triangles rooted at that node, then we can identify a triangle in $\triangle_3$ whose extension contains q on its right arm and is of minimum size by searching the $\mathcal{T}_{b\ell r}$ data structure in $O\left((\frac{\log n}{\log \log n})^2\right)$ time.

Theorem 4. *Given a set of n colored points, the smallest color spanning equilateral triangle of a fixed orientation containing query point q can be reported in $O\left((\frac{\log n}{\log \log n})^2\right)$ time using a data structure of size $O\left(n(\frac{\log n}{\log \log n})\right)$, built in $O(n \log n)$ time.*

References

1. Abellanas, M., et al.: Smallest color-spanning objects. In: auf der Heide, F.M. (ed.) ESA 2001. LNCS, vol. 2161, pp. 278–289. Springer, Heidelberg (2001). https://doi.org/10.1007/3-540-44676-1_23
2. Augustine, J., Das, S., Maheshwari, A., Nandy, S.C., Roy, S., Sarvattomananda, S.: Localized geometric query problems. Comput. Geom. **46**(3), 340–357 (2013)
3. Augustine, J., Das, S., Maheshwari, A., Nandy, S.C., Roy, S., Sarvattomananda, S.: Recognizing the largest empty circle and axis-parallel rectangle in a desired location. arXiv preprint arXiv:1004.0558 (2010)
4. Augustine, J., Putnam, B., Roy, S.: Largest empty circle centered on a query line. J. Discret. Algorithms **8**(2), 143–153 (2010)

5. de Berg, M., Cheong, O., van Kreveld, M., Overmars, M.: Computational Geometry Algorithms and Applications, 3rd edn. Springer, Heidelberg (2008). https://doi.org/10.1007/978-3-540-77974-2
6. Bint, G., Maheshwari, A., Nandy, S.C., Smid, M.: Partial enclosure range searching. Preprint
7. Chan, T.M.: Optimal partition trees. Discret. Comput. Geom. **47**(4), 661–690 (2012)
8. Chen, D.Z., Misiołek, E.: Algorithms for interval structures with applications. Theoret. Comput. Sci. **508**, 41–53 (2013)
9. Das, S., Goswami, P.P., Nandy, S.C.: Smallest color-spanning object revisited. Int. J. Comput. Geom. Appl. **19**(5), 457–478 (2009)
10. Gester, M., Hähnle, N., Schneider, J.: Largest empty square queries in rectilinear polygons. In: Gervasi, O., et al. (eds.) ICCSA 2015. LNCS, vol. 9155, pp. 267–282. Springer, Cham (2015). https://doi.org/10.1007/978-3-319-21404-7_20
11. Hasheminejad, J., Khanteimouri, P., Mohades, A.: Computing the smallest color spanning equilateral triangle. In: Proceedings of 31st EuroCG, pp. 32–35 (2015)
12. Huttenlocher, D.P., Kedem, K., Sharir, M.: The upper envelope of voronoi surfaces and its applications. Discret. Comput. Geom. **9**, 267–291 (1993)
13. JaJa, J., Mortensen, C.W., Shi, Q.: Space-efficient and fast algorithms for multidimensional dominance reporting and counting. In: Fleischer, R., Trippen, G. (eds.) ISAAC 2004. LNCS, vol. 3341, pp. 558–568. Springer, Heidelberg (2004). https://doi.org/10.1007/978-3-540-30551-4_49
14. Jiang, M., Wang, H.: Shortest color-spanning intervals. Theoret. Comput. Sci. **609**, 561–568 (2016)
15. Kaminker, T., Sharir, M.: Finding the largest disk containing a query point in logarithmic time with linear storage. JoCG **6**(2), 3–18 (2015)
16. Kaplan, H., Sharir, M.: Finding the maximal empty rectangle containing a query point. arXiv preprint arXiv:1106.3628 (2011)
17. Khanteimouri, P., Mohades, A., Abam, M.A., Kazemi, M.R.: Computing the smallest color-spanning axis-parallel square. In: Cai, L., Cheng, S.-W., Lam, T.-W. (eds.) ISAAC 2013. LNCS, vol. 8283, pp. 634–643. Springer, Heidelberg (2013). https://doi.org/10.1007/978-3-642-45030-3_59
18. Khanteimouri, P., Mohades, A., Abam, M.A., Kazemi, M.R.: Spanning colored points with intervals. In: CCCG (2013)

A Lower Bound for the Radio Number of Graphs

Devsi Bantva[✉]

Lukhdhirji Engineering College, Morvi 363 642, Gujarat, India
devsi.bantva@gmail.com

Abstract. A radio labeling of a graph G is a mapping $\varphi : V(G) \to \{0, 1, 2, \ldots\}$ such that $|\varphi(u) - \varphi(v)| \geq \mathrm{diam}(G) + 1 - d(u, v)$ for every pair of distinct vertices u, v of G, where $\mathrm{diam}(G)$ and $d(u, v)$ are the diameter of G and distance between u and v in G, respectively. The radio number $\mathrm{rn}(G)$ of G is the smallest number k such that G has radio labeling with $\max\{\varphi(v) : v \in V(G)\} = k$. In this paper, we slightly improve the lower bound for the radio number of graphs given by Das *et al.* in [5] and, give necessary and sufficient condition to achieve the lower bound. Using this result, we determine the radio number for cartesian product of paths P_n and the Peterson graph P. We give a short proof for the radio number of cartesian product of paths P_n and complete graphs K_m given by Kim *et al.* in [6].

Keywords: Radio labeling · Radio number ·
Peterson graph · Cartesian product of graphs

1 Introduction

A radio labeling is a distance constrained graph labeling problem originated from well known channel assignment problem. In channel assignment problem, a set of radio stations is given and the task is to assign channels to each radio station such that interference is minimum with optimum use of spectrum. It is known that the interference constraint relies on the distance between two radio stations. The interference between radio stations increases as distance between them decreases and vice-versa. This problem is modeled by graphs: Radio stations are represented by vertices of graphs and interference level is related with distance between them. The assignment of channels is converted into graph labeling problem. Motivated through this, Chartrand *et al.* introduced the concept of radio labeling in [3,4] as follows:

Definition 1. A *radio labeling* of a graph G is a mapping $\varphi : V(G) \to \{0, 1, 2, \ldots\}$ such that for every pair of distinct vertices u, v of G,

$$d(u, v) + |\varphi(u) - \varphi(v)| \geq \mathrm{diam}(G) + 1. \tag{1}$$

© Springer Nature Switzerland AG 2019
S. P. Pal and A. Vijayakumar (Eds.): CALDAM 2019, LNCS 11394, pp. 161–173, 2019.
https://doi.org/10.1007/978-3-030-11509-8_14

The integer $\varphi(u)$ is called the *label* of u under φ, and the *span* of φ is defined as $\mathrm{span}(\varphi) = \max\{|\varphi(u) - \varphi(v)| : u, v \in V(G)\}$. The *radio number* of G is defined as

$$\mathrm{rn}(G) := \min_{\varphi}\{\mathrm{span}(\varphi)\}$$

with minimum taken over all radio labelings φ of G. A radio labeling φ of G is *optimal* if $\mathrm{span}(\varphi) = \mathrm{rn}(G)$.

Note that an optimal radio labeling always assign 0 to some vertex and in this case, the span of φ is the maximum integer assign by φ. A radio labeling is a one-to-one integral function on $V(G)$ to the set of non-negative integers and hence it induces an ordering $x_0, x_1, \ldots, x_{p-1}$ ($p = |V(G)|$) of $V(G)$ such that $0 = \varphi(x_0) < \varphi(x_1) < \ldots < \varphi(x_{p-1}) = \mathrm{span}(\varphi)$. It is clear that if φ is an optimal radio labeling of graph G and ψ is any other radio labeling of G then $\mathrm{span}(\varphi) \leq \mathrm{span}(\psi)$.

A radio labeling problem is recognized as one of the tough graph labeling problems. In most of the research papers, the trend is to determine the radio number for specific graph families. A very few research papers are on general cases which gives lower bound for the radio number of trees and arbitrary graphs. These are as follows: In [7], Liu gave a lower bound for the radio number of trees and presented a class of trees, namely spiders, achieving this lower bound. In [1,2], Bantva et al. presented this lower bound using different notations and gave a necessary and sufficient condition to achieve this lower bound. Using this result they determined the radio number for banana trees, firecrackers trees and a special class of caterpillars. Recently, in [5], Das et al. gave a technique to find a lower bound for the radio number of any graphs.

In this paper, our focus is on a lower bound for the radio number of graphs. We slightly improve the technique to find a lower for the radio number of graphs given by Das et al. in [5] and, give a necessary and sufficient condition to achieve the lower bound. Our results are also useful to determine the radio number of graphs when it is slightly more than the lower bound for the radio number of graphs (see case of cartesian product of paths P_n with the Peterson graph P and complete graphs K_m when n is odd). We determine the radio number for cartesian product of paths P_n and the Peterson graph P and, give a short proof for the radio number of cartesian product of paths P_n and complete graphs K_m given by Kim et al. in [6].

2   A Lower Bound for the Radio Number of Graphs

In this section, we slightly improve the technique to find a lower bound for the radio number of graphs given by Das et al. in [5] and make it more effective (more detail is given in concluding remarks). We also give a necessary and sufficient condition to achieve the lower bound.

Let $G = (V, E)$ be a simple connected graph without loops and multiple edges. We denote the vertex set of G by $V(G)$. We assume $|V(G)| = p$ throughout this paper. The distance between two vertices u and v, denoted by $d(u, v)$, is the least length of a path joining u and v. The diameter of a graph G, denoted by $\text{diam}(G)$ (or simply d to use in equations), is $\max\{d(u, v) : u, v \in V(G)\}$. Let S be a induced subgraph of G, then for any $v \in V(G), d(v, S) = \min\{d(v, w) : w \in S\}$ and $\text{diam}(S) = \max\{d(u, v) : u, v \in S\}$. Denote $[0, n] = \{0, 1, 2, \ldots, n\}$. We follow [8] for standard graph theoretic definition and notation which are not defined here.

Let H be an induced subgraph of connected graph G. The choice of induced subgraph H in G is very crucial in our discussion. In fact, the choice of H plays an important role and key idea of our philosophy to improve a lower bound for the radio number of graphs. But at this moment, we provide only the following information about H and postpone the detail discussion about it till the end. We choose a subgraph H of G such that $\text{diam}(H) = k$. We set $L_0 = V(H)$. Let $N(L_0)$ denote the set of vertices which are adjacent to vertices of L_0. Set $L_1 = N(L_0) \setminus L_0$. Recursively define $L_{i+1} = N(L_i) \setminus (L_0 \cup \ldots \cup L_i)$. Assume that the maximum value of index i for L_i is h known as maximum level. Since G is connected it is clear that $L_s \neq \phi$ for $0 \leq s \leq h$ and $L_t = \phi$ for $t > h$. We fix these sets for rest of discussion.

Let φ be any radio labeling of G with $\text{span}(\varphi) = n$. Note that the function φ is injective but not surjective. Since φ is injective, it induces an ordering $x_0, x_1, \ldots, x_{p-1}$ of $V(G)$ with $0 = \varphi(x_0) < \varphi(x_1) < \ldots < \varphi(x_{p-1})$. Assume that the assign labels are $a_0, a_1, \ldots, a_{p-1}$ such that $\varphi(x_i) = a_i, 0 \leq i \leq p-1$ then $0 = a_0 < a_1 < \ldots < a_{p-1} = \text{span}(\varphi) = n$. Since φ is not surjective, it is clear that $\varphi(V(G)) = \{a_0, a_1, \ldots, a_{p-1}\} \subset [0, n]$. The labels assigned by φ to vertices of G are called *used labels* and the labels $[0, n] \setminus \{a_0, a_1, \ldots, a_{p-1}\}$ are called *unused labels*. So to give a lower bound, our aim is to count both the used and unused labels.

The number of unused labels are the sum of $a_{t+1} - a_t - 1$, where t varies from 0 to $p - 2$. Using definition of radio labeling and triangle inequality twice for $d(x_{t+1}, x_t)$ in G, we obtain

$$a_{t+1} - a_t - 1 \geq d + 1 - d(x_{t+1}, x_t) - 1 \tag{2}$$
$$\geq d + 1 - d(x_{t+1}, L_0) - d(x_t, L_0) - \text{diam}(L_0) - 1 \tag{3}$$
$$= d + 1 - d(x_{t+1}, L_0) - d(x_t, L_0) - k - 1.$$

Summing this latter inequality for 0 to $p - 2$, we obtain the total number of unused labels. Thus the number of unused labels is at least

$$\sum_{t=0}^{p-2} (a_{t+1} - a_t - 1) \geq \sum_{t=0}^{p-2} (d + 1 - d(x_{t+1}, L_0) - d(x_t, L_0) - k - 1)$$

$$= (p-1)(d-k) - 2 \sum_{t=0}^{p-2} d(x_t, L_0) + d(x_0, L_0) + d(x_{p-1}, L_0)$$

$$= (p-1)(d-k) + d(x_0, L_0) + d(x_{p-1}, L_0) - 2 \sum_{i=0}^{h} |L_i| i.$$

Note that as the label set for radio labeling includes 0 as well, the used labels have an additive factor of -1. Hence, the sum of used and unused labels is at least as follows.

$$\mathrm{span}(\varphi) \geq p - 1 + (p-1)(d-k) + d(x_0, L_0) + d(x_{p-1}, L_0) - 2\sum_{i=0}^{h} |L_i| i$$

$$\geq (p-1)(d-k+1) + d(x_0, L_0) + d(x_{p-1}, L_0) - 2\sum_{i=0}^{h} |L_i| i.$$

Note that $d(x_0, L_0) + d(x_{p-1}, L_0)$ has minimum value if $x_0, x_{p-1} \in L_0$ when $|L_0| \geq 2$ and $x_0 \in L_0, x_{p-1} \in L_1$ when $|L_0| = 1$. Define $\delta = 0$ if $|L_0| \geq 2$ and 1 if $|L_0| = 1$. Hence, we obtain

$$\mathrm{rn}(G) \geq (p-1)(d-k+1) + \delta - 2\sum_{i=0}^{h} |L_i| i.$$

We now come to the selection of H as an induced subgraph of G. We choose an induced subgraph H in G such that the set of vertices $V(G) \setminus V(H)$ can be partitioned into distinct sets $V_1, V_2, \ldots, V_m (m \geq 2)$ and when we fix $V(H)$ as L_0 then it possible to order the vertices of G as $x_0, x_1, \ldots, x_{p-1}$ such that $d(x_i, x_{i+1})$ satisfies the equation $d(x_i, x_{i+1}) = d(x_i, L_0) + d(x_{i+1}, L_0) + \mathrm{diam}(L_0)$, where $x_i \in V_i, x_{i+1} \in V_j, i \neq j$ or, one or both of x_i, x_{i+1} is in $V(H)$. We also keep in mind that such an ordering $x_0, x_1, \ldots, x_{p-1}$ satisfies conditions $d(x_0, L_0) = 0$, $d(x_{p-1}, L_0) = 1$ when $|L_0| = 1$ and $d(x_0, L_0) = d(x_{p-1}, L_0) = 0$ when $|L_0| \geq 2$. We also inform the readers that in case of trees, the set of weight center(s) $W(T)$ (see [7] and [2] for definition and detail about weight center) is always a good choice as L_0 and more useful results are given in [7] and [1,2] to determine the radio number of trees than the technique discussed above. We advised the readers to refer [7] and [1,2] for the radio number of trees.

Finally, from above discussion, we summarize our result as follows.

Theorem 1. *Let G be a simple connected graph of order p, diameter d and L_i's, δ are defined as earlier. Denote $\mathrm{diam}(L_0) = k$. Then*

$$\mathrm{rn}(G) \geq (p-1)(d-k+1) + \delta - 2\sum_{i=0}^{h} |L_i| i \qquad (4)$$

Theorem 2. *Let G be a simple connected graph of order p, diameter d and L_i's, δ are defined as earlier. Denote $\mathrm{diam}(L_0) = k$. Then*

$$\mathrm{rn}(G) = (p-1)(d-k+1) + \delta - 2\sum_{i=0}^{h} |L_i| i \qquad (5)$$

holds if and only if there exist a radio labeling φ with $0 = \varphi(x_0) < \varphi(x_1) < \ldots < \varphi(x_{p-1}) = \mathrm{span}(\varphi) = \mathrm{rn}(G)$ such that all the following hold for $0 \leq i \leq p - 1$:

(a) $d(x_i, x_{i+1}) = d(x_i, L_0) + d(x_{i+1}, L_0) + k$,

(b) $x_0, x_{p-1} \in L_0$ *if* $|L_0| \geq 2$ *and* $x_0 \in L_0, x_{p-1} \in L_1$ *if* $|L_0| = 1$,

(c) $\varphi(x_0) = 0$ *and* $\varphi(x_{i+1}) = \varphi(x_i) + d + 1 - d(x_i, L_0) - d(x_{i+1}, L_0) - k$.

Proof. **Necessity:** Suppose that (5) holds. Then there exist an optimal radio label-ing φ of G with $\text{span}(\varphi) = (p-1)(d-k+1) + \delta - 2\sum_{i=0}^{h} |L_i|i$. Let $x_0, x_1, \ldots, x_{p-1}$ with $0 = \varphi(x_0) < \varphi(x_1) < \cdots < \varphi(x_{p-1}) = \text{span}(\varphi)$ is an ordering of $V(G)$ induced by φ. Note that $\text{span}(\varphi) = (p-1)(d-k+1) + \delta - 2\sum_{i=0}^{h} |L_i|i$ is pos-sible if equalities hold in (2) and (3) together with $x_0, x_{p-1} \in L_0$ when $|L_0| \geq 2$ and, $x_0 \in L_0, x_{p-1} \in L_1$ when $|L_0| = 1$. Note that equalities in (2) and (3) gives $d(x_i, x_{i+1}) = d(x_i, L_0) + d(x_{i+1}, L_0) + k$. These all together turn the definition of radio labeling (1) as $\varphi(x_0) = 0$ and $\varphi(x_{i+1}) = \varphi(x_i) + d + 1 - L(x_i) - L(x_{i+1}) - k$.

Sufficiency: Suppose that there exist a radio labeling φ with $0 = \varphi(x_0) < \varphi(x_1) < \cdots < \varphi(x_{p-1}) = \text{span}(\varphi) = \text{rn}(G)$ such that (a), (b) and (c) holds. It is enough to prove that $\text{span}(\varphi) = (p-1)(d-k+1) + \delta - 2\sum_{i=0}^{h} |L_i|i$. From (b) and (c), we have

$$
\begin{aligned}
\text{span}(\varphi) &= \varphi(x_{p-1}) - \varphi(x_0) \\
&= \sum_{t=0}^{p-2} \left(\varphi(x_{t+1}) - \varphi(x_t) \right) \\
&= \sum_{t=0}^{p-2} \left(d + 1 - d(x_{t+1}, L_0) - d(x_t, L_0) - k \right) \\
&= (p-1)(d-k+1) - 2\sum_{t=0}^{p-2} d(x_t, L_0) + L(x_0, L_0) + d(x_{p-1}, L_0) \\
&= (p-1)(d-k+1) + \delta - 2\sum_{i=0}^{h} |L_i|i
\end{aligned}
$$

which completes the proof.

Remark 1. As a consequence of above Theorem 2, we obtain that if one or more conditions of Theorem 2 does not hold then

$$
\text{rn}(G) > (p-1)(d-k+1) + \delta - 2\sum_{i=0}^{h} |L_i|i \tag{6}
$$

3   Radio Number for Some Cartesian Product of Two Graphs

In this section, we continue to use the terminology and notation defined in previous section. We determine the radio number for cartesian product of paths P_n and the Peterson graph P using results of previous section. We present a short proof for the radio number of cartesian product of paths P_n and complete graphs K_m given by Kim *et al.* in [6] using our results approach.

Let $G = (V(G), E(G))$ and $H = (V(H), E(H))$ be two graphs. The cartesian product of G and H, denoted by $G \square H$, is the graph $G_\square = (V(G_\square), E(G_\square))$ where $V(G_\square) = V(G) \times V(H)$ and two vertices (a, b) and (c, d) are adjacent if $a = c$ and $(b, d) \in E(H)$ or $b = d$ and $(a, c) \in E(G)$.

3.1 Radio Number for $P_n \square P$

The peterson graph, denoted by P, is the complement of the line graph of complete graph K_5. The peterson graph and, the cartesian product of a path P_5 and the Peterson graph P is shown in Figs. 1 and 2, respectively. Note that $|P_n \square P| = |P_n| \times |P| = 10n$ and $\mathrm{diam}(P_n \square P) = n + 1$. We denote the vertex set of P_n by $V(P_n) = \{u_1, u_2, \ldots, u_n\}$ with $(u_i, u_{i+1}) \in E(P_n), 1 \le i \le n-1$ and the vertex set of P by $V(P) = \{v_1, v_2, \ldots, v_{10}\}$ with $E(P) = \{v_i v_{i+1}, v_1 v_6, v_1 v_8, v_2 v_7, v_3 v_9, v_4 v_8, v_5 v_7, v_6 v_9, v_7 v_{10}, v_8 v_{10}, v_9 v_{10} : 1 \le i \le 5\}$.

Fig. 1. Peterson graph P.   **Fig. 2.** Cartesian product $P_5 \square P$.

Theorem 3. *Let $n \ge 3$ be an integer. Then*

$$\mathrm{rn}(P_n \square P) := \begin{cases} 5n^2 - n + 1, & \text{if } n \text{ is even,} \\ 5n^2 - n + 6, & \text{if } n \text{ is odd.} \end{cases} \tag{7}$$

Proof. We consider the following two cases.

Case-1: n is even. In this case, we set the subgraph induced by vertex set $\{(u_{n/2}, v_1), (u_{n/2}, v_2), \ldots, (u_{n/2}, v_{10}), (u_{n/2+1}, v_1), (u_{n/2+1}, v_2), \ldots, (u_{n/2+1}, v_{10})\}$ of $P_n \square P$ as L_0 then $\mathrm{diam}(L_0) = k = 3$ and the maximum level in $P_n \square P$ is $h = n/2 - 1$. Note that $p = 10n$ and $\sum_{i=0}^{h} |L_i| i = 5n(n - 2)/2$. Substituting these all in (4), we obtain $\mathrm{rn}(P_n \square P) \ge 5n^2 - n + 1$.

We now prove that this lower bound is tight. Note that for this purpose, it suffices to give a radio labeling φ of $P_n \square P$ with span equal to this lower bound and for this, it is enough to give a radio labeling satisfying conditions of Theorem 2. We first order the vertices of $P_n \square P$ and define recursive formula of radio labeling φ on it. Let $\alpha = \left(\begin{smallmatrix} 1 & 2 & 3 & 4 & 5 & 6 & 7 & 8 & 9 & 10 \\ 1 & 8 & 3 & 7 & 2 & 10 & 5 & 4 & 6 & 9 \end{smallmatrix} \right)$, $\beta = \left(\begin{smallmatrix} 1 & 2 & 3 & 4 & 5 & 6 & 7 & 8 & 9 & 10 \\ 9 & 1 & 10 & 3 & 7 & 2 & 4 & 6 & 5 & 8 \end{smallmatrix} \right)$, $\sigma = \left(\begin{smallmatrix} 1 & 2 & 3 & 4 & 5 & 6 & 7 & 8 & 9 & 10 \\ 2 & 9 & 1 & 8 & 3 & 7 & 6 & 5 & 4 & 10 \end{smallmatrix} \right)$ and $\tau = \left(\begin{smallmatrix} 1 & 2 & 3 & 4 & 5 & 6 & 7 & 8 & 9 & 10 \\ 7 & 2 & 8 & 1 & 10 & 3 & 5 & 4 & 6 & 9 \end{smallmatrix} \right)$ be four permutations.

Using these four permutations, we first rename $(u_i, v_j)(1 \leq i \leq n, 1 \leq j \leq 10)$ as (a_r, b_s) as follows:

$$(a_r, b_s) := \begin{cases} (u_i, v_{\alpha(j)}), & \text{if } 1 \leq i \leq n/2 \text{ and } (n/2 - i) \equiv 0 \ (\text{mod } 2), \\ (u_i, v_{\beta(j)}), & \text{if } 1 \leq i \leq n/2 \text{ and } (n/2 - i) \equiv 1 \ (\text{mod } 2), \\ (u_i, v_{\sigma(j)}), & \text{if } n/2 + 1 \leq i \leq n \text{ and } (n - i) \equiv 0 \ (\text{mod } 2), \\ (u_i, v_{\tau(j)}), & \text{if } n/2 + 1 \leq i \leq n \text{ and } (n - i) \equiv 1 \ (\text{mod } 2). \end{cases}$$

We now define an ordering $x_0, x_1, \ldots, x_{p-1}$ as follows: Let $x_t := (a_r, b_s)$, where

$$t := \begin{cases} (n/2 - r)20 + 2(s - 1), & \text{if } 1 \leq r \leq n/2, \\ (n - r)20 + 2s - 1, & \text{if } n/2 + 1 \leq r \leq n. \end{cases}$$

Then note that $x_0, x_{p-1} \in L_0$ and for $0 \leq i \leq p - 2$, $d(x_i, x_{i+1}) = d(x_i, L_0) + d(x_{i+1}, L_0) + k$. Define φ as $\varphi(x_0) = 0$, $\varphi(x_{i+1}) = \varphi(x_i) + d + 1 - d(x_i, L_0) - d(x_{i+1}, L_0) - k$.

Claim-1: φ is a radio labeling with $\text{span}(\varphi) = 5n^2 - n + 1$.

Let x_i and $x_j, 0 \leq i < j \leq p - 1$ be two arbitrary vertices. If $j = i + 1$ then $\varphi(x_j) - \varphi(x_i) = d + 1 - d(x_i, L_0) - d(x_{i+1}, L_0) - k = d + 1 - d(x_i, x_{i+1})$. If $j \geq i + 4$ then $\varphi(x_j) - \varphi(x_i) \geq (j - i)(d - k + 1) - \sum_{t=i+1}^{j-1} d(x_t, L_0) - d(x_i, L_0) - d(x_j, L_0) \geq 4(n - 1) - (n - 2)/2 - n/2 - (n - 2)/2 - n/2 > n + 2 > n + 2 - d(x_i, x_j) = d + 1 - d(x_i, x_j)$. If $j = i + 3$ then $\varphi(x_j) - \varphi(x_i) = (j - i)(d - k + 1) - \sum_{t=i+1}^{j-1} d(x_t, L_0) - d(x_i, L_0) - d(x_j, L_0) \geq 3(n - 1) - n/2 - (n - 2)/2 - (n - 2)/2 = (3n - 2)/2 \geq n + 1 \geq n + 2 - d(x_i, x_j) = d + 1 - d(x_i, x_j)$ as $d(x_i, x_j) \geq 1$. If $j = i + 2$ then $\varphi(x_j) - \varphi(x_i) = (j - i)(d - k + 1) - d(x_i, L_0) - 2d(x_{i+1}, L_0) - d(x_{i+2}, L_0)$. If (1) $d(x_i, L_0) + 2d(x_{i+1}, L_0) + d(x_{i+2}, L_0) = n - 1$ then $d(x_i, x_j) = 3$ and hence $\varphi(x_j) - \varphi(x_i) = 2(n - 1) - (n - 1) = (n - 1) = n + 2 - d(x_i, x_j) = d + 1 - d(x_i, x_j)$. (2) $d(x_i, L_0) + 2d(x_{i+1}, L_0) + d(x_{i+2}, L_0) = n - 2$ then $d(x_i, x_j) = 2$ and hence $\varphi(x_j) - \varphi(x_i) = 2(n - 1) - (n - 2) = n = n + 2 - d(x_i, x_j) = d + 1 - d(x_i, x_j)$. Hence, φ is a radio labeling. The span of φ is $\text{span}(\varphi) = \varphi(x_{p-1}) - \varphi(x_0) = \sum_{t=0}^{p-1}(\varphi(x_{t+1}) - \varphi(x_t)) = (p - 1)(d - k + 1) - 2\sum_{t=0}^{p-1} d(x_t, L_0) = (p - 1)(d - k + 1) - 2\sum_{i=0}^{h} |L_i| i$ which is equal to $5n^2 - n + 1$ in the present case.

Case-2: n is odd. In this case, we set the subgraph induced by vertex set $\{(u_{(n+1)/2}, v_1), (u_{(n+1)/2}, v_2), \ldots, (u_{(n+1)/2}, v_{10})\}$ of $P_n \Box P$ as L_0 then $\text{diam}(L_0) = k = 2$ and the maximum level in $P_n \Box P$ is $h = n/2 - 1$. Note that $p = 10n$ and $\sum_{i=0}^{h} |L_i| i = 5(n^2 - 1)/2$. Substituting these all in (4), we obtain $\text{rn}(P_n \Box P) \geq 5n^2 - n + 5$. Now if possible then assume that $\text{rn}(P_n \Box P) = 5n^2 - n + 5$ then there exist a radio labeling φ of $P_n \Box P$ with $\text{span}(\varphi) = 5n^2 - n + 5$. By Theorem 2, φ induces an ordering $x_0, x_1, \ldots, x_{p-1}$ of $V(P_n \Box P)$ with $0 = \varphi(x_0) < \varphi(x_1) < \cdots < \varphi(x_{p-1}) = \text{span}(\varphi)$ which satisfies (a), (b) and (c) of Theorem 2. Let $L = \{(u_1, v_1), (u_1, v_2), \ldots, (u_1, v_{10})\}$, $C = \{(u_{(n+1)/2}, v_1), (u_{(n+1)/2}, v_2), \ldots, (u_{(n+1)/2}, v_{10})\}$ and $R = \{(u_n, v_1), (u_n, v_2), \ldots, (u_n, v_{10})\}$. Since $|L| = |R| = |C|$ and φ satisfies conditions (a), (b) and (c) of Theorem 2, there exist a vertex $x_t \in L$ or R such that $d(x_{t-1}, L_0) + d(x_t, L_0) > (n - 1)/2$ and $d(x_t, L_0) + d(x_{t+1}, L_0) > (n - 1)/2$. Without loss of generality, assume that $d(x_{t-1}, L_0) + d(x_t, L_0) \geq d(x_t, L_0) + d(x_{t+1}, L_0)$. Since an ordering $x_0, x_1, \ldots, x_{p-1}$ of $V(P_n \square P)$ satisfies condition (a) of Theorem 2, it is clear that $d(x_{t-1}, x_{t+1}) = d(x_{t-1}, L_0) - d(x_{t+1}, L_0) + 2$. Now consider $\varphi(x_{t+1}) - \varphi(x_{t-1}) = \varphi(x_{t+1}) - \varphi(x_t) + \varphi(x_t) - \varphi(x_{t-1}) = n + 2 - d(x_{t+1}, L_0) - d(x_t, L_0) - 2 + n + 2 - d(x_t, L_0) - d(x_{t-1}, L_0) - 2 = 2n - (d(x_{t-1}, L_0) - d(x_{t+1}, L_0) + 2) - 2(d(x_t, L_0) + d(x_{t+1}, L_0) - 1) \leq 2n - d(x_{t-1}, x_{t+1}) - 2((n+1)/2 - 1) = n + 1 - d(x_{t-1}, x_{t+1}) < n + 2 - d(x_{t-1}, x_{t+1})$, a contradiction with φ is a radio labeling. Hence, $rn(P_n \square P) \geq 5n^2 - n + 6$. We now prove that this lower bound is the actual value for $rn(P_n \square P)$. Note that for this purpose, it is enough to give a radio labeling φ of $P_n \square P$ with $span(\varphi) = 5n^2 - n + 6$. We order the vertices of $P_n \square P$ and define recursive formula on this ordering for φ. We consider the following two cases.

Subcase-2.1: $n \equiv 1 \pmod 4$.

Let $\alpha = \left(\begin{smallmatrix} 1 & 2 & 3 & 4 & 5 & 6 & 7 & 8 & 9 & 10 \\ 1 & 4 & 3 & 6 & 2 & 7 & 9 & 8 & 10 & 5 \end{smallmatrix} \right)$, $\beta = \left(\begin{smallmatrix} 1 & 2 & 3 & 4 & 5 & 6 & 7 & 8 & 9 & 10 \\ 2 & 7 & 1 & 5 & 3 & 6 & 8 & 10 & 9 & 4 \end{smallmatrix} \right)$, $\sigma = \left(\begin{smallmatrix} 1 & 2 & 3 & 4 & 5 & 6 & 7 & 8 & 9 & 10 \\ 2 & 3 & 1 & 7 & 4 & 5 & 6 & 9 & 10 & 8 \end{smallmatrix} \right)$ be three permutations. Using these three permutations, we first rename (u_i, v_j), $(1 \leq i \leq n, 1 \leq j \leq 10)$ as (a_r, b_s) as follows:

$$(a_r, b_s) := \begin{cases} (u_i, v_{\alpha(j)}), & \text{if } i = (n+1)/2, \\ (u_i, v_{\sigma^{2(n-i)}\beta(j)}), & \text{if } (n+1)/2 < i \leq n, \\ (u_i, v_{\sigma^{2((n+1)/2-i)+1}\beta(j)}), & \text{if } 1 \leq i < (n+1)/2. \end{cases}$$

We now define an ordering $x_0, x_1, \ldots, x_{p-1}$ as follows: Set $x_0 = (a_{(n+1)/2}, b_1)$, $x_{p-1} = (a_{(n+1)/2}, b_{10})$ and for $1 \leq t \leq p - 2$, let $x_t := (a_r, b_s)$, where

$$t := \begin{cases} (n+1-2r) + n(s-1), & \text{if } 1 \leq r \leq (n+1)/2 \text{ and } 1 \leq s \leq 7, \\ 2(n-r) + n(s-1) + 1, & \text{if } (n+1)/2 < r \leq n \text{ and } 1 \leq s \leq 7, \\ (n+1-2r) + n(s-1) - 1, & \text{if } 1 \leq r < (n+1)/2 \text{ and } 8 \leq s \leq 10, \\ 2(n-r) + n(s-1), & \text{if } (n+1)/2 < r \leq n \text{ and } 8 \leq s \leq 10, \\ ns - 1, & \text{if } r = (n+1)/2 \text{ and } 8 \leq s \leq 10. \end{cases}$$

Then note that $x_0, x_{p-1} \in L_0$ and for $0 \leq i \leq p - 2$, $d(x_i, x_{i+1}) = d(x_i, L_0) + d(x_{i+1}, L_0) + k$. Define φ as follows: $\varphi(x_0) = 0$ and $\varphi(x_{i+1}) = \varphi(x_i) + d + 1 - d(x_i, L_0) - d(x_{i+1}, L_0) - k$ for $0 \leq i \leq p - 2, i \neq p - 3n - 1$ and $\varphi(x_{p-3n}) = \varphi(x_{p-3n-1}) + d + 1 - d(x_i, L_0) - d(x_{i+1}, L_0) - k + 1$.

Claim-2: φ is a radio labeling with $span(\varphi) = 5n^2 - n + 6$.

Let x_i and x_j, $0 \leq i < j \leq p - 1$ be two arbitrary vertices. If $j = i + 1$ then it is clear that $\varphi(x_j) - \varphi(x_i) \geq d + 1 - d(x_i, L_0) - d(x_j, L_0) - k = d + 1 - d(x_i, x_j)$. If $j \geq i + 3$ then if (1) $0 \leq i \leq p - 3n - 4$ or $p - 3n \leq i \leq p - 4$ then $\varphi(x_j) - \varphi(x_i) \geq (j-i)(d - k + 1) - 2 \sum_{t=i+1}^{j-1} d(x_t, L_0) - d(x_i, L_0) - d(x_j, L_0) = 3n - (d(x_i, L_0) + d(x_{i+1}, L_0)) + (d(x_{i+1}, L_0) + d(x_{i+2}, L_0)) + (d(x_{i+2}, L_0) + d(x_{i+3}, L_0)) \geq 3n - (n+1)/2 - (n-1)/2 - (n+1)/2 = (3n-1)/2 > n + 1 > n + 2 - d(x_i, x_j) = d + 1 - d(x_i, x_j)$ as $d(x_i, x_j) \geq 1$; (2) $i \in \{p - 3n - 3, p - 3n - 2, p - 3n - 1\}$ then $\varphi(x_j) - \varphi(x_i) \geq (j-i)(d-k+1) - 2 \sum_{t=i+1}^{j-1} d(x_t, L_0) - d(x_i, L_0) - d(x_j, L_0) + 1 = 3n - (d(x_i, L_0) + d(x_{i+1}, L_0)) + (d(x_{i+1}, L_0) + d(x_{i+2}, L_0)) + (d(x_{i+2}, L_0) + d(x_{i+3}, L_0)) + 1 \geq 3n - (n+1)/2 - (n-1) - (n+1)/2 + 1 = n + 1 > n + 2 - d(x_i, x_j) = d + 1 - d(x_i, x_j)$

as $d(x_i, x_j) \geq 1$. If $j = i + 2$ then if (1) $0 \leq i \leq p - 3n - 3$ or $p - 3n \leq i \leq p - 3$ then $\varphi(x_j) - \varphi(x_i) = (j - i)(d - k + 1) - (d(x_i, L_0) + d(x_{i+1}, L_0)) - (d(x_{i+1}, L_0) + d(x_{i+1}, L_0)) \geq 2n - (n+1)/2 - (n-1)/2 = n \geq n + 2 - d(x_i, x_j) = d + 1 - d(x_i, x_j)$ as $d(x_i, x_j) \geq 2$; (2) $i \in \{p - 3n - 2, p - 3n - 1\}$ then it is easy to verify $\varphi(x_j) - \varphi(x_i) \geq n + 2 - d(x_i, x_j) = d + 1 - d(x_i, x_j)$. Hence, φ is a radio labeling. The span of φ is $\text{span}(\varphi) = (p - 1)(d - k + 1) - 2\sum_{i=0}^{h} |L_i| i + 1$ which is equal to $5n^2 - n + 6$ in the present case.

Subcase-2.2: $n \equiv 3 \pmod 4$.

Let $\alpha = \begin{pmatrix} 1 & 2 & 3 & 4 & 5 & 6 & 7 & 8 & 9 & 10 \\ 1 & 7 & 2 & 9 & 3 & 8 & 5 & 6 & 4 & 10 \end{pmatrix}$ and $\sigma = \begin{pmatrix} 1 & 2 & 3 & 4 & 5 & 6 & 7 & 8 & 9 & 10 \\ 3 & 1 & 2 & 6 & 4 & 5 & 8 & 9 & 10 & 7 \end{pmatrix}$ be two permutations. Using these two permutations, we first rename $(u_i, v_j), (1 \leq i \leq n, 1 \leq j \leq 10)$ as (a_r, b_s) as follows:

$$(a_r, b_s) := \begin{cases} (u_i, v_{\alpha(j)}), & \text{if } i = (n+1)/2, \\ (u_i, v_{\sigma^{2(n-i)+1}\alpha(j)}), & \text{if } (n+1)/2 < i \leq n, \\ (u_i, v_{\sigma^{2((n+1)/2-i)-1}\alpha(j)}), & \text{if } 1 \leq i < (n+1)/2. \end{cases}$$

We now define an ordering $x_0, x_1, \ldots, x_{p-1}$ as follows: Set $x_0 = (a_{(n+1)/2}, b_1)$, $x_{p-1} = (a_{(n+1)/2}, b_{10})$ and for $1 \leq t \leq p - 2$, let $x_t := (a_r, b_s)$, where

$$t := \begin{cases} (n + 1 - 2r) + n(s - 1), & \text{if } 1 \leq r \leq (n+1)/2 \text{ and } 1 \leq s \leq 9, \\ 2(n - r) + n(s - 1) + 1, & \text{if } (n+1)/2 < r \leq n \text{ and } 1 \leq s \leq 9, \\ (n + 1 - 2r) + n(s - 1) - 1, & \text{if } 1 \leq r < (n+1)/2 \text{ and } s = 10, \\ 2(n - r) + n(s - 1), & \text{if } (n+1)/2 < r \leq n \text{ and } s = 10. \end{cases}$$

Then note that $x_0, x_{p-1} \in L_0$ and for $0 \leq i \leq p - 2$, $d(x_i, x_{i+1}) = d(x_i, L_0) + d(x_{i+1}, L_0) + k$. Define φ as follows: $\varphi(x_0) = 0$, $\varphi(x_{i+1}) = \varphi(x_i) + d + 1 - d(x_i, L_0) - d(x_{i+1}, L_0) - k$ for $0 \leq i \leq p - 2, i \neq p - n - 1$ and $\varphi(x_{p-n}) = \varphi(x_{p-n-1}) + d + 1 - d(x_i, L_0) - d(x_{i+1}, L_0) - k + 1$.

Claim-3: φ is a radio labeling with $\text{span}(\varphi) = 5n^2 - n + 6$.

Let x_i and $x_j, 0 \leq i < j \leq p - 1$ be two arbitrary vertices. If $j = i + 1$ then it is clear that $\varphi(x_j) - \varphi(x_i) \geq d + 1 - d(x_i, L_0) - d(x_j, L_0) - k = d + 1 - d(x_i, x_j)$. If $j \geq i + 3$ then if (1) $0 \leq i \leq p - n - 4$ or $p - n \leq i \leq p - 4$ then $\varphi(x_j) - \varphi(x_i) \geq (j - i)(d - k + 1) - 2\sum_{t=i+1}^{j-1} d(x_t, L_0) - d(x_i, L_0) - d(x_j, L_0) = 3n - (d(x_i, L_0) + d(x_{i+1}, L_0)) + (d(x_{i+1}, L_0) + d(x_{i+2}, L_0)) + (d(x_{i+2}, L_0) + d(x_{i+3}, L_0)) \geq 3n - (n + 1)/2 - (n-1)/2 - (n+1)/2 = (3n-1)/2 > n + 1 > n + 2 - d(x_i, x_j) = d + 1 - d(x_i, x_j)$ as $d(x_i, x_j) \geq 1$; (2) $i \in \{p - n - 3, p - n - 2, p - n - 1\}$ then $\varphi(x_j) - \varphi(x_i) \geq (j-i)(d-k+1) - 2\sum_{t=i+1}^{j-1} d(x_t, L_0) - d(x_i, L_0) - d(x_j, L_0) + 1 = 3n - (d(x_i, L_0) + d(x_{i+1}, L_0)) + (d(x_{i+1}, L_0) + d(x_{i+2}, L_0)) + (d(x_{i+2}, L_0) + d(x_{i+3}, L_0)) + 1 \geq 3n - (n+1)/2 - (n-1) - (n+1)/2 + 1 = n + 1 > n + 2 - d(x_i, x_j) = d + 1 - d(x_i, x_j)$ as $d(x_i, x_j) \geq 1$. If $j = i + 2$ then if (1) $0 \leq i \leq p - n - 3$ or $p - n \leq i \leq p - 3$ then $\varphi(x_j) - \varphi(x_i) = (j - i)(d - k + 1) - (d(x_i, L_0) + d(x_{i+1}, L_0)) - (d(x_{i+1}, L_0) + d(x_{i+1}, L_0)) \geq 2n - (n+1)/2 - (n-1)/2 = n \geq n + 2 - d(x_i, x_j) = d + 1 - d(x_i, x_j)$ as $d(x_i, x_j) \geq 2$; (2) $i \in \{p - n - 2, p - n - 1\}$ then it is easy to verify $\varphi(x_j) - \varphi(x_i) \geq n + 2 - d(x_i, x_j) = d + 1 - d(x_i, x_j)$. Hence, φ is a radio labeling. The span of φ

is span$(\varphi) = (p-1)(d-k+1) - 2\sum_{i=0}^{h}|L_i|i + 1$ which is equal to $5n^2 - n + 6$ in the present case.

Example 1. In Table 1, an ordering and the corresponding optimal radio labeling of vertices of $P_6 \square P$ is shown.

Table 1. An ordering and optimal radio labeling for vertices of $P_6 \square P$.

$(u_i, v_j)\frac{i\rightarrow}{j\downarrow}$	1		2		3		4		5		6	
1	x_{40}	118	x_{36}	107	$\underline{x_0}$	$\underline{0}$	x_{43}	127	x_{33}	98	x_3	9
2	x_{54}	160	x_{20}	59	x_{14}	42	x_{57}	169	x_{23}	68	x_{17}	51
3	x_{44}	130	x_{38}	113	x_4	12	x_{41}	121	x_{35}	104	x_1	3
4	x_{52}	154	x_{24}	71	x_{12}	36	x_{55}	163	x_{21}	62	x_{15}	45
5	x_{42}	124	x_{32}	95	x_2	6	x_{45}	133	x_{39}	116	x_5	15
6	x_{58}	172	x_{22}	65	x_{18}	54	x_{53}	157	x_{25}	74	x_{13}	39
7	x_{48}	142	x_{26}	77	x_8	24	x_{51}	151	x_{29}	86	x_{11}	33
8	x_{46}	136	x_{30}	89	x_6	18	x_{49}	145	x_{27}	80	x_9	27
9	x_{50}	148	x_{28}	83	x_{10}	30	x_{47}	139	x_{31}	92	x_7	21
10	x_{56}	166	x_{34}	101	x_{16}	48	$\underline{x_{59}}$	$\underline{\mathbf{175}}$	x_{37}	110	x_{19}	57

Example 2. In Table 2, an ordering and the corresponding optimal radio labeling of vertices of $P_5 \square P$ is shown.

Table 2. An ordering and optimal radio labeling for vertices of $P_5 \square P$.

$(u_i, v_j)\frac{i\rightarrow}{j\downarrow}$	1		2		3		4		5	
1	x_9	23	x_{12}	31	$\underline{x_0}$	$\underline{0}$	x_3	8	x_6	16
2	x_{19}	49	x_{27}	70	x_{15}	39	x_{23}	60	x_{31}	81
3	x_4	10	x_7	18	x_{10}	26	x_{13}	34	x_1	3
4	x_{29}	75	x_{17}	44	x_{25}	65	x_{33}	86	x_{21}	55
5	x_{14}	36	x_2	5	x_5	13	x_8	21	x_{11}	29
6	x_{34}	88	x_{22}	57	x_{30}	78	x_{18}	47	x_{26}	68
7	x_{38}	97	x_{41}	105	x_{44}	113	x_{47}	121	$\underline{x_{35}}$	$\underline{90}$
8	x_{48}	123	x_{36}	92	x_{39}	100	x_{42}	108	x_{45}	116
9	x_{43}	110	x_{46}	118	$\underline{x_{49}}$	$\underline{\mathbf{126}}$	x_{37}	95	x_{40}	103
10	x_{24}	62	x_{32}	83	x_{20}	52	x_{28}	73	x_{16}	42

Example 3. In Table 3, an ordering and the corresponding optimal radio labeling of vertices of $P_7 \square P$ is shown.

Table 3. An ordering and optimal radio labeling for vertices of $P_7\Box P$.

$(u_i, v_j)\xrightarrow[j\downarrow]{i\to}$	1		2		3		4		5		6		7	
1	x_6	21	x_{18}	64	x_9	32	$\underline{x_0}$	$\mathbf{0}$	x_{12}	43	x_3	11	x_{15}	54
2	x_{62}	221	x_{46}	164	x_{58}	207	x_{42}	150	x_{54}	193	x_{65}	230	x_{50}	179
3	x_{13}	46	x_4	14	x_{16}	57	x_7	25	x_{19}	68	x_{10}	36	x_1	4
4	x_{48}	171	x_{60}	214	x_{44}	157	x_{56}	200	x_{67}	237	x_{52}	186	$\underline{x_{63}}$	$\underline{223}$
5	x_{20}	71	x_{11}	39	x_2	7	x_{14}	50	x_5	18	x_{17}	61	x_8	29
6	x_{68}	240	x_{53}	189	x_{64}	226	x_{49}	175	x_{61}	218	x_{45}	161	x_{57}	204
7	x_{34}	121	x_{25}	89	x_{37}	132	x_{28}	100	x_{40}	143	x_{31}	111	x_{22}	79
8	x_{41}	146	x_{32}	114	x_{23}	82	x_{35}	125	x_{26}	93	x_{38}	136	x_{29}	104
9	x_{27}	96	x_{39}	139	x_{30}	107	x_{21}	75	x_{33}	118	x_{24}	86	x_{36}	129
10	x_{55}	196	x_{66}	233	x_{51}	182	$\underline{x_{69}}$	$\mathbf{244}$	x_{47}	168	x_{59}	211	x_{43}	154

3.2   Radio Number for $P_n\Box K_m$

In this section, using Theorems 1 and 2, we give a short proof for the radio number of $P_n\Box K_m$ given by Kim *et al.* in [6].

We assume that $m \geq 3$ and $n \geq 4$. Note that $|P_n\Box K_m| = |P_n| \times |K_m| = nm$ and $\mathrm{diam}(P_n\Box K_m) = n$. We denote the vertex set of P_n by $V(P_n) = \{u_1, u_2, \ldots, u_n\}$ with $(u_i, u_{i+1}) \in E(P_n)$, $1 \leq i \leq n-1$ and the vertex set of K_m by $V(K_m) = \{v_1, v_2, \ldots, v_m\}$ with $(v_i, v_j) \in E(K_m), 1 \leq i, j \leq m, i \neq j$ then the vertex set of $P_n\Box K_m$ is $V(P_n\Box K_m) = \{(u_i, v_j) : 1 \leq i \leq n, 1 \leq j \leq m\}$.

Theorem 4. *Let $m \geq 3$ and $n \geq 4$ be integers. Then*

$$\mathrm{rn}(P_n\Box K_m) := \begin{cases} \frac{mn^2 - 2n + 2}{2}, & \text{if } n \text{ is even,} \\ \frac{mn^2 - 2n + m + 2}{2}, & \text{if } n \text{ is odd.} \end{cases} \tag{8}$$

Proof. We consider the following two cases.

Case-1: n is even. In this case, we set the subgraph induced by vertex set $\{(u_{n/2}, v_1), (u_{n/2}, v_2), \ldots, (u_{n/2}, v_m), (u_{n/2+1}, v_1), (u_{n/2+1}, v_2), \ldots, (u_{n/2+1}, v_m)\}$ of $P_n\Box K_m$ as L_0 then $\mathrm{diam}(L_0) = k = 2$ and the maximum level in $P_n\Box K_m$ is $h = n/2 - 1$. Note that $p = mn$ and $\sum_{i=0}^{h} |L_i| i = mn(n-2)/2$. Substituting these all in (4), we obtain $\mathrm{rn}(P_n\Box K_m) \geq (mn^2 - 2n + 2)/2$. In fact, this lower bound is the actual value for $\mathrm{rn}(P_n\Box K_m)$ and for that, it is enough to give a radio labeling with span equal to this lower bound. Note that the radio labeling given by Kim *et al.* in [6] serve this purpose (the readers are required to understand and adjust with notation matter) which complete the proof.

Case-2: n is odd. In this case, we set the subgraph induced by vertex set $\{(u_{(n+1)/2}, v_1), (u_{(n+1)/2}, v_2), \ldots, (u_{(n+1)/2}, v_m)\}$ in $P_n\Box K_m$ as L_0 then $\mathrm{diam}(L_0) = k = 1$ and the maximum level in $P_n\Box K_m$ is $h = n/2 - 1$. Note that $p = nm$ and $\sum_{i=0}^{h} |L_i| i = m(n^2 - 1)/4$. Substituting these all in (4), we obtain $\mathrm{rn}(P_n\Box K_m) \geq (mn^2 - 2n + m)/2$. Now if possible then assume that $\mathrm{rn}(P_n \Box K_m) = (mn^2 - 2n + m)/2$ then there exist a radio labeling φ of $P_n \Box K_m$ with $\mathrm{span}(\varphi) = (mn^2 - 2n + m)/2$. By Theorem 2, φ induces an ordering $x_0, x_1, \ldots, x_{p-1}$ of $V(P_n \Box P)$ with $0 = \varphi(x_0) < \varphi(x_1) < \cdots < \varphi(x_{p-1}) = \mathrm{span}(\varphi)$ which satisfies (a), (b) and (c) of Theorem 2. Let $L = \{(u_1, v_1), (u_1, v_2), \ldots, (u_1, v_m)\}$, $C = \{(u_{(n+1)/2}, v_1), (u_{(n+1)/2}, v_2), \ldots, (u_{(n+1)/2}, v_m)\}$ and $R = \{(u_n, v_1), (u_n, v_2), \ldots, (u_n, v_m)\}$. Since $|L| = |R| = |C|$ and φ satisfies conditions (a), (b) and (c) of Theorem 2, there exist a vertex $x_t \in L$ or R such that $d(x_{t-1}, L_0) + d(x_t, L_0) > (n-1)/2$ and $d(x_t, L_0) + d(x_{t+1}, L_0) > (n-1)/2$. Without loss of generality assume that $d(x_{t-1}, L_0) + d(x_t, L_0) \geq d(x_t, L_0) + d(x_{t-1}, L_0)$. Since an ordering $x_0, x_1, \ldots, x_{p-1}$ of $V(P_n \Box K_m)$ satisfies condition (a) of Theorem 2, it is clear that $d(x_{t-1}, x_{t+1}) = d(x_{t-1}, L_0) - d(x_{t+1}, L_0) + 1$. Now consider $\varphi(x_{t+1}) - \varphi(x_{t-1}) = \varphi(x_{t+1}) - \varphi(x_t) + \varphi(x_t) - \varphi(x_{t-1}) = n + 1 - d(x_{t+1}, L_0) - d(x_t, L_0) - 1 + n + 1 - d(x_t, L_0) - d(x_{t-1}, L_0) - 1 = 2n - (d(x_{t-1}, L_0) - d(x_{t+1}, L_0) + 1) - 2(d(x_t, L_0) + d(x_{t+1}, L_0) - 1/2) \leq 2n - d(x_{t-1}, x_{t+1}) - 2(n/2 + 1 - 1/2) = n - 1 - d(x_{t-1}, x_{t+1}) < n + 1 - d(x_{t-1}, x_{t+1})$, a contradiction with φ is a radio labeling. Hence, $\mathrm{rn}(P_n \Box K_m) \geq (mn^2 - 2n + m + 2)/2$. In fact, this lower bound is the actual value for $\mathrm{rn}(P_n \Box K_m)$ and for that, it is enough to give a radio labeling with span equal to this lower bound. Again note that the radio labeling given by Kim et al. in [6] serve this purpose (the readers are required to understand and adjust with notation matter) which complete the proof.

4   Concluding Remarks

In [5], Das et al. gave a technique to find a lower bound for the radio k-coloring of graphs which also cover the case of radio labeling when $k = \mathrm{diam}(G)$. In [5], authors fixed a vertex as L_0 when $\mathrm{diam}(G)$ is even and a maximal clique C of G as L_0 when $\mathrm{diam}(G)$ is odd. We remark that our approach is more useful to find a better lower bound for the radio number of graphs than one given by Das et al. in [5] and this can be realize for the graph $P_n \Box P$. Note that in case of $P_n \Box P$, if we fix a vertex or a maximal clique then there is a large gap between a lower bound for the radio number of $P_n \Box P$ and the actual value of radio number of $P_n \Box P$. Moreover, a necessary and sufficient condition to achieve the lower is useful to determine the exact radio number of graphs. It is also possible to determine the existing radio number for complete graph K_n, wheel graph W_n, n-gear graph G_n, paths P_n using Theorems 1 and 2.

Finally, we suggest the following further work in direction of present research work.

1. Find graphs G such that $\mathrm{rn}(P_n \Box G)$ can be determine using Theorems 1 and 2 (we suggest star graph, wheel graph etc. as G).
2. Find graphs G_1 and G_2 such that $\mathrm{rn}(G_1 \Box G_2)$ can be determine using Theorems 1 and 2.
3. More generally, find graphs G other than trees whose radio number can be determine using Theorems 1 and 2.

Acknowledgements. I want to express my deep gratitude to anonymous referees for kind comments and constructive suggestions.

References

1. Bantva, D., Vaidya, S., Zhou, S.: Radio number of trees. Electron. Notes Discret. Math. **48**, 135–141 (2015)
2. Bantva, D., Vaidya, S., Zhou, S.: Radio number of trees. Discret. Appl. Math. **217**, 110–122 (2017)
3. Chartrand, G., Erwin, D., Harary, F., Zhang, P.: Radio labelings of graphs. Bull. Inst. Combin. Appl. **33**, 77–85 (2001)
4. Chartrand, G., Erwin, D., Zhang, P.: A graph labeling suggested by FM channel restrictions. Bull. Inst. Combin. Appl. **43**, 43–57 (2005)
5. Das, S., Ghosh, S., Nandi, S., Sen, S.: A lower bound technique for radio k-coloring. Discret. Math. **340**, 855–861 (2017)
6. Kim, B.M., Hwang, W., Song, B.C.: Radio number for the product of a path and a complete graph. J. Comb. Optim. **30**(1), 139–149 (2015)
7. Liu, D.: Radio number for trees. Discret. Math. **308**, 1153–1164 (2008)
8. West, D.B.: Introduction to Graph Theory. Prentice-Hall of India, New Delhi (2001)

Improved Descriptional Complexity Results on Generalized Forbidding Grammars

Henning Fernau[1], Lakshmanan Kuppusamy[2], Rufus O. Oladele[3], and Indhumathi Raman[4(✉)]

[1] Abteilung Informatikwissenschaften, CIRT, Fachbereich 4, Universität Trier, 54286 Trier, Germany
fernau@uni-trier.de

[2] School of Computer Science and Engineering, VIT, Vellore 632014, India
klakshma@vit.ac.in

[3] Department of Computer Science, University of Ilorin, P.M.B. 1515, Ilorin, Nigeria
roladele@unilorin.edu.ng

[4] Department of Applied Mathematics and Computational Sciences, PSG College of Technology, Coimbatore 641004, India
indhumathi.raman@gmail.com

Abstract. In formal language theory, if a grammar consists of rules of the type (r, F_r), where r is a context-free rule and F_r is an associated finite set of strings called the forbidding set, such that r can be applied to a string only if none of the strings in F_r is present in the string, then the grammar is said to be a generalized forbidding (GF) grammar. There are four main parameters that describe the size of a GF grammar, namely, (i) d, the maximum length of strings in the forbidding sets, (ii) i, the maximum cardinality of the forbidding sets, (iii) n, the number of nonterminals used in the grammar, and (iv) c, the number of rules with nonempty forbidding set. The family of languages described by a GF grammar of size (at most) (d, i, n, c) is denoted by $\mathrm{GF}(d, i, n, c)$. We study for what sizes of generalized forbidding grammars one can obtain the computational power of Turing machines. We specifically show this result for sizes $(2, 6, 8, 6)$, $(2, 5, 8, 8)$, $(2, 4, 9, 9)$ and $(2, 3, 20, 18)$. These results improve on previous results on the descriptional complexity of generalized forbidding grammars.

Keywords: Descriptional complexity in formal languages ·
Semi-conditional grammars · Generalized forbidding grammars ·
Computational completeness

1 Introduction

Generalized forbidding (GF) grammars were first introduced and investigated by Meduna [12]. These grammars are based on context-free rules, each of which may

© Springer Nature Switzerland AG 2019
S. P. Pal and A. Vijayakumar (Eds.): CALDAM 2019, LNCS 11394, pp. 174–188, 2019.
https://doi.org/10.1007/978-3-030-11509-8_15

be regulated with finitely many forbidding strings. A nonterminal is rewritten by a rule only if none of the forbidding strings of the rule occur in the current sentential form; other than this, these grammars work just like context-free grammars. Although these grammars were suggested rather with a mathematical motivation, as a further variant of regulated rewriting, there are recent papers that underpin the linguistic interest of finding and dealing with forbidden substrings for language processing; see [14,16].

In this paper, we are looking into descriptional complexity issues of GF grammars. Recall that the class of recursively enumerable (RE) languages is the class of languages that can be accepted by Turing machines. It was proved in [12] that every RE language can be generated by some GF grammar whose forbidding strings have length at most two; the corresponding language class is also denoted as GF(2). Notice that GF(1) is also known as the class of languages described by forbidden random context grammars, equivalent to ordered grammars, known to be properly included in RE [3,4]. Similar considerations concerning different grammatical formalisms are abundant.

Besides an upper-bound on the length of the forbidding strings, known as the *degree* of a GF grammar, its *index*, which is an upper-bound on the number of forbidding strings per rule, was also considered in [10,11]. Further, the number of nonterminals and the number of conditional rules are considered as measures of descriptional complexity of GF grammars, where a rule is said to be *conditional* if the set of forbidding strings associated to this rule is non-empty. $GF(d, i, n, c)$ denotes the class of languages that can be described by GF grammars of degree d, of index i, with n nonterminals and at most c conditional rules. We only mention [1–3,5–7,9,15,19] as classical and more recent examples on this line of research in descriptional complexity, which can be traced back to Shannon [17] who discussed the question of how many states a Turing machine should have in order to describe RE. Notice that GF grammars can be considered as a special case of (simple) semi-conditional, (S)SC, grammars of degree (i, j) with $i = 0$. This relates and embeds our results into the existing literature in that area.

The usual ambition within descriptional complexity is to see how small these four parameters (in our case) could be for a class of GF grammars, maintaining the property that the corresponding language class still coincides with RE. This ambition fits into a general research line in combinatorial mathematics that tries to understand and describe the frontiers of certain formalisms. We summarize the previously known and new results in Table 1. Notice that in two cases marked by (∗), we were only able to describe the class of all recursively enumerable languages that contain the empty word, which we denote by RE^λ. In the literature on formal languages ([3, p. 12] is one example), this specialty is often neglected. Our improvements are easily read off from Table 1. Notice that never before, characterizations of RE with GF grammars were obtained with index 5 or 3, or with only 7 nonterminals or only 6 conditional rules.

Since our results rely on specific properties of two types of normal forms originally derived by Geffert [8], we first state these normal forms and also another normal form of type-0 grammars proposed by Masopust and Meduna (in [10]) in the next section. Then, we state and prove our main results concerning the descriptional complexity of GF grammars, as already discussed above. Finally, we give some prospects on future research.

Table 1. RE results on GF grammars: existing ones on the left, new ones on the right

Degree d	Index i	# Nonterminals n	# Cond. rules c	Reference (Year)	Degree d	Index i	# Nonterminals n	# Cond. rules c	Reference
2				[12] (1990)	2	6	8	6	Theorem 1
2		15	13	[13] (2003)	2	5	8	8	Theorem 2
2	9	10	8	[11] (2007)	2	5	7	10	*Theorem 3
2	6	9	10	[10] (2007)	2	4	9	9	Theorem 4
2	4	10	11	[10] (2007)	2	4	8	11	*Theorem 5
2		8	9	[10] (2007)	2	3	20	18	Theorem 6

2   Preliminaries and Definitions

It is assumed in this paper that the reader is familiar with the fundamentals of language theory. Let Σ^* denote the free monoid generated by a set Σ (called alphabet) under the operation of concatenation, where λ denotes the unit of Σ^*, also called the empty string. Any element of Σ^* is also called a *word* or *string* over Σ. Recall that $\Sigma^* = \bigcup_{m \geq 0} \Sigma^m$ and if $w \in \Sigma^*$, then there is a unique m such that $w \in \Sigma^m$; this number m is also known as the *length* of w and is also written as $|w|$. String y is a *subword* of w if there are strings x, z such that $w = xyz$. Let $sub(w) = \{y \mid y \text{ is a subword of } w\}$.

2.1   Generalized Forbidding Grammars

We are now introducing the central concept of this paper, GF grammars.

Definition 1. *A generalized forbidding (GF) grammar is a quadruple $G = (V, T, P, S)$; V is the total alphabet, $T \subset V$ is the terminal alphabet, $S \in V \setminus T$ is the start symbol and P is a set of rules of the form $(A \rightarrow x, F)$, where $A \in V \setminus T$, $x \in V^*$, $F \subseteq V^+$; $|F|, |P| < \infty$.*

We partition the rules set P of a GF grammar into P_u and P_c such that $P_u = \{(A \rightarrow x, F) \in P \mid F = \emptyset\}$ and $P_c = \{(A \rightarrow x, F) \in P \mid F \neq \emptyset\}$. We call P_u the set of *unconditional* rules and P_c the set of *conditional* rules. For simplicity, we also write $(A \rightarrow x)$ instead of $(A \rightarrow x, \emptyset)$. A rule $p = (A \rightarrow x, F)$ *can be applied* (or, is *applicable*) to a word w if and only if $F \cap sub(w) = \emptyset$ and $A \in sub(w)$. If p is applicable to w, then there are strings $r, s \in V^*$ such that $w = rAs$. Moreover, the result of applying p to w is $w' = rxs$. We also say that w (directly) derives w' and denote this by $w \Rightarrow w'$ or $w \Rightarrow_p w'$ or $w \Rightarrow_G w'$ if we want to make the rule p or the grammar G that is used explicit. If rule p is applied $k \geq 2$ times on a string w to obtain w'', then we denote this by $w \Rightarrow_p^k w''$. As usual, the reflexive transitive closure of $\Rightarrow$ is written $\Rightarrow^*$. The language of G, denoted $L(G)$, is defined by $L(G) = \{w \in T^* \mid S \Rightarrow^* w\}$.

For integers $d, i, n, c \geq 0$, the language family $\mathrm{GF}(d, i, n, c)$ is defined as follows: $L \in \mathrm{GF}(d, i, n, c)$ if and only if there is a generalized forbidding grammar $G = (V, T, P, S)$ that satisfies all of the following: (i) $L = L(G)$, (ii) (degree) $d \geq d(G) := \max\limits_{(A \to x, F) \in P} \max\limits_{f \in F} |f|$, (iii) (index) $i \geq i(G) := \max\limits_{(A \to x, F) \in P} |F|$, (iv) (number of nonterminals) $n \geq n(G) := |V \setminus T|$, (v) (number of conditional rules) $c \geq c(G) := |P_c|$.

As an example, we show how the non-context-free cross-dependency language $L_{cd} = \{a^m b^n c^m d^n \mid n, m \geq 1\}$ can be generated by a GF grammar.

Example 1. $L_{cd} = \{a^m b^n c^m d^n \mid n, m \geq 1\} \in \mathrm{GF}(1, 2, 9, 12)$.

Proof. Consider $G = (\{S, \#, \$, \hat{\#}, \hat{\$}, \#', \$', \#'', \$'', a, b, c, d\}, \{a, b, c, d\}, P_u \cup P_c, S)$ where P_u contains the unconditional rule $(S \to \#\$)$ and P_c contains the following twelve rules.

1: $(\# \to a\#', \{\$'\})$ 5: $(\#' \to \hat{\#}, \{\$\})$ 9: $(\#'' \to \hat{\#}, \{\hat{\$}\})$
2: $(\$ \to c\$', \{\#\})$ 6: $(\$' \to \hat{\$}, \{\#, \#'\})$ 10: $(\$'' \to \hat{\$}, \{\#''\})$
3: $(\#' \to \#, \{\$\})$ 7: $(\hat{\#} \to b\#'', \{\$', \$''\})$ 11: $(\#'' \to \lambda, \{\hat{\$}\})$
4: $(\$' \to \$, \{\#', \hat{\#}\})$ 8: $(\hat{\$} \to d\$'', \{\hat{\#}\})$ 12: $(\$'' \to \lambda, \{\hat{\#}, \#''\})$

The reader is encouraged to verify that $L(G) = L_{cd}$.

2.2 Geffert Normal Forms

Recall that the class RE can be characterized by so-called type-0 grammars, being the bottom line of the well-known Chomsky hierarchy. As with GF grammars, a quadruple $G = (N, T, P, S)$ can be used to describe such a grammar; the main difference to GF grammars are the rules collected in P, which are now of the form $\alpha \to \beta$, with $\alpha, \beta \in V^*$, where $V = N \cup T$. This means that in one derivation step, one can replace the substring α by β. In 1991, Geffert introduced several normal forms for type-0 grammars [8], each one with arbitrarily many context-free rules but only a few non-context-free rules and with a few nonterminals. The best known of these normal forms, often referred to as *Geffert normal form* (GNF), uses five non-terminals S, A, B, C, D and two erasing non-context-free rules $AB \to \lambda$ and $CD \to \lambda$. We will call this normal form as $(5, 2)$-GNF, highlighting the number of nonterminals and the number of non-context-free rules in this way, because they are characteristic for the normal forms that Geffert found. The derivation of a type-0 grammar in $(5, 2)$-GNF proceeds in two phases, where the first phase splits into two stages. In phase one, stage one, rules of the form $S \to uSa$ are used, with $u \in \{A, C\}^*$, $a \in T$. In stage two of phase one, rules of the form $S \to uSv$ are used, with $u \in \{A, C\}^*$

and $v \in \{B, D\}^*$. Also, rules of the form $S \to uv$ are available (see [8]) that prepare the transition into phase two, where the two erasing non-context-free rules $AB \to \lambda$ and $CD \to \lambda$ are used exclusively until a terminal string is derived.

Accordingly, a type-0 grammar is said to be in $(4,1)$-GNF if it has exactly four nonterminals S, A, B, C and a single non-context-free erasing rule of the form $ABC \to \lambda$. Geffert has shown in [8] that this normal form is obtained from $(5,2)$-GNF by applying the morphism $A \mapsto AB$, $B \mapsto C$, $C \mapsto A$ and $D \mapsto BC$ to all context-free rules. This implies:

Proposition 1. *The following properties hold for $(4,1)$-GNF grammars:*

1. *If $S \Rightarrow^* w$, then $w \in \{A, AB\}^*\{S, ABC, \lambda\}(\{BC, C\} \cup T)^*$.*
2. *No sentential form derivable in the grammar contains substrings BBB or CA (in fact, after any C no A can occur).*
3. *If $S \Rightarrow^* w$, with $w = w't$, where $w' \in \{A, B, C\}^+$ and $t \in (T\{BC, C\}^*)^*$, then w' contains exactly one occurrence from $\{ABC, AC, ABBC\}$ as a substring. We refer to this substring as the* central part *of w. Notice that only with ABC as central part, possibly $w' \Rightarrow^* \lambda$ as intended in a derivation that yields a terminal string. Hence, only with ABC as central part and with $t \in T^*$, a terminal string could be produced.*
4. *Again, the derivation proceeds in two phases, the first one split into two stages. Only in phase two, a central part will appear. Till then, S separates the central part. To the right of S, some suffix $t \in (T\{BC, C\}^*)^*$ is situated.*

Masopust and Meduna [10] stated another normal form of type-0 grammars derived from GNF, which we call MMNF. We state it in the following proposition for easy reference.

Proposition 2. *Let L be a recursively enumerable language with $L = L(\tilde{G})$, where $\tilde{G} = (\tilde{N}, T, \tilde{P}, \tilde{S})$ is a grammar in $(5,2)$-GNF with $\tilde{N} = \{S, A, B, C, D\}$. Then, there is a grammar $G = (\{S, 0, 1, \$\}, T, P_u \cup \{0\$0 \to \$, 1\$1 \to \$, \$ \to \lambda\}, S)$, with P_u containing only unconditional rules of the form*

- $S \to h(u)Sa$ if $S \to uSa \in \tilde{P}$,
- $S \to h(u)Sh(v)$ if $S \to uSv \in \tilde{P}$,
- $S \to h(u)\$h(v)$ if $S \to uv \in \tilde{P}$,

where $h : \{A, B, C, D\}^* \to \{0, 1\}^*$ is a homomorphism defined by $h(A) = h(B) = 00$, $h(C) = 01$, and $h(D) = 10$, such that $L(G) = L(\tilde{G})$.

Clearly, the derivation of an MMNF grammar also proceeds in two phases: the first phase generates a string $uSvt$, with $u \in \{00, 01\}^*$, $v \in \{00, 10\}^*$ and $t \in T^*$, and is ended by replacing S by $\$$, and the second phase uses $0\$0 \to \$$, $1\$1 \to \$$, terminated by $\$ \to \lambda$. Moreover, if $S \Rightarrow^* w$, then $w \in \{00, 01\}^*(\{S, \$\} \cup \{\xi\$\eta \mid \xi, \eta \in \{0, 1\}\})(\{00, 10\} \cup T)^*$; a string $x\$y$, $x, y \in \{0, 1\}$ is the *central part*, with $x = y$ being intended.

3   Main Results

This section presents the main results of this paper. We present them in an order that first shows an RE characterization with relatively many forbidden strings associated to each conditional rule, but with relatively few nonterminals and conditional rules, then moving on to less and less forbidden strings associated to each conditional rule at the expense of more and more nonterminals and conditional rules. Parts of the proofs of some results are omitted due to space constraints.

Theorem 1. RE $= \mathrm{GF}(2, 6, 8, 6)$.

Proof. Let L be a recursively enumerable language described by a type-0 grammar $G = (\{S, A, B, C\}, T, P_u \cup \{ABC \to \lambda\}, S)$ that is in $(4, 1)$-GNF. Construct the grammar $G' = (\{S', S, \sigma, A, B, C, \#, \$\}, T, P'_u \cup P', S')$, where S' is the start symbol, P'_u contains an unconditional rule $(S' \to \sigma S \sigma)$, plus all rules $(S \to g(x))$ when $S \to x \in P_u$, where g maps all terminal symbols a to $\sigma a \sigma$ and acts as identity on all nonterminals; P' contains the following six conditional rules:

1: $(B \to \$\ ,\ \{S, AC, BB, \$, \#\})$ 3: $(C \to \$\#,\ \{S, AC, BB, \$\#, C\#, \#C\})$ 5: $(\# \to \lambda,\ \{\$, \#A, C\#\})$
2: $(A \to \#\$,\ \{S, AC, BB, \#\})$ 4: $(\$ \to \lambda\ ,\ \{A\$, \$C, \$\sigma, \sigma\$, B\$, \$B\})$ 6: $(\sigma \to \lambda,\ \{A, B, C, \#, \$, S\})$

Let us first show that $L(G) \subseteq L(G')$ by exhibiting how each rule from G can be simulated by some rule(s) from G'. Recall that in Phase one, only context-free rules are used in G which clearly correspond to the listed unconditional rules, such that $S \Rightarrow^n_G \alpha S \beta$ implies that $S' \Rightarrow^{n+1}_{G'} \sigma \alpha S g(\beta) \sigma$. Then, we apply $S \to uv$ in G to get rid of the nonterminal S. This way, we enter Phase two of the Geffert normal form grammar. This can be trivially simulated by applying $(S \to uv)$ in G'. Now, applying $ABC \to \lambda$ on $\alpha ABC \beta t$, $\alpha \in \{A, AB\}^*$, $\beta \in \{BC, C\}^*$, is simulated as follows in G':

$$\sigma \alpha ABC \beta g(t) \sigma \Rightarrow_1 \sigma \alpha A\$C \beta g(t) \sigma \Rightarrow_2 \sigma \alpha \#\$\$C \beta g(t) \sigma \Rightarrow_3$$
$$\sigma \alpha \#\$\$\$\# \beta g(t) \sigma \Rightarrow^3_4 \sigma \alpha \#\# \beta g(t) \sigma \Rightarrow^2_5 \sigma \alpha \beta g(t) \sigma.$$

This shows that each rule of G can be simulated by a certain sequence of rule applications of G'. Finally, after having derived a string over $(T \cup \{\sigma\})^*$ that corresponds to a terminal string $w \in L(G)$, this terminal string can be obtained by applying Rule 6 repeatedly. By induction on the number of derivation steps of G, $L(G) \subseteq L(G')$ follows.

Now, we proceed to show the reverse inclusion, namely $L(G') \subseteq L(G)$. Consider a string w that is derivable from S' that contains an occurrence of the nonterminal S. By induction, it is easy to see that $w = \sigma w' \sigma$. Moreover, after applying $(S' \to \sigma S \sigma)$ to the start symbol S' of G', only context-free rules of the original grammar G are simulated in a straightforward manner. Let $g' : (T \cup \{\sigma\})^* \to T^*$ be the morphism defined as the identity on T and mapping σ to λ, i.e., it is somewhat the inverse of g. Hence, more precisely, the variation $g'(w') = g'(w)$ is derivable in G, where we tacitly extended g' to act as identity on further nonterminal symbols. By Proposition 1, it follows that $w' = \alpha S \beta g'(t)$, for some $\alpha \in \{A, AB\}^*$, $\beta \in \{B, BC\}^*$, $t \in (\{\sigma\}T\{\sigma\}\{BC, C\}^*)^*$. However, when G is assumed to produce a terminal string, actually $t \in (\{\sigma\}T\{\sigma\})^*$. Notice that we cannot erase σ prematurely, as S is always present, nor can any other rule from P' be applied. If we decide to apply $(S \to uv)$, we arrive at a string of the form $w = \sigma\alpha uv\beta t$ in G', while $g'(w)$ is the corresponding sentential form in G. This corresponds to the transition between Phase 1 and Phase 2 of the original grammar given in $(4,1)$-GNF. Now, the rules of P' become applicable.

Consider any string $w = \sigma\alpha\beta t$ with $\alpha \in \{A, AB\}^*$, $\beta \in \{BC, C\}^*$, $t \in (\{\sigma\}T\{\sigma\}\{BC, C\}^*)^*\{\sigma\}$, that is derivable in G'. By induction, these are exactly the type of strings that occur in derivations of G' after applying $(S \to uv)$ and before applying Rule 6. By induction again, we can assume that $\alpha\beta g'(t)$ is derivable in G. Moreover, actually $t \in (\{\sigma\}T\{\sigma\})^*\{\sigma\}$ if this sentential form should ever yield a terminal string, as we will see. All this is clearly true for strings that show up immediately after applying $(S \to uv)$, which proves the induction basis. In the long version, we discuss several cases of α, β and t. To simplify these arguments, we state and prove some claims.

Claim (I): If $w \Rightarrow_3 \hat{w}$ or $w \Rightarrow_1 w' \Rightarrow_3 \hat{w}$, then there is no $x \in T^*$ with $\hat{w} \Rightarrow^*_{G'} x$.

Namely, consider $w \Rightarrow_3 \hat{w}$ first. As $\$, \# \in sub(\hat{w})$, Rules 1, 2, 5 and 6 are blocked. Further Rule 3 cannot be applied on $\hat{w}$ since $\$\#$ blocks its application. $\hat{w} \Rightarrow_4 \tilde{w}$ is only possible if to the left of the occurrence of C that was replaced by $\$\#$, none of the symbols A, B, σ appears. As $w = \sigma\alpha\beta t$ with $\alpha \in \{A, AB\}^*$, $\beta \in \{BC, C\}^*$, $t \in (\{\sigma\}T\{\sigma\}\{BC, C\}^*)^*\{\sigma\}$, this means that $C\$\# \in sub(\hat{w})$; hence $C\# \in sub(\tilde{w})$. This blocks all rules to be applied in $\tilde{w}$, certifying the claim.

If $w \Rightarrow_1 w' \Rightarrow_3 \hat{w}$, then either actually the conversion from w to $\hat{w}$ can be explained by a replacement of the substring BC by $\$\$\#$, or not. If not, then the argument given in the previous paragraph literally applies, it might only happen that Rule 4 blocks even earlier, as we have now two occurrences of $\$$ that might form forbidden contexts. If $\$\$\# \in sub(\hat{w})$, then again quite a similar analysis applies, with the only difference that Rule 4 can be applied twice.

Claim (II): If $w \Rightarrow_2 \hat{w}$ or $w \Rightarrow_1 w' \Rightarrow_2 \hat{w}$, then there is no $x \in T^*$ such that $\hat{w} \Rightarrow^*_{G'} x$, assuming that Rule 3 does not apply on $\hat{w}$.

The argument for this claim is very similar to the one of Claim (I), exchanging the roles of A and C, as well as of left and right. So, if $w \Rightarrow_3 \hat{w} \Rightarrow \tilde{w}$, then the last derivation step is possible only if $\#\$A \in sub(\hat{w})$ and hence $\#A \in sub(\tilde{w})$. This blocks all rules as claimed.

The claims are essential to an inductive argument showing $L(G') \subseteq L(G)$.   □

While the variation of normal forms introduced by Geffert himself are rather standard when it comes to construct proofs showing computational completeness in this area of research, alternatives like the normal form designed by Masopust and Meduna are far less employed. Each of these normal forms has its own pros and cons, for instance, MMNF has a central part clearly marked by the special symbol $\$$. To give an idea how this helps getting our results, we selected the following theorem to be presented with all details in the main text body.

Theorem 2. RE $= \mathrm{GF}(2, 5, 8, 8)$.

Proof. Let L be a recursively enumerable language. Then, by Proposition 2 there is a grammar $G = (\{S, 0, 1, \$\}, T, P_u \cup \{0\$0 \to \$, 1\$1 \to \$, \$ \to \lambda\}, S)$ in MMNF such that $L = L(G)$. Construct the grammar $G' = (\{S', S, 0, 1, \sigma, \$, \#, \dagger\}, T, P'_u \cup P', S')$. Here, P'_u contains rules of the forms $(S' \to \sigma S \sigma)$ and $(S \to g(x))$ if $S \to x \in P_u$, where $g : T^* \to (T \cup \{\sigma\})^*$ is the morphism given by $a \mapsto \sigma a \sigma$ for $a \in T$, slightly extended by assuming that it acts as the identity on the nonterminals. Moreover, P' contains the following eight conditional rules:

1: $(0 \to \#,\ \{\$1, \#, \dagger, \$\sigma, S\})$ 4: $(\dagger \to \lambda,\ \{\#, \sigma\dagger, S\})$ 7: $(\$ \to \lambda,\ \{0, 1, \dagger, S\})$
2: $(0 \to \dagger,\ \{1\$, 0\$, \dagger, \sigma\$, S\})$ 5: $(1 \to \#,\ \{\$0, \#, \dagger, \$\sigma, S\})$ 8: $(\sigma \to \lambda,\ \{\dagger, \$, S\})$
3: $(\# \to \lambda,\ \{\$0, \$1, \$\#, \dagger\#, S\})$ 6: $(1 \to \dagger,\ \{1\$, 0\$, \dagger, \sigma\$, S\})$

The intended derivation is as follows. We simulate an application of $0\$0 \to \$$, yielding $u0\$0vt \Rightarrow uvt$, $u \in \{0, 1\}^*$, $v \in \{0, 1\}^*$, $t \in T^*$, by

$$\sigma u0\$0vg(t)\sigma \Rightarrow_1 \sigma u\#\$0vg(t)\sigma \Rightarrow_2 \sigma u\#\$\dagger vg(t)\sigma \Rightarrow_3 \sigma u\$\dagger vg(t)\sigma \Rightarrow_4 \sigma u\$vg(t)\sigma.$$

where $g : T^* \to (T \cup \{\sigma\})^*$ is the morphism given by $a \mapsto \sigma a$, so that the string $g(t)\sigma$ always starts with σ. Similarly, we simulate $1\$1 \to \$$ as

$$\sigma u1\$1vg(t)\sigma \Rightarrow_5 \sigma u\#\$1vg(t)\sigma \Rightarrow_6 \sigma u\#\$\dagger vg(t)\sigma \Rightarrow_3 \sigma u\$\dagger vg(t)\sigma \Rightarrow_4 \sigma u\$vg(t)\sigma.$$

After matching all 0's and 1's in this way, we are left with a string $\sigma\$g(t)\sigma$ (via G') that corresponds to $\$t$ in G, finally leading to $t \in T^*$. As this string does not contain $0, 1, \dagger$ nor $\#$, we can (now) delete $\$$ and then all occurrences of σ. This proves $L(G) \subseteq L(G')$ by an easy induction argument.

We are now arguing why $L(G) \supseteq L(G')$. The correctness of the simulation of Phase 1 is pretty straightforward. If we decide to apply $(S \to u\$v)$, we arrive at a string of the form $\sigma u\$v\sigma$, while $u\$g'(v)$ is the corresponding sentential form in G. This corresponds to the transition between Phase 1 and Phase 2 of the original grammar given in MMNF. Now, rules from P' become applicable and we have to study this situation in more details.

Consider any string $\sigma\alpha\$\beta t$ with $\alpha \in \{0, 1\}^*$, $\beta \in \{0, 1\}^*$, $t \in (T \cup \{\sigma, 00, 10\})^*$, that is derivable in G'. By induction, these are exactly the type of strings that occur in derivations of G' after applying $(S \to u\$v)$ and before applying Rule 7. By induction again, we can assume that $\alpha\$\beta g'(t)$ is derivable in G. Moreover, we can assume by induction that $t \in (\{\sigma\}T\{\sigma\} \cup \{00, 10\})^*\{\sigma\}$ such that t starts with σ. All this is clearly true for strings that show up immediately after applying $(S \to u\$v)$, which proves the induction basis.

There are a couple of cases that could happen. If $\alpha = \beta = \lambda$ and if t contains no symbols from $\{0, 1\}$, then only Rule 7 is applicable, leading to t. Recall that t might still contain some nonterminals, occurrences of σ. These can be deleted by using Rule 8, possibly repeatedly, hence finally arriving at the terminal string $g'(t)$ that could be likewise derived from $\$g'(t)$ by applying $\$ \to \lambda$ from G.

We analyze how a derivation is stuck up when u or v is empty in the following. Consider now a string $w = \sigma\$vt$, $v \in \{0, 1\}^*$, $g'(t) \in (T \cup \{0, 1\})^*$, that might be generated by G' such that $\$vg'(t)$ can be generated by G and such that $g'(vt)$ contains at least one occurrence from $\{0, 1\}$. As $\sigma\$$ is forbidden in Rules 2 and 6, it is possible only to apply either Rule 1 or 5 on w. If we apply the former, then some 0 in vt is changed into $\#$ upon $w \Rightarrow_1 w'$. Also, $\$$ is followed by v and the first symbol of v can be one of $\{0, \#\}$ now, as $\$1$ and $\$\sigma$ were checked upon applying Rule 1. Hence w' contains one of $\{\$0, \$\#\}$ as a subword and since both of these are forbidden in Rule 3, the symbol $\#$ cannot be deleted and hence the derivation is stuck, as no other rule is applicable either. If we want to apply Rule 5 on w, then v has to begin with 1, since $\$0$ and $\$\#$ are forbidden in Rule 5. By a symmetric argument, $w \Rightarrow_5 w'$ means that w' contains one of $\{\$1, \$\#\}$ as a subword and since both of these are forbidden in Rule 3, the symbol $\#$ cannot be deleted and hence the derivation is stuck. Notice that also the corresponding sentential form $\$vg'(t)$ of G would not allow for any further derivation steps.

Similarly, we can discuss $w = \sigma u \$t$ for $u \in \{0,1\}^+$, $t \in (T \cup \{\sigma, 00, 10\})^*$, that is derivable in G'. Recall that t starts with σ: the presence of $\$\sigma$ in w blocks the application of Rules 1 and 5. As u ends with 0 or 1, w contains $0\$$ or $1\$$ as a substring, blocking Rules 2, 6, 7 and 8. Trivially, Rules 3 and 4 are not applicable, either, so that the derivation is blocked.

We now discuss $w = \sigma u \$vt$ for $u, v \in \{0,1\}^+$, $t \in (T \cup \{\sigma, 00, 10\})^*$, that is derivable in G'. Recall that t starts and ends with σ. Notice that Rules 3, 4, 7 and 8 are never applicable on w.

<u>Case 1:</u> $u = u'0$, $v = 0v'$. This blocks Rules 2, 5 and 6. Hence, we have to apply Rule 1. If we apply it to u' or $v't$, then none of the formerly blocking situations is resolved, and moreover Rule 1 becomes inapplicable, so that no further continuation is possible. First consider $w \Rightarrow w' = \sigma u \$\# v't$. The substrings $\$\#$ and $0\$$ block any further rule applications. Alternatively, consider $w \Rightarrow w' = \sigma u' \# \$vt$. Although Rule 1 is disabled now, Rules 2 and 6 are no longer blocked. However, if we apply any of these to replace a symbol from $u'v't$ by $\dagger$, then the derivation is blocked. Therefore, $w' \Rightarrow w''$ enforces applying Rule 2 to the occurrence of 0 next to $\$$, i.e., $w'' = \sigma u' \# \$ \dagger v't$. Now, Rule 3 is the only applicable rule, leading us to $w''' = \sigma u' \$ \dagger v't$. The presence of $\dagger$ blocks Rules 1, 2, 5, 6, 7, 8. As Rule 3 is clearly not applicable either, Rule 4 must be applied, leading us to $w'''' = \sigma u' \$v't$. Clearly, $u' \$v'$ belongs to $\{0,1\}^* \{\$\} \{0,1\}^*$, and also $w \Rightarrow_G w''''$ by using the rule $0\$0 \to \$$; this proves the induction step for this subcase.
<u>Case 2:</u> $u = u'1$, $v = 1v'$. By symmetry, this case is very similar to Case 1.
<u>Case 3:</u> $u = u'0$, $v = 1v'$. Applying Rule 5 is enforced, but then we are stuck.
<u>Case 4:</u> $u = u'1$, $v = 0v'$ is symmetric to Case 3. $\square$

In the following theorem, we use a special trick by re-using the terminal companion symbol σ as a start symbol, but otherwise keeping the construction of the previous theorem. Of course, this idea runs into the problem that we might re-start the grammar, which is not intended. This is why the start rule is now conditional. Yet, it might be still possible that we start (unintendedly) with other rules with left-hand side σ. As we cannot avoid this with forbidden context alone, we make sure that then the empty word is generated. This explains why we have (only) a characterization of RE^λ, i.e., disregarding the empty word. More precisely, instead of the unconditional rule $(S' \to \sigma S \sigma)$ (in P_u), we introduce the conditional rule $(\sigma \to \sigma S\sigma, \{\sigma\sigma, S, \$, \dagger\})$. Moreover, we replace Rule 8 by $8' : (\sigma \to \dagger\#, \{0, 1, \$, S\})$ and $9' : (\# \to \lambda, \{\sigma\})$. For details, refer to the long version.

Theorem 3. $RE^{\lambda} = GF(2, 5, 7, 10)$.   □

Theorem 4. $RE = GF(2, 4, 9, 9)$.

Proof. Let $L \in RE$ with $L \subseteq T^*$ be generated by a type-0 grammar in MMNF $G = (\{S, 0, 1, \$\}, T, P_u \cup P_c, S)$ where $P_c = \{0\$0 \to \$, 1\$1 \to \$, \$ \to \lambda\}$ such that $L = L(G)$ and P_u contains rules of the form as shown in Proposition 2. Construct the grammar $G' = (\{S', S, 0, 1, \sigma, \$, \#, \#', \dagger\}, T, P'_u \cup P', S')$. Here, P'_u contains rules of the forms $(S' \to \sigma S\sigma)$ and $(S \to g(x))$ if $S \to x \in P_u$, where $g : T^* \to (T \cup \{\sigma\})^*$ is the morphism given by $a \mapsto \sigma a\sigma$ for $a \in T$, slightly extended by assuming that it acts as the identity on the nonterminals. Moreover, P' contains the following nine conditional rules:

1: $(0 \to \#', \{\$1, \#, \#', S\})$ 4: $(\dagger \to \lambda, \{\#, \#', \sigma\dagger, \sigma\$\})$ 7: $(\$ \to \lambda, \{0, 1, \#', \dagger\})$
2: $(0 \to \dagger, \{1\$, 0\$, \dagger, S\})$ 5: $(1 \to \#', \{\$0, \#, \#', S\})$ 8: $(\sigma \to \lambda, \{\dagger, \$, S\})$
3: $(\# \to \lambda, \{\$0, \$1, \$\#, \dagger\#\})$ 6: $(1 \to \dagger, \{0\$, 1\$, \dagger, S\})$ 9: $(\#' \to \#, \{\#, \dagger, \sigma\$, \$\sigma\})$

For proof details, we refer to the long version.   □

Similar to Theorem 3, if we impose the idea of introducing the conditional rule $(\sigma \to \sigma S\sigma, \{\sigma\sigma, S, \$, \dagger\})$ instead of the unconditional rule $(S' \to \sigma S\sigma)$ of Theorem 4 and then replace Rule 8 of Theorem 4 by the two conditional rules $8' : (\sigma \to \dagger\#, \{0, 1, \$, S\})$ and $9' : (\# \to \lambda, \{\sigma\})$, we get the following theorem. Again, we arrive at RE^{λ}, not at RE, as we cannot avoid generating λ.

Theorem 5. $RE^{\lambda} = GF(2, 4, 8, 11)$.   □

Theorem 6. $RE = GF(2, 3, 20, 18)$.

Let L be an RE language described by a type-0 grammar G, where G is in $(5, 2)$-GNF. Construct $G' = (\{S', S, Z, Z', A, B, C, D, A', B', C', D', A'', B'', C'', D'', \rhd, \lhd, \blacktriangleleft, \blacktriangleright\}, T, P'_u \cup P', S')$, where P'_u contains an unconditional rule $(S' \to ZSZZ')$ plus all rules $(S \to g(x))$ when $S \to x \in P_u$, where g maps all terminal symbols a to ZaZ and acts as identity on all nonterminals; P' contains the following eighteen conditional rules. In addition, we assume that if $\lambda \in L(G)$, then the empty string is generated by the (additional) rule $S \to Z$ only and that otherwise we are in fact considering the grammar that generates $L(G) \setminus \{\lambda\}$.

1: $(B \to B' \blacktriangleleft B'', \{\blacktriangleleft, ZB, S\})$ 7: $(D'' \to \lambda, \{AD', A'D', CD'\})$ 13: $(\rhd \to \lambda, \{\rhd B', \rhd D', \rhd \blacktriangleleft\})$
2: $(A \to A'' \blacktriangleright A', \{\blacktriangleright, AZ, S\})$ 8: $(C'' \to \lambda, \{C'B, C'B', C'D\})$ 14: $(\lhd \to \lambda, \{A'\lhd, C'\lhd, \rhd\})$
3: $(D \to D' \blacktriangleleft D'', \{\blacktriangleleft, ZD, S\})$ 9: $(A' \to \rhd, \{A'', A'A, A'C\})$ 15: $(\blacktriangleright \to \lambda, \{A', C', \lhd\})$
4: $(C \to C'' \blacktriangleright C', \{\blacktriangleright, CZ, S\})$ 10: $(B' \to \lhd, \{B'', BB', DB'\})$ 16: $(\blacktriangleleft \to \lambda, \{B', D', \blacktriangleright\})$
5: $(B'' \to \lambda, \{C'B', CB', AB'\})$ 11: $(C' \to \rhd, \{C'', C'C, C'A\})$ 17: $(Z \to \lambda, \{S, A, C\})$
6: $(A'' \to \lambda, \{A'D, A'D', A'B\})$ 12: $(D' \to \lhd, \{D'', DD', BD'\})$ 18: $(Z' \to \lambda, \{Z, B, D\})$

The intended simulation of $AB \to \lambda$ works as follows in G' on $uABvt$ of G:

$$ZuABvg(t)ZZ' \Rightarrow_1 uAB' \blacktriangleleft B''vg(t)ZZ' \Rightarrow_2 ZuA'' \blacktriangleright A'B' \blacktriangleleft B''vg(t)ZZ' \Rightarrow_5$$
$$ZuA'' \blacktriangleright A'B' \blacktriangleleft vg(t)ZZ' \Rightarrow_6 Zu \blacktriangleright A'B' \blacktriangleleft vg(t)ZZ' \Rightarrow_9 Zu \blacktriangleright \triangleright B' \blacktriangleleft vg(t)ZZ' \Rightarrow_{10} \cdots$$

A similar derivation simulating $CD \to \lambda$ on $uCDvt$ (of G) within G' is as follows:

$$ZuCDvg(t)ZZ' \Rightarrow_3 ZuCD' \blacktriangleleft D''vg(t)ZZ' \Rightarrow_4 ZuC'' \blacktriangleright C'D' \blacktriangleleft D''vg(t)ZZ' \Rightarrow_7$$
$$ZuC'' \blacktriangleright C'D' \blacktriangleleft vg(t)ZZ' \Rightarrow_8 Zu \blacktriangleright C'D' \blacktriangleleft vg(t)ZZ' \Rightarrow_{11} Zu \blacktriangleright \triangleright D' \blacktriangleleft vg(t)ZZ' \Rightarrow_{12} \cdots$$

From this point on, both simulations continue in the same way as follows:

$$Zu \blacktriangleright \triangleright \triangleleft \blacktriangleleft vg(t)ZZ' \Rightarrow_{13} Zu \blacktriangleright \triangleleft \blacktriangleleft vg(t)ZZ' \Rightarrow_{14}$$
$$Zu \blacktriangleright \quad \blacktriangleleft vg(t)ZZ' \Rightarrow_{15} Zu \blacktriangleleft vg(t)ZZ' \Rightarrow_{16} Zuvg(t)ZZ'.$$

This shows that $L(G) \subseteq L(G')$.

The forbidden context in Rule 6 and 8 avoids wrong matchings, i.e., that A matches with D (or its primed versions) and C matches with B (or its primed versions). Also, the forbidden context $A'B$ in Rule 6 (and $C'D$ in Rule 8) ensures that B is encoded to its primed version (D is encoded to its primed version, respectively) before A'' (or C'') is deleted. Otherwise, it would be the case that A (or C) alone is encoded and deleted without changing B (or D), ending up with many A's (or C's) can be deleted without deleting the corresponding B (or D). Also, the forbidden context $A'A$ ($C'C$) in Rule 9 (or Rule 11) ensures that only the central A (or C) is encoded and not some other A (or C). Likewise, the forbidden contexts in Rule 10 (and Rule 12) ensure that the central B (or D) is only encoded and not some other B (or D).

To show the reverse inclusion, we consider a string w that is derivable from S' that contains an occurrence of the nonterminal S. By induction, it is easy to see that $w = Zw'ZZ'$. Moreover, after applying $(S' \to ZSZZ')$ to the start symbol S' of G', only context-free rules of the original grammar G are simulated in a straightforward manner. Let $g' : (T \cup \{Z\})^*Z' \to T^*$ be the morphism defined as the identity on T and mapping σ to λ, i.e., it is somewhat the inverse of g. Hence, more precisely, the variation $g'(w') = g'(w)$ is derivable in G, where we tacitly extended g' to act as identity on further nonterminal symbols. Notice that we cannot erase Z prematurely, as S is always present, nor can any other rule from P' be applied.

Consider any string $w = Z\alpha\beta t$ with $\alpha \in \{A,C\}^*$, $\beta \in \{B,D\}^*$, $t \in (T \cup \{Z,B,D\})^*Z'$, that is derivable in G'. By induction, these are exactly the type of strings that occur in derivations of G' between the application of rule $(S \to uv)$ and Rule 8. By induction again, we can assume that $\alpha\beta g'(t)$ is derivable in G. Moreover, we can assume by induction that $t \in (\{Z\}T\{Z\} \cup \{B,D\})^*\{ZZ'\}$ and that t starts with Z. All this is clearly true for strings that show up immediately after applying $(S \to uv)$, which proves the induction basis. The discussion of cases when α or β is empty is suppressed due to space constraints. We now discuss only when both α, β are nonempty.

Let $\alpha \in \{A,C\}^+$ and $\beta \in \{B,D\}^+$ in $w = Z\alpha\beta t$. The central part could be one of AB, AD, CB, CD. Note that Rules 17 and 18 are not at all applicable, since α and β contains at least one of A, B, C, D. In principle, it could be that t contains occurrences of B or D, but as argued before, because Z will be to the left of such maximal substrings, although we might start touching these, finally we will run into a situation where we have substrings that block any further applications of Rule 1 or Rule 3, this way preventing us from termination. Hence, we can assume in the following case analysis that t contains no occurrences of B or D, i.e., $t \in (\{Z\}T\{Z\})^*\{ZZ'\}$.

<u>Case 1.1</u>: If a non-central nonterminal A in w is changed to $A'' \blacktriangleright A'$ by Rule 2, then we note that this A' is followed by either A or C and hence $A'A$ and $A'C$ is a substring of the resulting string w', which forbids application of Rule 9 to eliminate A' (or to change A' to $\triangleright$) and hence this A' stays forever. More precisely, if $A'A$ was a substring of w', then in order to make Rule 9 applicable again, we would have to change A into A' by Rule 2, which is not possible as $\blacktriangleright$ is present in w', but $\blacktriangleright$ cannot be eliminated using Rule 15, because A' occurs in w'. A similar reasoning applies to the case when $A'C$ is a substring of w'.

<u>Case 1.2</u>: If a non-central nonterminal B in w is changed to $B' \blacktriangleleft B''$ by Rule 1, then we note that this B' is preceded by either B or D and hence BB' and DB' is a substring of the resulting string w' which forbids application of Rule 10 to eliminate B' (or to change B' to $\triangleleft$) and hence this B' stays forever, as a case analysis similar to Case 1.1 applies.

<u>Case 1.3</u>: When a non-central nonterminal C in w is changed to $C'' \blacktriangleright C'$ by Rule 4, the case is similar to Case 1.1 (see the forbidden contexts $C'C$, $C'A$ in Rule 11).

<u>Case 1.4</u>: When a non-central nonterminal D in w is changed to $D' \blacktriangleleft D''$ by Rule 3, the case is similar to Case 1.2 (see the forbidden contexts $D'D$, BD' in Rule 12).

Also, note that if Rule 1 is applied, then Rule 3 cannot be applied (until $\blacktriangleleft$ is present) and vice versa. Similarly, if Rule 2 is applied, then Rule 4 cannot be applied (until $\blacktriangleright$ is present) and vice versa. Due to the above arguments, only the nonterminals of the central part can be replaced by Rules 1 to 4 so as to have a fruitful derivation. We omit the discussion when the central part is AD or CB due to space constraints; it may be easily noted that the derivation is stuck in both the cases. Hence, we discuss in the following when the central part is AB or CD.

When the central part is AB, $w = Z\alpha\beta t$ implies $w = Z\alpha'AB\beta't$. We could apply Rules 1 and 2 and their order of application does not really matter here. Let us clarify first why this is the only potentially fruitful derivation. If we apply Rule 1 on w first, we arrive at $w' = Z\alpha'AB' \blacktriangleleft B''\beta't$. Then Rules 4 and 3 are applicable only on non-central C and D and by an analysis similar to Cases 1.3 and 1.4, respectively, the derivation is stuck. The presence of AB' in the central part forbids the application of Rule 5. The presence of B'' and B' forbids applying Rules 10 and 16, respectively. Other rules are clearly inapplicable. The only applicable rule is hence Rule 2. If we do not apply Rule 2 to the central part, we run into similar problems as discussed above. If we apply Rule 2 on

w first, then we arrive at $Z\alpha'A'' \blacktriangleright A'B\beta't$. By a similar reasoning as above, Rules 3 and 4 lead to blocked derivation. The presence of $A'B$ in the central part forbids applying Rule 6. The presence of A'' and A' further forbid applying Rules 9 and 15. Other rules are inapplicable since there their left-hand sides are absent. The only applicable rule is hence Rule 2 and the application must take place in the central part.

Hence on applying Rules 1 and 2 (irrespective of the order) to w, we obtain $w_2 = Z\alpha'A'' \blacktriangleright A'B' \blacktriangleleft B''\beta't$. Now, Rules 1 through 4 and Rules 7 through 18 are not applicable, either due to forbidden contexts (black triangles or primed and double-primed nonterminals) or because the left-side nonterminals are absent. Thus, the rules that are applicable on w_2 are Rules 5 and 6. The following application of rules (discussed in case A and B) could be done in a parallel or in a sequential manner. However, both has to be done. Other rules are inapplicable.

<u>Case A</u>: If we apply Rule 6 on w_2, then A'' is deleted yielding $w_2' = Z\alpha' \blacktriangleright A'B' \blacktriangleleft B''\beta't$ and then on applying Rule 9 (all others except Rule 5 are inapplicable, which we defer to the next case), A' is replaced by $\triangleright$ thereby resulting in $w_2'' = Z\alpha' \blacktriangleright \triangleright B' \blacktriangleleft B''\beta't$. If the sequence of derivations discussed in this case is applied first, then we continue with the sequence of derivations listed in case B which yields $w_5 = Z\alpha' \blacktriangleright \triangleright\triangleleft \blacktriangleleft \beta't$.

<u>Case B</u>: Applying Rule 5 on w_2 deletes B'', resulting in $w_3 = Z\alpha'A'' \blacktriangleright A'B' \blacktriangleleft \beta't$. All rules except Rules 6 and 10 are inapplicable. The former is discussed in the previous case A. Applying Rule 10, we have $w_4 = Z\alpha'A'' \blacktriangleright A'\triangleleft \blacktriangleleft \beta't$. If the sequence of derivations discussed in this case is applied first, then we continue with case A which yields $w_5 = Z\alpha' \blacktriangleright \triangleright \triangleleft \blacktriangleleft \beta't$.

Hence, Cases A and B have converged again, inevitably leading to w_5. Now, it is easy to see that Rules 1 through 12 are inapplicable. The symbol $\triangleright$ or $\blacktriangleright$ forbids applying Rules 14, 15 and 16. The only applicable rule is number 13 which eliminates $\triangleright$, yielding $w_6 = Z\alpha' \blacktriangleright \triangleleft \blacktriangleleft \beta't$. The only rule that is applicable is Rule 14 which eliminates $\triangleleft$ yielding $w_7 = Z\alpha' \blacktriangleright\blacktriangleleft \beta't$. Proceeding in a more deterministic way, we are forced to apply Rules 15 and 16 (in order) which deletes the symbols $\blacktriangleright$ and $\blacktriangleleft$, respectively, yielding $w_8 = Z\alpha'\beta't$. The sequence of derivations that transforms w to $w_8 = Z\alpha'\beta't$ corresponds to applying the erasing rule $AB \to \lambda$.

When the central part is CD, the case follows in a symmetric way to the above case, thereby simulating $CD \to \lambda$ correctly.

Our inductive arguments show that we have to simulate the derivation of G in the intended manner, so that $L(G') \subseteq L(G)$ follows.   □

4   Future Research

From a mathematical point of view, it is of course interesting to investigate further possibilities to shrink the resources like the number of nonterminals or the number of conditional rules while maintaining characterizations of RE. Yet, from a more applied point of view, it would be at least as interesting to find nice classes of formal languages that can be described, for instance, by generalized forbidding grammars with certain restrictions, and that have good parsing properties. For a proper understanding, also developing combinatorial lemmas as presented in [18] for $d = 1$ (disallowing erasing rules) would be helpful for small parameter settings. Also, looking at Theorem 6, one concrete generalized question would be $GF(2, 2, *, *) \neq RE$? Notice that in formal language theory, proofs that show that certain languages cannot be described by some formalism often have quite a combinatorial flavor. Possibly, proof patterns as found in extremal combinatorics could help getting stronger results.

References

1. Csuhaj-Varjú, E., Kelemenová, A.: Descriptional complexity of context-free grammar forms. Theoret. Comput. Sci. **112**(2), 277–289 (1993)
2. Csuhaj-Varjú, E., Păun, G., Vaszil, G.: PC grammar systems with five context-free components generate all recursively enumerable languages. Theoret. Comput. Sci. **299**(1–3), 785–794 (2003)
3. Dassow, J., Păun, G.: Regulated Rewriting in Formal Language Theory. EATCS Monographs in Theoretical Computer Science. Springer, Heidelberg (1989)
4. Fernau, H.: Closure properties of ordered languages. EATCS Bull. **58**, 159–162 (1996)
5. Fernau, H., Freund, R., Oswald, M., Reinhardt, K.: Refining the nonterminal complexity of graph-controlled, programmed, and matrix grammars. J. Autom. Lang. Combin. **12**(1/2), 117–138 (2007)
6. Fernau, H., Kuppusamy, L., Oladele, R.O.: New nonterminal complexity results for semi-conditional grammars. In: Manea, F., Miller, R.G., Nowotka, D. (eds.) CiE 2018. LNCS, vol. 10936, pp. 172–182. Springer, Cham (2018). https://doi.org/10.1007/978-3-319-94418-0_18
7. Fernau, H., Kuppusamy, L., Oladele, R.O., Raman, I.: Minimizing rules and nonterminals in semi-conditional grammars: non-trivial for the simple case. In: Durand-Lose, J., Verlan, S. (eds.) MCU 2018. LNCS, vol. 10881, pp. 88–104. Springer, Cham (2018). https://doi.org/10.1007/978-3-319-92402-1_5
8. Geffert, V.: Normal forms for phrase-structure grammars. RAIRO Informatique théorique et Applications **25**, 473–498 (1991)
9. Masopust, T.: Simple restriction in context-free rewriting. J. Comput. Syst. Sci. **76**(8), 837–846 (2010)
10. Masopust, T., Meduna, A.: Descriptional complexity of generalized forbidding grammars. In: Geert, V., Pighizzini, G.: (eds.) 9th International Workshop on Descriptional Complexity of Formal Systems - DCFS, pp. 170–177. University of Kosice, Slovakia (2007)
11. Masopust, T., Meduna, A.: Descriptional complexity of grammars regulated by context conditions. In: Loos, R., Fazekas, S.Z., Martín-Vide, C. (eds.) LATA 2007. Proceedings of the 1st International Conference on Language and Automata Theory and Applications, volume report 35/07, pp. 403–412. Research Group on Mathematical Linguistics, Universitat Rovira i Virgili, Tarragona (2007)
12. Meduna, A.: Generalized forbidding grammars. Intern. J. Comput. Math. **36**, 31–39 (1990)

13. Meduna, A., Svec, M.: Descriptional complexity of generalized forbidding grammars. Intern. J. Comput. Math. **80**(1), 11–17 (2003)
14. Meduna, A., Zemek, P.: Regulated Grammars and Automata. Springer, New York (2014). https://doi.org/10.1007/978-1-4939-0369-6
15. Okubo, F.: A note on the descriptional complexity of semi-conditional grammars. Inform. Process. Lett. **110**(1), 36–40 (2009)
16. Rogers, J., Lambert, D.: Extracting forbidden factors from regular stringsets. In: Kanazawa, M., de Groote, P., Sadrzadeh, M. (eds.) Proceedings of the 15th Meeting on the Mathematics of Language, MOL, pp. 36–46. ACL (2017)
17. Shannon, C.E.: A universal Turing machine with two internal states. In: Shannon, C.E., McCarthy, J. (eds.) Automata Studies. Annals of Mathematics Studies, vol. 34, pp. 157–165. Princeton University Press, Princeton (1956)
18. van der Walt, A.P.J., Ewert, S.: A shrinking lemma for random forbidding context languages. Theoret. Comput. Sci. **237**(1–2), 149–158 (2000)
19. Vaszil, G.: On the descriptional complexity of some rewriting mechanisms regulated by context conditions. Theoret. Comput. Sci. **330**, 361–373 (2005)

On Selecting Leaves with Disjoint Neighborhoods in Embedded Trees

Kolja Junginger, Ioannis Mantas, and Evanthia Papadopoulou[(✉)]

Faculty of Informatics, USI Università della Svizzera italiana, Lugano, Switzerland
{kolja.junginger,ioannis.mantas,evanthia.papadopoulou}@usi.ch

Abstract. We present a generalization of a combinatorial result from Aggarwal, Guibas, Saxe and Shor [1] on selecting a fraction of leaves, with pairwise disjoint neighborhoods, in a tree embedded in the plane. This result has been used by linear-time algorithms to compute certain tree-like Voronoi diagrams, such as the Voronoi diagram of points in convex position. Our generalization allows that only a fraction of the tree leaves is considered: Given is a plane tree T of n leaves, m of which have been *marked*. Each marked leaf is associated with a *neighborhood* (a subtree of T) and any topologically consecutive marked leaves have disjoint neighborhoods. We show how to select in linear time a constant fraction of the marked leaves that have pairwise disjoint neighborhoods.

Keywords: Tree · Linear-time algorithm · Neighborhood · Voronoi diagram

1 Introduction

In 1987, Aggarwal, Guibas, Saxe and Shor [1] introduced a linear-time approach to compute the Voronoi diagram of points in convex position, which can also be used to compute additional tree-like Voronoi diagrams of points, such as: (1) updating a nearest-neighbor Voronoi diagram of points after deletion of one site; (2) computing the farthest-point Voronoi diagram, after the convex hull of the points is available; (3) computing an order-k Voronoi diagram of points, given its order-(k-1) counterpart. Since then, this framework has been used in various ways to tackle linear-time Voronoi constructions, including the medial axis of a simple polygon by Chin et al. [4]. Other examples include the *Hamiltonian abstract* Voronoi diagram by Klein and Lingas [7], and more recently, some *forest-like abstract* Voronoi diagrams by Bohler et al. [3]. If the framework of [1] is applied to the iterative construction of *order-k* Voronoi diagrams of point-sites [8], then the construction is directly improved by a logarithmic factor to $O(nk^2 + n \log n)$. This in turn can be used in various scenarios, e.g., So and Ye [10] provide algorithms to treat *coverage problems* in wireless networks.

Research supported in part by the Swiss National Science Foundation, project SNF 200021E-154387.

© Springer Nature Switzerland AG 2019
S. P. Pal and A. Vijayakumar (Eds.): CALDAM 2019, LNCS 11394, pp. 189–200, 2019.
https://doi.org/10.1007/978-3-030-11509-8_16

The linear-time approach of Aggarwal et al. [1] relies on the following combinatorial result on a binary tree embedded in the plane. Thus, this theorem is inherently used by all other results that follow this linear-time framework.

Theorem 1 ([1]). *Let T be an unrooted binary tree embedded in the plane. Each leaf of T is associated with a neighborhood, which is a subtree of T rooted at that leaf; leaves that are adjacent in the topological ordering of T have disjoint neighborhoods. Then there exists a fixed fraction of leaves whose neighborhoods are of constant size and disjoint, such that no tree edge has its endpoints in two different neighborhoods. Such a set of leaves can be found in time linear in the complexity of T.*

For generalized sites, other than points in the plane, or for abstract Voronoi diagrams, linear-time constructions for the counterparts of problems (1)–(3) have since remained open, even in the case of simple geometric sites such as line segments or circles in the Euclidean plane. The main reason is disconnected Voronoi regions whose presence complicates the problem considerably. The underlying diagrams remain tree-like, however, their regions are disconnected [2,5,6,9]. To potentially extend the linear-time framework of Aggarwal et al. [1], in these cases, we first need a generalized version of this combinatorial result. In particular, not all leaves in the tree may be available to be selected. In addition, the unavailable leaves may be spread along the topological ordering of the tree in any order. Still we need to compute in linear time a constant fraction of the leaves such that their neighborhoods are pairwise disjoint.

In this paper we generalize Theorem 1 when an arbitrary fraction of the leaves of T can only be considered. In particular, we prove the following theorem.

Theorem 2. *Let T be an unrooted binary tree embedded in the plane with n leaves, m of which have been marked. Each marked leaf of T is associated with a neighborhood, which is a subtree of T rooted at that leaf; consecutive marked leaves in the topological ordering of T have disjoint neighborhoods. Then, there exist at least $\frac{1}{10}m$ marked leaves in T with pairwise disjoint neighborhoods, such that no tree edge has its endpoints in two different neighborhoods. We can select at least $\frac{p}{10}m$ marked leaves in time $O(\frac{1}{1-p}n)$, for any $p \in (0,1)$.*

The main contribution in this paper is the algorithm to compute a constant fraction of the m marked leaves in $O(n)$ time. The algorithm allows for a trade-off between the fraction of the returned marked leaves and its time complexity using a parameter $p \in (0,1)$. We expect this result to be used in other linear-time Voronoi constructions, as in problems (1)–(3) for generalized sites. Recently, we have considered randomized linear-time constructions for some of these problems, in particular for the farthest segment Voronoi diagram, given the sequence of its faces at infinity [6], and for abstract Voronoi diagrams [5] (problems (1), (2)). In [5], Junginger and Papadopoulou introduced *Voronoi-like* diagrams, a relaxed Voronoi construct that is easier to compute and can achieve a randomized linear construction. To potentially use these Voronoi-like structures in the deterministic linear-time framework of [1], we first need the generalized Theorem 2. Although we are inspired by Voronoi diagrams in this paper, Theorem 2 is purely combinatorial, and thus, it can find applications in different contexts.

2 Preliminaries

Throughout this work, we consider an unrooted binary tree T embedded in the plane with n leaves and the following additional properties:

- Out of the n leaves, m are *marked* and the remaining $n - m$ are *unmarked*.
- Every marked leaf ℓ is associated with a *neighborhood*, denoted $nh(\ell)$, which is a binary subtree of T rooted at ℓ.
- Every two consecutive marked leaves in the topological ordering of T have disjoint neighborhoods. See Fig. 4 for an illustration.

For brevity we call such a tree T a *marked tree*. The following definition is isolated from the proof of Theorem 1 by Aggarwal et al. [1]. Throughout this paper, whenever possible, we use the terminology of [1].

Definition 1 ([1]). *Let T be a binary tree and let T^* be the tree obtained after deleting all leaves from T. A node u is called:*

(a) Leaf *or* L-node *if $deg(u) = 1$ in T^*. So, u neighbors two leaves in T.*
(b) Comb *or* C-node *if $deg(u) = 2$ in T^*. So, u neighbors one leaf in T.*
(c) Junction *or* J-node *if $deg(u) = 3$ in T^*. So, u neighbors no leaves in T.*

A spine *is a maximal sequence of consecutive C-nodes. Thus, each spine is delimited by J- or L-nodes.*

In Fig. 1, the middle and the right tree illustrate the various nodes of Definition 1. The tree contains two spines, one with five and one with six C-nodes.

Given a marked tree T, consider a *labeling* of its nodes defined as follows. Let T_{del} be the tree resulting from T by deleting all its unmarked leaves and contracting the resulting degree-2 vertices. Let T_{del}^* be the tree resulting from T_{del} by applying Definition 1. The vertices in T_{del}^* are labeled as L-, C- and J-nodes. The labeling of nodes in T_{del}^* is then carried back to the corresponding vertices of T. We call T, augmented with these labels, a *labeled tree*. The transformations of a marked tree T to T_{del} and T_{del}^* are illustrated in Fig. 1. The resulting labeled tree is shown in Fig. 2. Note that not all vertices of T need to receive a label, e.g., node u is unlabeled in Fig. 2.

Definition 2. *We define two types of* components *in a labeled marked tree T.*

(a) L-component: *For an L-node λ we define its L-component to be the node λ union the two subtrees of T, which are incident to λ and contain no labeled node. See e.g. the L-component K_2 in Fig. 3.*

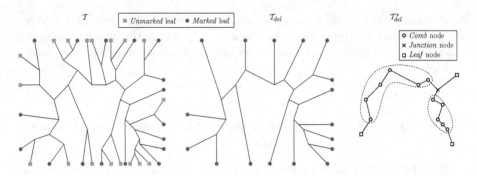

Fig. 1. A marked tree $\mathcal{T}$, $\mathcal{T}_{del}$ and $\mathcal{T}_{del}^*$. The dashed curves illustrate the two spines.

(b) 5-*component: A spine is divided in groups of five consecutive C-nodes. For a group of five consecutive C-nodes $c_i, \ldots, c_{i+4}$, let $P_{c_i:c_{i+4}}$ be the path from c_i to c_{i+4} in $\mathcal{T}$. We define the 5-component of this group to be the path $P_{c_i:c_{i+4}}$ union all the subtrees of $\mathcal{T}$, which are incident to the nodes of $P_{c_i:c_{i+4}}$ and contain no labeled node. See e.g. K_1 in Fig. 3.*

The tree $\mathcal{T}$ of Fig. 3 has three L-components and two 5-components that are indicated with shaded domains. The L-node λ defines the L-component K_2. The 5-component K_1 contains the path from c_1 to c_5 which is shown in thick black lines and denoted $P_{c_1:c_5}$. Note that this path $P_{c_1:c_5}$ contains unlabeled nodes, such as u.

Each spine, and thus each 5-component, has two *sides* in which its marked leaves lie, depending on their topological ordering. Figure 3 illustrates a spine which consists of the C-nodes $c_1, c_2, c_3, c_4, c_5, c_6$. It is delimited by the L-node λ' and the J-node ι. It has five marked leaves from one side and one from the other.

From now on, we assume that $\mathcal{T}$ has been labeled and its components are available. The following observations can be easily derived from the definitions.

Observation 1. *All components are pairwise disjoint subtrees of $\mathcal{T}$.*

Observation 2. *Every L-component contains exactly two marked leaves and every 5-component contains exactly five marked leaves.*

Every L-node must belong to a component of $\mathcal{T}$, whereas for C-nodes this is not true, as there may be up to four *ungrouped* C-nodes in each spine. For example, in Fig. 3, C-node c_6 is not part of any 5-component. Moreover, there may exist subtrees consisting solely of unmarked leaves and unlabeled vertices that do not belong to any component. As an illustration, in Fig. 3, node u' is an unlabeled node and the gray dotted subtree incident to it consists solely of unmarked leaves and unlabeled vertices that do not belong to any component.

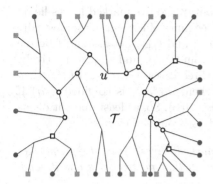

Fig. 2. The labeled marked tree $\mathcal{T}$ corresponding to T of Fig. 1.

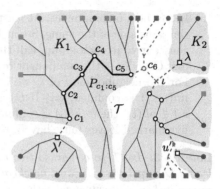

Fig. 3. The L- and 5-components of the labeled marked tree $\mathcal{T}$.

3 Existence of Leaves with Disjoint Neighborhoods

For the proof of Theorem 1, Aggarwal et al. [1] showed that for every eight ungrouped C-nodes there exists at least one L-node. This is a combinatorial argument which also holds for our labeled trees, since the labeling is done after the unmarked leaves have been deleted. We adapt this proposition to our setting, in the following lemma (see [1] for its proof).

Lemma 1. *For every 8 ungrouped C-nodes there exists at least one L-component.*

Recall that each marked leaf ℓ is associated with a neighborhood $nh(\ell)$. We say that a neighborhood $nh(\ell)$ is *confined to a component K* if it is a subtree of K, that is, $nh(\ell) \subseteq K$.

The following two lemmas establish the existence of a constant fraction of marked leaves with pairwise disjoint neighborhoods. The counting arguments used are similar to those of [1] and we follow them closely. The key difference is that, here, only marked leaves have neighborhoods whereas unmarked leaves do not have, in contrast to [1] where each leaf has a neighborhood. Between two marked leaves there can be an arbitrary number of unmarked leaves, which we have to consider when searching for confined neighborhoods.

Lemma 2. *In every component K, there exists a marked leaf $\ell \in K$ whose neighborhood is confined to K.*

Proof.(a) Let K be an L-component, where s is the corresponding L-node and let ℓ_i and ℓ_{i+1} be the two marked leaves of K. For $nh(\ell_i)$ we consider two cases: (i) If $s \in nh(\ell_i)$ then $nh(\ell_{i+1})$ is confined to K and (ii) if $s \in nh(\ell_{i+1})$ then $nh(\ell_i)$ is confined to K; see Fig. 4(a). So, at least one of $nh(\ell_i)$ and $nh(\ell_{i+1})$ must be confined to K.

(b) Let K be a 5-component. At least three out of the five marked leaves lie on the same side of the spine. So, at least three marked leaves are topologically consecutive in each group, call them ℓ_{i-1}, ℓ_i and ℓ_{i+1}. Let r, s and t be the first C-nodes in the component reachable from ℓ_{i-1}, ℓ_i and ℓ_{i+1}, respectively. For the neighborhood of ℓ_i we consider three cases: (i) If $r \in nh(\ell_i)$ then $nh(\ell_{i-1})$ is confined; (ii) If $t \in nh(\ell_i)$ then $nh(\ell_{i+1})$ is confined; (iii) If $r, t \notin nh(\ell_i)$ then $nh(\ell_i)$ is confined; see Fig. 4(b). So, at least one out of the five marked leaves must have a neighborhood confined to K.   □

Lemma 3. *Let T be a marked tree with m marked leaves. At least $\frac{1}{10}m$ marked leaves have pairwise disjoint neighborhoods, such that no tree edge has its endpoints in two different neighborhoods.*

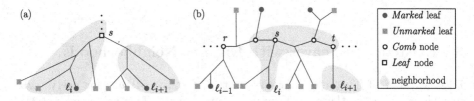

Fig. 4. Marked leaves with their neighborhoods. Cases analysis of Lemma 2: (a) L-component. (b) 5-component. Due to the extent of $nh(\ell_i)$, over s in case (a) and over t in case (b) the neighborhoods $nh(\ell_{i+1})$ are confined to the components.

Proof. Every spine of T has up to four ungrouped C-nodes. From Lemma 1 there exists at least one L-component for every eight ungrouped C-nodes. From Lemma 2 every component has at least one marked leaf with a neighborhood confined to that component. So, overall, at least $\frac{1}{5}$ of the marked leaves from each group of 5 C-nodes and at least $\frac{1}{10}$ marked leaves of the remaining nodes have pairwise disjoint neighborhoods.

All components are disjoint subtrees of T (see Observation 1) and any neighborhood confined to a component is a binary subtree of that component. It follows that the confined neighborhoods are pairwise disjoint and that no tree edge has its endpoints in two different confined neighborhoods.

We remark that the disjoint neighborhoods guaranteed by Lemma 3 may not be of constant complexity, which was the case in [1], but may have $O(n)$ size.

4 Selecting Leaves with Disjoint Neighborhoods

Lemma 3 established the existence of $\frac{1}{10}m$ marked leaves with pairwise disjoint neighborhoods, such that no edge has its endpoints in two different neighborhoods. This is the same constant fraction as in [1], despite the presence of the unmarked leaves. In this section, we present an algorithm to select $\frac{p}{10}m$ marked leaves with pairwise disjoint neighborhoods in time $O(\frac{1}{1-p}n)$, where $0 < p < 1$.

The main challenge in finding these neighborhoods is due to the presence of unmarked leaves that are arbitrarily distributed among the marked leaves. This may result in confined neighborhoods of large size, which may each require $\Theta(n)$ time to be identified. If we spend less time for each component, we cannot guarantee to output all $\frac{m}{10}$ leaves with confined neighborhoods of Lemma 3. If we spend $O(n)$ for each component, the overall time complexity could be up to $O(n^2)$. To tackle this issue we take into account the ratio between unmarked and marked leaves. By introducing a parameter $p \in (0,1)$, we guarantee to find at least a p-fraction of the possible $\frac{m}{10}$ marked leaves in time $O(\frac{1}{1-p}n)$.

Let $\ell_1, \ldots, \ell_m$ be the topological ordering of the marked leaves counterclockwise around $\mathcal{T}$. Let M denote the set of all marked leaves of $\mathcal{T}$. For two consecutive marked leaves ℓ_i and ℓ_{i+1}, let the *interval* (ℓ_i, ℓ_{i+1}) denote the list of the unmarked leaves between ℓ_i and ℓ_{i+1}. The number of these unmarked leaves is denoted as $|(\ell_i, \ell_{i+1})|$ and it is called the *interval size*. We show the following *pigeonhole lemma* for unmarked leaves and intervals.

Lemma 4. *Let $\mathcal{T}$ be a marked tree and let $c = \lceil \frac{n-m}{m} \rceil$ be the ratio between unmarked and marked leaves. For any $x \in \mathbb{N}$ with $x \leq n - m$, let $M_x = \{\ell_i \in M \mid |(\ell_i, \ell_{i+1})| \leq x\}$, be the set of marked leaves whose intervals have at most x unmarked leaves. Then, $|M_x| \geq \frac{x-c+1}{x+1}m$.*

Proof. Let $\overline{M}_x = \{\ell_i \in M \mid \ell_i \notin M_x\}$ be the complement of M_x. The unmarked leaves in $\overline{M}_x$ are equal to at most all the unmarked leaves in $\mathcal{T}$. Thus:

$$|\overline{M}_x|(x+1) \leq cm \;\Rightarrow\; |\overline{M}_x| \leq \frac{cm}{x+1} \tag{1}$$

$$|M_x| = m - |\overline{M}_x| \overset{(1)}{\geq} m - \frac{cm}{x+1} = m\left(1 - \frac{c}{x+1}\right) \tag{2}$$

$$(2) \Rightarrow |M_x| \geq \frac{x-c+1}{x+1}m$$

□

Each interval (ℓ_i, ℓ_{i+1}) is associated with a subtree $T_{(\ell_i, \ell_{i+1})}$, which is the minimal subtree of $\mathcal{T}$ containing ℓ_i, ℓ_{i+1}, and all the unmarked leaves within (ℓ_i, ℓ_{i+1}). See Fig. 5(b) for an example. If $T_{(\ell_i, \ell_{i+1})}$ has a common edge with a component K, we say that the interval (ℓ_i, ℓ_{i+1}) is *related* to K. The set of all *related intervals* of a component K is denoted I_K.

In the case of an L-component, it is easy to see that, there are three intervals whose associated subtrees can share an edge with the component, see Fig. 5(a).

In the case of a 5-component, the marked leaves as well as the related intervals of the component lie on two different sides of the component. The number of related intervals at each side is equal to the number of marked leaves on that side plus one. So, there are seven related intervals for each 5-component (seven marked leaves plus one from each side), see Fig. 5(b). We summarize in the following observation.

Observation 3. *If K is an L-component it has three related intervals, $|I_K| = 3$. If K is a 5-component it has seven related intervals, $|I_K| = 7$.*

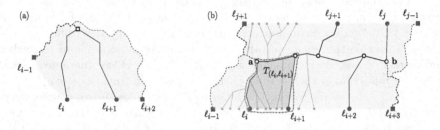

Fig. 5. (a) An L-component and (b) a 5-component and their related intervals, shaded. The dark shaded area in the 5-component highlights the subtree $T_{(\ell_i, \ell_{i+1})}$. Squared leaves denote topologically consecutive marked leaves not belonging to the components.

The following lemma gives a relation between the set of the related intervals of a component K, which is denoted as I_K, and the size (number of nodes) of a neighborhood confined to K.

Lemma 5. *Let K be a component, let ℓ_i be a marked leaf of K such that $nh(\ell_i)$ is confined to K. Let $\delta_K = \max_{(\ell_i, \ell_{i+1}) \in I_K} |(\ell_i, \ell_{i+1})|$, be the maximum size of the intervals that are related to K.*

(a) If K is an L-component, then $nh(\ell_i)$ intersects the associated subtrees of at most two related intervals and has at most $4\delta_K - 1$ nodes.

(b) If K is a 5-component, then $nh(\ell_i)$ intersects the associated subtrees of at most five subtrees of related intervals and has at most $10\delta_K - 1$ nodes.

Proof. Neighborhood $nh(\ell_i)$ is confined to K, so $nh(\ell_i)$ is a binary subtree of K. If K is an L-component, $nh(\ell_i)$ does not contain the L-node as it is a binary subtree. Therefore, $nh(\ell_i)$ intersects the associated subtrees of at most two related intervals. Namely, if ℓ_i is the first topological marked leaf of K, these are (ℓ_{i-1}, ℓ_i) and (ℓ_i, ℓ_{i+1}); otherwise these are (ℓ_i, ℓ_{i+1}) and (ℓ_{i+1}, ℓ_{i+2}). See Fig. 6(a).

If K is a 5-component let a, b be the first and last C-nodes in that component. Let ℓ_a be the marked leaf in K for which there is a path from a to ℓ_a that does not include any other C-node. Analogously we define ℓ_b, see Fig. 6(b),(c). First observe, that if $\ell_i \in \{\ell_a, \ell_b\}$, then $nh(\ell_i)$ can share an edge with at most two subtrees, namely $T_{(\ell_{i-1},\ell_i)}, T_{(\ell_i,\ell_{i+1})}$. If ℓ_i is a marked leaf in K different from ℓ_a and ℓ_b, then $nh(\ell_i)$ may intersect the subtrees belonging to at most five related intervals. By Observation 3 there are at most seven related intervals. Further, notice that $nh(\ell_i)$ cannot extend beyond the vertices a and b, because it is confined to K. Thus, we can exclude the two related intervals which are incident to ℓ_a and ℓ_b and whose associated subtrees do not contain any C-nodes of K apart from a and b. So, we obtain the upper bound of five related intervals.

Trivially, no subtree associated with an interval which is not related to K may be intersected by $nh(\ell_i)$. For each interval the confined neighborhood $nh(\ell_i)$ intersects at most the entire subtree of this interval. Since $nh(\ell_i)$ is a binary tree and δ_K is the size of the biggest related interval, the claim follows.   □

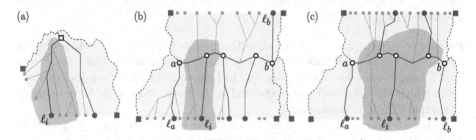

Fig. 6. Different component configurations where the confined neighborhoods $nh(\ell_i)$ (shown highlighted) intersect subtrees of two (a), three (b) and five (c) related intervals.

By considering the proof of Lemma 2, we define a *representative leaf* and at most two *delimiting nodes* for each component. These are used by our algorithm to identify the marked leaves with neighborhoods confined to that component.

Definition 3. *(a)* If K is an L-component, its representative leaf *is the first of its marked leaves, in the topological ordering; its* delimiting node *is the L-node. See Fig. 4(a), where ℓ_i is the representative leaf and s is the delimiting node.*
(b) If K is a 5-component, consider the side of the component containing at least three marked leaves. The *representative leaf is the topologically second among these leaves. The first C-nodes that are reachable from the first and third of these leaves are the* delimiting nodes. *See Fig. 4(b), where ℓ_i is the representative leaf and r, t are the delimiting nodes.*

Our algorithm takes as input a marked tree T and a parameter p, and returns a fraction of the marked leaves with pairwise disjoint neighborhoods. A pseudocode description is given in Algorithm 1. The algorithm iterates over all components and for each one it adds at most one marked leaf to the solution. After identifying the representative leaf and the delimiting nodes of the component, it *traces* the neighborhood of the representative leaf. Procedure `trace()` can be thought as a depth-first search, starting on the representative leaf and extending over the entire neighborhood. While tracing, the size (number of nodes) of the neighborhood is counted. If a pre-fixed limit on the neighborhood size is exceeded, we abandon K and proceed with the next component. If not, we decide which marked leaf to select as indicated in the proof of Lemma 2. In particular, we check which delimiting nodes belong to the neighborhood of the representative leaf. If the neighborhood of the representative leaf does not contain the delimiting nodes, then we add the representative leaf to the solution. Otherwise, one neighbor of the representative leaf must have a confined neighborhood and we add that leaf to the solution. After a leaf is selected, we proceed to the next component.

Lemma 6. *Algorithm 1 returns at least $\frac{p}{10}m$ marked leaves with pairwise disjoint neighborhoods, such that no tree edge has its endpoints in two different neighborhoods.*

Proof. For each component K the algorithm selects a marked leaf to the solution if some conditions are satisfied within a certain number of steps. The selected leaf has its neighborhood confined to K due to Lemma 2. The fixed number of steps up to which the neighborhood is checked is: $4z$ steps, if K is an L-component; $10z$ steps, if K is a 5-component.

Lemma 5 guarantees that such a marked leaf will be found if $z \geq \delta_K$, where δ_K is the maximum size of the intervals related to K. Note that one interval may be related to more than one component. However, no unmarked leaf of such an interval may be considered by two components, because the neighborhood, which the algorithm checks, is confined within the delimiting nodes of each component.

By Lemma 4, the number $|M_z|$ of marked leaves whose intervals in T have size at most z, is at least $\frac{z-c+1}{z+1}m$. For the value z we have $z = \left\lceil \frac{10c}{1-p} \right\rceil - 1$, so:

$$|M_z| \geq \frac{z-c+1}{z+1}m \geq \frac{9+p}{10}m \tag{3}$$

$$|\overline{M}_z| = m - |M_z| \overset{(3)}{\Rightarrow} |\overline{M}_z| \leq \frac{1-p}{10}m$$

Thus, for at most $|\overline{M}_z|$ components, the algorithm will not find a marked leaf. By Lemma 1, there exist at least $\frac{1}{10}m$ components in T and no edge has its endpoints in two different components, as established in Lemma 3. Therefore, the algorithm will select at least $\frac{1}{10}m - |\overline{M}_z| \geq p\frac{m}{10}$ marked leaves with pairwise disjoint neighborhoods. $\square$

Algorithm 1: Selecting leaves with pairwise disjoint neighborhoods.

Input : A marked tree T with n leaves and a parameter $p \in (0,1)$.
Output: At least $\frac{p}{10}m$ leaves with pairwise disjoint neighborhoods.

1 Obtain the labeling of T.
2 Partition T into components as indicated in Definition2.
3 $sol \leftarrow \emptyset, \qquad c \leftarrow \left\lceil \frac{n-m}{m} \right\rceil, \qquad z \leftarrow \left\lceil \frac{10c}{1-p} \right\rceil - 1$
4 **for** each *component K of T* **do**
5   **if** K *is an L-component* **then**
6           $\ell_i \leftarrow$ representative leaf, $s \leftarrow$ delimiting node
7   **while** $\text{trace}(nh(\ell_i))$ **and** $nh(\ell_i) < 4z$
8   **if** $s \in nh(\ell_i)$ **then**
9                   $sol \leftarrow sol \cup \{\ell_{i+1}\}$
10   **else if** $s \notin nh(\ell_i)$ **then**
11                  $sol \leftarrow sol \cup \{\ell_i\}$
12   **else if** K *is a 5-component* **then**
13          $\ell_i \leftarrow$ representative leaf, $r,t \leftarrow$ delimiting nodes
14   **while** $\text{trace}(nh(\ell_i))$ **and** $nh(\ell_i) < 10z$
15   **if** $r \in nh(\ell_i)$ **then**
16                  $sol \leftarrow sol \cup \{\ell_{i-1}\}$
17   **else if** $t \in nh(\ell_i)$ **then**
18                  $sol \leftarrow sol \cup \{\ell_{i+1}\}$
19   **else if** $r \notin nh(\ell_i)$ **and** $t \notin nh(\ell_i)$ **then**
20                  $sol \leftarrow sol \cup \{\ell_i\}$
21 **return** sol

Lemma 7. *Algorithm 1 has time complexity $O(\frac{1}{1-p}n)$.*

Proof. The first step of the algorithm is to obtain the labeling and the components. This can be done in $O(n)$ time by traversing the tree T. Then, for each component we consider the neighborhood of the representative leaf. There are at least $\frac{1}{10}m = \Theta(m)$ components and we examine the neighborhood of each representative leaf up to a size of at most $10z = \Theta(\frac{c}{1-p})$. So, we have $O(\frac{c}{1-p} \cdot m)$ time complexity. Recall that $c = \left\lceil \frac{n-m}{m} \right\rceil$. If $m = \Theta(n)$, then $c = \Theta(1)$, so $cm = \Theta(n)$. If $m = o(n)$, then again $cm = \Theta(n)$. In any case, the time complexity of the algorithm is $O(\frac{1}{1-p}n)$. $\qquad\square$

By combining Lemmas 3, 6 and 7 we establish Theorem 2, which we re-state here for completeness.

Theorem 2. *Let T be a marked tree with n leaves, m of which have been marked. Then there exist at least $\frac{1}{10}m$ leaves in T with pairwise disjoint neighborhoods, such that no tree edge has its endpoints in two different neighborhoods. Further, we can select at least $\frac{p}{10}m$ marked leaves in time $O(\frac{1}{1-p}n)$, for any $p \in (0,1)$.*

We remark that if a constant fraction of the marked leaves is required, it suffices to give a constant value to the parameter $p \in (0,1)$, then the algorithm has linear time complexity $O(n)$. Moreover, if $m = \Theta(n)$, then the selected neighborhoods are of constant size, as in [1].

References

1. Aggarwal, A., Guibas, L., Saxe, J., Shor, P.: A linear-time algorithm for computing the Voronoi diagram of a convex polygon. Discret. Comput. Geom. **4**, 591–604 (1989)
2. Bohler, C., Cheilaris, P., Klein, R., Liu, C.H., Papadopoulou, E., Zavershynskyi, M.: On the complexity of higher order abstract Voronoi diagrams. Comput. Geom.: Theory Appl. **48**(8), 539–551 (2015)
3. Bohler, C., Klein, R., Lingas, A., Liu, C.-H.: Forest-like abstract Voronoi diagrams in linear time. Comput. Geom. **68**, 134–145 (2018)
4. Chin, F., Snoeyink, J., Wang, C.A.: Finding the medial axis of a simple polygon in linear time. Discret. Comput. Geom. **21**(3), 405–420 (1999)
5. Junginger, K., Papadopoulou, E.: Deletion in abstract Voronoi diagrams in expected linear time. In: Proceedings of 34th International Symposium on Computational Geometry (SoCG), LIPIcs, vol. 99, pp. 50:1–50:14 (2018)
6. Khramtcova, E., Papadopoulou, E.: An expected linear-time algorithm for the farthest-segment Voronoi diagram. arXiv:1411.2816v3 [cs.CG] (2017). Preliminary Version in Proceedings of 26th International Symposium on Algorithms and Computation (ISAAC), LNCS, vol. 9472, pp. 404–414. Springer, Heidelberg (2015).https://doi.org/10.1007/978-3-662-48971-0_35
7. Klein, R., Lingas, A.: Hamiltonian abstract Voronoi diagrams in linear time. In: Du, D.-Z., Zhang, X.-S. (eds.) ISAAC 1994. LNCS, vol. 834, pp. 11–19. Springer, Heidelberg (1994). https://doi.org/10.1007/3-540-58325-4_161
8. Lee, D.-T.: On k-nearest neighbor Voronoi diagrams in the plane. IEEE Trans. Comput. **100**(6), 478–487 (1982)
9. Papadopoulou, E., Zavershynskyi, M.: The higher-order Voronoi diagram of line segments. Algorithmica **74**(1), 415–439 (2016)
10. So, A.M.-C., Ye, Y.: On solving coverage problems in a wireless sensor network using Voronoi diagrams. In: Deng, X., Ye, Y. (eds.) WINE 2005. LNCS, vol. 3828, pp. 584–593. Springer, Heidelberg (2005). https://doi.org/10.1007/11600930_58

The Balanced Connected Subgraph Problem

Sujoy Bhore[1], Sourav Chakraborty[2], Satyabrata Jana[2], Joseph S. B. Mitchell[3], Supantha Pandit[3(✉)], and Sasanka Roy[2]

[1] Ben-Gurion University, Beer-Sheva, Israel
sujoy.bhore@gmail.com
[2] Indian Statistical Institute, Kolkata, India
chakraborty.sourav@gmail.com, satyamtma@gmail.com, sasanka.ro@gmail.com
[3] Stony Brook University, Stony Brook, NY, USA
joseph.mitchell@stonybrook.edu, pantha.pandit@gmail.com

Abstract. The problem of computing induced subgraphs that satisfy some specified restrictions arises in various applications of graph algorithms and has been well studied. In this paper, we consider the following *Balanced Connected Subgraph* (shortly, *BCS*) problem. The input is a graph $G = (V, E)$, with each vertex in the set V having an assigned color, "red" or "blue". We seek a maximum-cardinality subset $V' \subseteq V$ of vertices that is *color-balanced* (having exactly $|V'|/2$ red nodes and $|V'|/2$ blue nodes), such that the subgraph induced by the vertex set V' in G is connected. We show that the BCS problem is NP-hard, even for bipartite graphs G (with red/blue color assignment not necessarily being a proper 2-coloring). Further, we consider this problem for various classes of the input graph G, including, e.g., planar graphs, chordal graphs, trees, split graphs, bipartite graphs with a proper red/blue 2-coloring, and graphs with diameter 2. For each of these classes either we prove NP-hardness or design a polynomial time algorithm.

Keywords: Balanced connected subgraph · Trees · Split graphs · Chordal graphs · Planar graphs · Bipartite graphs · NP-hard · Color-balanced

1 Introduction

Several problems in graph theory and combinatorial optimization involve determining if a given graph G has a subgraph with certain properties. Examples

S. Bhore—Partially supported by the Lynn and William Frankel Center for Computer Science, Ben-Gurion University of the Negev, Israel.
J. S. B. Mitchell—Support from the National Science Foundation (CCF-1526406) and the US-Israel Binational Science Foundation (project 2016116).
S. Pandit—Partially supported by the Indo-US Science & Technology Forum (IUSSTF) under the SERB Indo-US Postdoctoral Fellowship scheme with grant number 2017/94, Department of Science and Technology, Government of India.

© Springer Nature Switzerland AG 2019
S. P. Pal and A. Vijayakumar (Eds.): CALDAM 2019, LNCS 11394, pp. 201–215, 2019.
https://doi.org/10.1007/978-3-030-11509-8_17

include seeking paths, cycles, trees, cliques, vertex covers, matching, independent sets, bipartite subgraphs, etc. Related optimization problems include finding a maximum clique, a maximum (connected) vertex cover, a maximum independent set, a minimum (connected) dominating set, etc. These well-studied problems have significant theoretical interest and many practical applications.

We study the problem in which we are given a simple connected graph $G = (V, E)$ whose vertex set V has each node being "red" or "blue" (note, the color assignment might not be a proper 2-coloring of the vertices, i.e., we allow nodes of the same color to be adjacent in G). We seek a maximum-cardinality subset $V' \subseteq V$ of the nodes such that V' is *color-balanced*, i.e. having same number of red and blue nodes in V', and such that the induced subgraph H by V' in G is connected. We refer to this as the *Balanced Connected Subgraph (BCS)* problem:

Balanced Connected Subgraph (BCS) Problem
Input: A graph $G = (V, E)$, with node set $V = V_R \cup V_B$ partitioned into red nodes (V_R) and blue nodes (V_B).
Goal: Find a maximum-cardinality color-balanced subset $V' \subseteq V$ that induces a connected subgraph H.

1.1   Connection with the Graph Motif Problem

Here we establish a connection between the BCS problem and the *Graph Motif* problem [7,14,20]. In the Graph Motif problem, we are given the input as a graph $G = (V, E)$, a color function $col : V \rightarrow C$ on the vertices, and a multiset M of colors of C; the objective is to find a subset $V' \subseteq V$ such that the induced subgraph on V' is connected and $col(V') = M$. We note that if $C = \{\text{red, blue}\}$ and the motif has same number of blues and reds, then the solution of the Graph Motif problem gives a balanced connected subgraph (not necessarily a maximum balanced connected subgraph).

Fellows et al. [14] showed that the Graph Motif problem is NP-complete for trees of maximum degree 3 where the given motif is a colorful set instead of a multiset (that is, no color occurs more than once). They also showed that the Graph Motif problem remains NP-hard for bipartite graphs of maximum degree 4 and the motif contains only two colors. It is easy to observe that a solution to the Graph Motif problem (essentially) gives a solution to the BCS problem, with an impact of a polynomial factor in the running time. On the other hand the NP-hardness result for the BCS problem on a particular graph class implies the NP-hardness result for the Graph Motif problem on the same class. We conclude that BCS problem is a special case of the Graph Motif problem. Note that much of the work on the Graph Motif problem (e.g., [7,14,20]) is addressing the parameterized complexity of the Graph Motif problem.

1.2 Motivation and Possible Applications

In [7], the authors mentioned that the vertex-colored graph problems have numerous applications in bioinformatics. See the references of [7] for more specific applications. However, the Graph Motif problem is motivated by the applications in biological network analysis [20]. This problem also has applications in social or technical networks [4,14] or in the context of mass spectrometry [6,14].

The *BCS* problem is closely related to the *Maximum Node Weight Connected Subgraph (MNWCS)* problem [12,16]. In the *MNWCS* problem, we are given a connected graph $G(V, E)$, with an integer weight associated with each node in V, and an integer bound B; the objective is to decide whether there exists a subset $V' \subseteq V$ such that the subgraph induced by V' is connected and the total weight of the vertices in V' is at least B. In the *MNWCS* problem, if the weight of each vertex is either $+1$ (red) or -1 (blue), and if we ask for a largest connected subgraph whose total weight is *exactly* zero, then it is equivalent to the *BCS* problem. The *MNWCS* problem along with its variations have numerous practical application in various fields (see [12] and the references therein). We believe some of these applications also serve well to motivate the *BCS* problem.

1.3 Related Work

Bichromatic input points, often referred to as "red-blue" input, has appeared extensively in numerous problems. For a detailed survey on geometric problems with red-blue points see [17]. In [5,10,11] colored points have been considered in the context of matching and partitioning problems. In [1], Aichholzer et al. considered the balanced island problem and devised polynomial algorithms for points in the plane. On the combinatorial side, Balanchandran et al. [2] studied the problem of unbiased representatives in a set of bicolorings. Kaneko et al. [18] considered the problem of balancing colored points on a line. Later on, Bereg et al. [3] studied balanced partitions of 3-colored geometric sets in the plane.

Finding a certain type of subgraph in a graph is a fundamental algorithmic question. In [13], Feige et al. studied the dense k-subgraph problem in which we are given a graph G and a parameter k, and the goal is to find a set of k vertices with maximum average degree in the subgraph induced by this set. Crowston et al. [8] considered parameterized algorithms for the balanced subgraph problem. Kierstead et al. [19] studied the problem of finding a colorful induced subgraph in a properly colored graph. In [9], Derhy and Picouleau considered the problem of finding induced trees in both weighted and unweighted graphs and obtained hardness and algorithmic results. They have studied bipartite graphs and triangle-free graphs; moreover, they have considered the case in which the number of prescribed vertices is bounded.

1.4 Our Results

In this paper, we consider the balanced connected subgraph problem on various graph families and present several hardness and algorithmic results.

On the hardness side, in Sect. 2, we prove that the *BCS* problem is NP-hard on general graphs, even for planar graphs, bipartite graphs (with a general red/blue color assignment, not necessarily a proper 2-coloring), and chordal graphs. Furthermore, we show that the existence of a balanced connected subgraph containing a specific vertex is NP-complete. In addition to that, we prove that finding the maximum balanced path in a graph is NP-hard. Note that, Fellows et al. [14] showed that the Graph Motif problem is NP-complete for bipartite graphs with two colors. However, their reduction does not imply that the *BCS* problem on bipartite graph is NP-hard since in their reduction the motif is not color-balanced (i.e., does not include the same number of blues and reds).

On the algorithmic side, in Sect. 3, we devise polynomial-time algorithms for trees (in $O(n^4)$ time), split graphs (in $O(n^2)$ time), bipartite graphs with a proper 2-coloring (in $O(n^2)$ time), and graphs with diameter 2 (in $O(n^2)$ time). Here, n is the number of vertices in the input graphs.

2   Hardness Results

2.1   BCS Problem on Bipartite Graphs

In this section we prove that the *BCS* problem is NP-hard for bipartite graphs with a general red/blue color assignment, not necessarily a proper 2-coloring. We give a reduction from the *Exact-Cover-by-3-Sets (EC3Set)* problem [15]. In this *EC3Set* problem, we are given a set U with $3k$ elements and a collection S of m subsets of U such that each $s_i \in S$ contains exactly 3 elements. The objective is to find an exact cover for U (if one exists), i.e., a sub-collection $S' \subseteq S$ such that every element of U occurs in exactly one member of S'. During the reduction, we generate an instance $G = (R \cup B, E)$ of the *BCS* problem from an instance $X(S, U)$ of the *EC3Set* problem as follows:

Reduction: For each set $s_i \in S$, we take a blue vertex $s_i \in B$. For each element $u_j \in U$, we take a red vertex $u_j \in R$. Now consider a set $s_i \in S$ containing three elements, u_α, u_β, and u_γ, and add the three edges $(s_i, u_\alpha), (s_i, u_\beta)$, and (s_i, u_γ) to the edge set E. Additionally, we consider a path of $5k$ blue vertices starting and ending with vertices b_1 and b_{5k}, respectively. Similarly, we consider a path of $3k$ red vertices starting and ending with vertices r_1 and r_{3k}, respectively. We connect these two paths by joining the vertices r_{3k} and b_1 by an edge. Finally, we add edges connecting each vertex s_i with b_{5k}. This completes the construction. See Fig. 1 for the complete construction. Clearly, the numbers of vertices and edges in G are polynomial in terms of the numbers of elements and sets in X; hence, the construction can be done in polynomial time. We now prove the following lemma.

Lemma 1. *The instance X of the EC3Set problem has a solution if and only if the instance G of the BCS problem has a connected balanced subgraph T with $12k$ vertices ($6k$ red and $6k$ blue).*

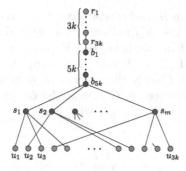

Fig. 1. Construction of the instance G of the BCS problem. (Color figure online)

Proof. Assume that $EC3Set$ problem has a solution. Let S^* be an optimal solution in it. We choose the corresponding vertices of S^* in T. Since this solution covers all u_j's. So we select all u_j's in T. Finally we select all the $5k$ blue and $3k$ red vertices in T, resulting in a total of $6k$ red and $6k$ blue vertices.

On the other hand, assume that there is a balanced tree T in G with $6k$ vertices of each color. The solution must pick the $5k$ blue vertices $b_1, \ldots, b_{5k}$. Otherwise, it exclude the $3k$ red vertices $r_1, \ldots, r_{3k}$, and reducing the size of the solution. Since the graph G has at most $6k$ red vertices, at most k vertices can be picked from the set $s_1, \ldots, s_m$ and need to cover all the $3k$ red vertices corresponding to u_j for $1 \leq j \leq 3k$. Hence, this k sets give an exact cover. $\square$

It is easy to see that the graph we constructed from the ($EC3Set$) problem in Fig. 1 is indeed a bipartite graph. Hence we conclude the following theorem.

Theorem 1. *The BCS problem is NP-hard for bipartite graphs.*

2.2 NP-Hardness: BCS Problem on Special Classes of Graphs

In this section, we show that the BCS problem is NP-hard even if we restrict the graph classes to be planar, or chordal graphs.

Planar Graphs: In this section we prove that BCS problem is NP-hard for planar graphs. We give a reduction from the *Steiner Tree problem in planar graphs (STPG)* [15]. In this problem, we are given a planar graph $G = (V, E)$, a subset $X \subseteq V$, and a positive integer $k \in \mathbb{N}$. The objective is to find a tree $T = (V', E')$ with at most k edges such that $X \subseteq V'$. Without loss of generality we assume that $k \geq |X| - 1$, otherwise the $STPG$ problem has no solution.

Reduction: We generate an instance $H = (R \cup B, E(H))$ for the BCS problem from an instance $G = (V, E)$ of the $STPG$ problem. We color all the vertices, V, in G as blue. We create a set of $|X|$ red vertices as follows: for each vertex $u_i \in X$, we create a red vertex u'_i in H, and we connect u'_i to u_i via an edge. Additionally, we take a set Z of $(k + 1 - |X|)$ red vertices in H and the edges (z_j, u'_1) into $E(H)$, for each $z_j \in Z$. Hence we have, $B = V$, and $R = Z \cup \{u'_i; 1 \le i \le |X|\}$. Note that $|R| < |B|$ and $|R| = (k + 1)$. This completes the construction. For an illustration see Fig. 2. Clearly the number of vertices and edges in H are polynomial in terms of vertices in G. Hence the construction can be done in polynomial time. We now prove the following lemma.

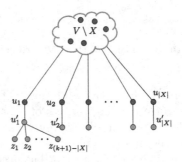

Fig. 2. Schematic construction for planar graphs. (Color figure online)

Lemma 2. *The STPG problem has a solution if and only if the instance H of the BCS problem has a balanced connected subgraph with $(k+1)$ vertices each of the two colors.*

Proof. Assume that $STPG$ has a solution. Let $T = (V', E')$ be the resulting Steiner tree, which contains at most k edges and $X \subseteq V'$. If $|V'| = (k+1)$ then the subgraph of H induced by $(V' \cup R)$ is connected and balanced with $(k+1)$ vertices of each color. If $|V'| < (k+1)$ then we take a set Y of $((k+1)-|V'|)$ many vertices from V such that the subgraph of G induced by $(V' \cup Y)$ is connected. Clearly $|V' \cup Y| = (k+1)$. Now the subgraph of H induced by $(V' \cup Y \cup R)$ is connected and balanced with $(k+1)$ vertices of each red and blue color.

On the other hand, assume that there is a balanced connected subgraph H' of H with $(k+1)$ vertices of each color. Note that, except vertex u'_1, in H all the red vertices are of degree 1 and connected to blue vertices. Let G' be the subgraph of G induced by all blue vertices in H'. Since H is connected and there is no edge between any two red vertices, G' is connected. Since G' contains $(k+1)$ vertices, any spanning tree T of H' contains k edges. So T is a solution of the $STPG$ problem. $\qquad\square$

Theorem 2. *The BCS problem is NP-hard for planar graphs.*

Chordal Graphs: We prove that the BCS problem is NP-hard where the input graph is a chordal graph. The hardness construction is similar to the construction in Sect. 2.1; we modify the construction so that the graph is chordal.

In particular, we add edges between s_i and s_j for each $i \neq j, 1 \leq i, j \leq m$. For this modified graph, it is easy to see that a lemma identical to Lemma 1 holds. Hence, we conclude that the BCS problem is NP-hard for chordal graphs.

2.3   NP-Hardness: BCS Problem with a Specific Vertex

In this section we prove that the existence of a balanced subgraph containing a specific vertex is NP-complete. We call this problem the BCS-existence problem. The reduction is similar to the reduction used in showing the NP-hardness of the BCS problem; we also use here a reduction from the $EC3Set$ problem (see Sect. 2.1 for the definition).

Reduction: Assume that we are given a $EC3Set$ problem instance $X = (U, S)$, where set U contains $3k$ elements and a collection S of m subsets of U such that each $s_i \in S$ contains exactly 3 elements. We generate an instance $G(R, B, E)$ of the BCS-existence problem from X as follows. The red vertices R are the elements $u_j \in U$; i.e., $R = U$. The blue vertices B are the 3-element sets $s_i \in S$; i.e., $B = S$. For each blue vertex $s_i = \{u_\alpha, u_\beta, u_\gamma\} \in S = B$, we add the 3 edges (s_i, u_α), (s_i, u_β), and (s_i, u_γ) to the set E of edges of G. We instantiate an additional set of $2k$ blue vertices, $\{b_1, \ldots, b_{2k}\}$, and add edges to E to link them into a path $(b_1, b_2, \ldots, b_{2k})$. Finally, we add an edge from b_{2k} to each of the blue vertices s_i. Refer to Fig. 3.

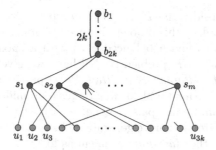

Fig. 3. Construction of the instance G of the BCS problem containing b_1. (Color figure online)

Clearly, the number of vertices and edges in G are polynomial in terms of number of elements and sets in the $EC3Set$ problem instance X, and hence the construction can be done in polynomial time. We now prove the following lemma.

Lemma 3. *The instance X of the EC3Set problem has a solution iff the instance G of the corresponding BCS existence problem has a balanced subgraph T containing the vertex b_1.*

Proof. Assume that the *EC3Set* problem has a solution, and let S^* be the collection of $k = |S^*|$ sets of S in the solution. Then, we obtain a balanced subgraph T that contains b_1 as follows: T is the induced subgraph of the $3k$ red vertices U, together with the k blue vertices S^* and the $2k$ blue vertices $b_1, \ldots, b_{2k}$. Note that T is balanced and connected and contains b_1.

Conversely, assume there is a balanced connected subgraph T containing b_1. Let t be the number of (blue) vertices of S within T. First, note that $t \leq k$. (Since T is balanced and contains at most $3k$ red vertices, it must contain at most $3k$ blue vertices, $2k$ of which must be $\{b_1, \ldots, b_{2k}\}$, in order that T is connected.) Next, we claim that, in fact, $t \geq k$. To see this, note that each of the t blue vertices of T that corresponds to a set in S is connected by edges to 3 red vertices; thus, T has at most $3t$ red vertices. Now, T has $2k + t$ blue vertices (since it has t vertices other than the path $(b_1, \ldots, b_{2k})$), and T is balanced; thus, T has exactly $2k + t$ red vertices, and we conclude that $2k + t \leq 3t$, implying $k \leq t$, as claimed. Therefore, we need to select exactly k blue vertices corresponding to the sets S, and these vertices connect to all $3k$ of the red vertices. The k sets corresponding to these k blue vertices is a solution for the *EC3Set* problem. $\square$

Clearly, the *BCS* existence problem is in NP. Hence, we conclude:

Theorem 3. *It is NP-complete to decide if there exists a connected balanced subgraph that contains a specific vertex.*

2.4   NP-Hardness: Balanced Connected Path Problem

In this section we consider the *Balanced Connected Path (BCP)* Problem and prove that it is NP-hard. In this problem instead of finding a balanced connected subgraph, our goal is to find a balanced path with a maximum cardinality of vertices. To prove the *BCP* problem is NP-hard we give a polynomial time reduction from the *Hamiltonian Path (Ham-Path)* problem which is known to be NP-complete [15]. In this problem, we are given an undirected graph Q, and the goal is to find a Hamiltonian path in Q i.e., a path which visits every vertex in Q exactly once. In the reduction we generate an instance G of the *BCP* problem from an instance Q of the *Ham-Path* problem as follows:

Reduction: We make a new graph Q' from Q. Let us assume that the graph Q contains m vertices. If m is even then $Q' = Q$. If m is odd, then we add a dummy vertex u in Q, connect to every other vertices in Q by edges with u and attach a path of length 2 to u. The resulting graph is our desired Q'. It is easy to observe that, Q has a Hamiltonian path if and only if Q' has a Hamiltonian path.

Now we have a *Ham-Path* instance Q' with even number of vertices, say n. We arbitrary choose any $n/2$ vertices in Q' and color them red and color the remaining $n/2$ vertices blue. Let G be the colored graph. This completes the construction. Clearly, this can be done in polynomial time.

Lemma 4. *Q' has a Hamiltonian path T if and only if G has a balanced path P with exactly n vertices.*

Proof. Assume that Q' has a Hamiltonian path T. This implies that, T visits every vertex in Q'. Since by the construction there are exactly half of the vertices in G is red and remaining are blue, the same path T is balanced with $n/2$ vertices of each color. On the other hand, assume that there is a balanced path P in G with exactly $n/2$ vertices of each color. Since, G has a total of n vertices, the path P visits every vertex in G. Hence, P is a Hamiltonian path.   □

Theorem 4. *The BCP problem is NP-hard for general graph.*

3   Algorithmic Results

In this section, we consider several graph families and devise polynomial time algorithms for the *BCS* problem. Notice that, if the graph is a path or cycle, the optimal solution is just a path. Hence, one can do brute-force search to obtain the maximum balanced path. In case of a complete graph K_n, we output a subgraph H of K_n induced by V, where $|V| = 2|B|$, $B \subset V$, and B is the set of all blue vertices in K_n (assuming that, the number of blue vertices is at most the number of red vertices in K_n). Clearly, H is the maximum-cardinality balanced connected subgraph in K_n. We consider trees, split graphs, bipartite graphs (properly colored), graphs of diameter 2, and present polynomial algorithms for each of them.

3.1   Trees

In this section we give a polynomial time algorithm for the *BCS* problem where the input graph is a tree. We first consider the following problem.

Problem 1: Given a tree $T = (V, E)$, and a root $t \in V$ where $V = V_R \cup V_B$. The vertices in V_R and V_B are colored red and blue,respectively. The objective is to find maximum balanced tree with root t.

We now design an algorithm to solve this problem. Let v be a vertex in G. We associate a set P_v of *pairs* of the form (r, b) to v, where r is the count of red vertices and b is the count of blue vertices. A single pair (r, b) associated with vertex v indicates that there is a subtree rooted at v having r red and b blue vertices. Note that r may not be equal to b. Now for any k pairs, the sum is also a pair which is defined as the element-wise sum of these k pairs. Let $A_1, A_2, \ldots, A_k$ be k sets. The Minkowski sum $^M \sum_{i=1}^{k} A_i$ denotes the set of sums of k elements one from each set A_i i.e., $^M \sum_{i=1}^{k} A_i = A_1 \oplus A_2 \oplus \ldots \oplus A_k$. We use $\oplus$ to denote Minkowski sum between sets. For example, for the Minkowski sum of the sets A and B, we write $A \oplus B$ and it means $A \oplus B = \{a + b \colon a \in A, b \in B\}$.

Now we are ready to describe the algorithm to solve Problem 1. In Algorithm 1, we describe how to get maximum balanced subtree with root t for a tree T rooted at t.

Algorithm 1. Construct red-blue pair-sets in a rooted tree.

Input : (i) A rooted tree $T = (B \cup R, E)$ with root t.
        (ii) B and R are colored blue and red respectively.
Output: A set of pairs at each node in T.

1 **if** v *is a leaf with red color* **then**
2 $\quad\big|\quad P_v = \{(0,0),(1,0)\}$;
3 **if** v *is a leaf with blue color* **then**
4 $\quad\big|\quad P_v = \{(0,0),(0,1)\}$;
5 **if** v *be a vertex with red color and v has k children $u_1, u_2, ..., u_k$ in T with root*
    at r, **then**
6 $\quad\big|\quad P_v = \{(0,0)\} \cup \{{}^M\sum_{i=1}^{k} P_{u_i} \oplus \{(1,0)\}\}$; // $\quad$ *Here $\oplus$ denotes Minkowski*
    Set sum.
7 **if** v *be a vertex with blue color and v has k children $u_1, u_2, ..., u_k$ in T with root*
    at r, **then**
8 $\quad\big|\quad P_v = \{(0,0)\} \cup \{{}^M\sum_{i=1}^{k} P_{u_i} \oplus \{(0,1)\}\}$;
9 **return** P_t

In Algorithm 1 we compute a finite set P_t of pairs $\{(r,b)\}$ at the root t in T. To do so, we recursively calculate the set of pairs from leaf to the root. For an internal vertex v, the set P_v is calculated as follows: let the color of v is red and it has k children $u_1, u_2, \ldots, u_k$. Then, $P_v = \{(0,0)\} \cup \{{}^M\sum_{i=1}^{k} P_{u_i} \oplus \{(1,0)\}\}$. We now prove the following lemma.

Lemma 5. *Let T be rooted tree with t as a root. Then Algorithm 1 produces all possible balanced subtrees rooted at t in $O(n^6)$ time.*

Proof. Notice that in Algorithm 1, at each node $v \in T$, we store a set P_v of pairs $\{(r_i, b_i)\}$, where each (r_i, b_i) indicates that there exists a subtree T' with root v such that number of red and blue vertices in T' are r_i and b_i, respectively. Note that r_i may not be the same as b_i. When we construct the set P_v, all the sets corresponding to its children are already calculated. Finally, in steps 6 and 8 of Algorithm 1 we calculate the set P_v based on the color of v. Hence, when Algorithm 1 terminates, we get the set P_t where t is the root of T.

Now we calculate the time taken by Algorithm 1. Clearly, steps 2 and 4 take $O(1)$ time to construct the P_v when v is a leaf. Note that, the size of P_v, for an internal node v is $O(n^2)$. Since there are at most n blue and red vertices in the subtree rooted at v. If v has k children then we have to take Minkowski sum of the sets corresponds to the children of v. To get the sum of two sets it takes $O(n^4)$ time. As there are at most n children of node v, so the time taken by steps 6 and 8 are $O(n^5)$. Finally, we traverse the tree from bottom to the root. Hence, the total time taken by the algorithm is $O(n^6)$. $\qquad\square$

We can now improve the time complexity by slightly modifying the Algorithm 1. For an internal vertex v, we actually do not need all the pairs to get the maximum balanced subtree. Suppose there are two pairs (a, b) and (c, d) in P_v, where $(b - a) = (d - c)$ and $a < c$. Then, instead of using the subtree with pair (a, b), it is better to use the subtree with pair (c, d), since it may help to construct a larger balance subtree. Therefore, in a set P_v if there are k pairs $\{(a_i, b_i); 1 \leq i \leq k\}$ such that $(b_i - a_i) = (b_j - a_j)$ whenever $i \neq j, 1 \leq i, j \leq k$. Then we remove the $(k-1)$ pairs and store only the pair which is largest among all these k pairs. We say (a_m, b_m) is largest when $a_m > a_i$ and $b_m > b_i$ for $1 \leq i \leq k, i \neq m$. So we reduce the size of P_v for each vertex $v \in T$ from $O(n^2)$ to $O(n)$. Let $T(n)$ be the time to compute red-blue pairset for the root vertex t in the tree T with size n. If r has k children $u_1, u_2, \ldots, u_k$ with size $n_1, n_2, \ldots, n_k$. Then the recurrence is $T(n) = T(n_1) + T(n_2) + \ldots + T(n_k) + O(\sum_{i=1}^{k-1}(n_1 + n_2 + \cdots + n_i)n_{i+1})$. Now $\sum_{i=1}^{k-1}(n_1 + n_2 + \ldots n_i)n_{i+1} \leq \sum_{i=1}^{k-1} n n_{i+1} = n \sum_{i=1}^{k-1} n_{i+1} \leq n^2$. which gives the solution that $T(n) = O(n^3)$. Hence, we conclude the following lemma.

Lemma 6. *Let T be rooted tree with t as a root. We can produces all possible balanced subtrees rooted at t in $O(n^3)$ time and $O(n^2)$ space complexity.*

Optimal Solution for BCS Problem in Tree
If there are n nodes in the tree T, then, for each node $v_i, 1 \leq i \leq n$, we consider T to be a tree rooted at v_i. We then apply Algorithm 1 to find maximum-cardinality balanced subtree rooted at v_i; let T_i be the resulting balanced subtree, having m_i vertices of each color. Then, to obtain an optimal solution for the $BCST$ problem in T we choose a balanced subtree that has $\max\{m_i; 1 \leq i \leq n\}$ vertices of each color. Now we can state the following theorem.

Theorem 5. *Let T be a tree whose n vertices are colored either red or blue. Then, in $O(n^4)$ time and $O(n^2)$ space, one can compute a maximum-cardinality balanced subtree of T.*

3.2 Split Graphs

A graph $G = (V, E)$ is defined to be a split graph if there is a partition of V into two sets S and K such that S is an independent set and K is a complete graph. There is no restriction on edges between vertices of S and K. Here we give a polynomial time algorithm for the BCS problem where the input graph $G = (V, E)$ is a split graph. Let V be partitioned into S and K where S and K induce an independent set and a clique respectively in G. Also, let S_B and S_R be the sets of blue and red vertices in S, respectively. Similarly, let K_B and K_R be the sets of blue and red vertices in K, respectively. We argue that there exists a balanced connected subgraph in G, having $\min\{|S_B \cup K_B|, |S_R \cup K_R|\}$ vertices of each color.

Note that if $|S_B \cup K_B| = |S_R \cup K_R|$ then G itself is balanced. Now, w.l.o.g., we can assume that $|S_B \cup K_B| < |S_R \cup K_R|$. We will find a connected balanced subgraph H of G, where the number of vertices in H is exactly $2|S_B \cup K_B|$. To do so, we first modify the graph $G = (V, E)$ to a graph $G' = (V, E')$. Then, from G', we will find the desired balanced subgraph with $|S_B \cup K_B|$ many vertices of each color. Moreover, this process is done in two steps.

Step 1: Construct $G' = (V, E')$ from $G = (V, E)$.

For each $u \in S_B$, if u is adjacent to at least a vertex u' in K_R, then remove all adjacent edges with u except the edge (u, u'). Similarly, for each $v \in S_R$, if v is adjacent to at least a vertex v' in K_B, then remove all adjacent edges with v except the edge (v, v').

Step 2: Delete $|S_R \cup K_R| - |S_B \cup K_B|$ vertices from G'.

Let $k = |S_R \cup K_R| - |S_B \cup K_B|$. Now we have following cases.

Case 1: $|S_R| \geq k$. We remove k vertices from S_R in G'. Clearly, after this modification, G' is connected, and we get a balanced subgraph having $|S_B \cup K_B|$ vertices of each color.

Case 2: $|S_R| < k$. Then we know, $|K_R| > |K_B \cup S_B|$. Let $S'_B \subseteq S_B$ be the set of vertices in G' such that each vertex of S'_B has exactly one neighbor in K_R. Then, we take a set $X \subset K_R$ with cardinality $|K_B \cup S_B|$ such that X contains all adjacent vertices of S'_B. Now we take the subgraph H of G' induced by $(S_B \cup K_B \cup X)$. H is optimal and balanced.

Running Time: Step 1 takes $O(|E|)$ time to construct G' from G. Now in step 2, both Case 1 and Case 2 take $O(|V|)$ time to delete $|S_R \cup K_R| - |S_B \cup K_B|$ vertices from G'. Hence, the total time taken is $O(n^2)$, where n is the number of vertices in G. We conclude in the following theorem.

Theorem 6. *Given a split graph G of n vertices, with r red and b blue ($n = r+b$) vertices, then, in $O(n^2)$ time we can find a balanced connected subgraph of G having $\min\{b, r\}$ vertices of each color.*

3.3   Bipartite Graphs, Properly Colored

In this section, we describe a polynomial-time algorithm for the *BCS* problem where the input graph is a bipartite graph whose nodes are colored red/blue according to proper 2-coloring of vertices in a graph. We show that there is a balanced connected subgraph of G having $\min\{b, r\}$ vertices of each color where G contains r red vertices and b blue vertices. Note that we earlier showed that the *BCS* problem is NP-hard in bipartite graphs whose vertices are colored red/blue arbitrarily; here, we insist on the coloring being a proper coloring (the construction in the hardness proof had adjacent pairs of vertices of the same color). We begin with the following lemma.

Lemma 7. *Consider a tree T (which is necessarily bipartite) and a proper 2-coloring of its nodes, with r red nodes and b blue nodes. If $r < b$, then T has at least one blue leaf.*

Proof. We prove it by contradiction. Let there is no blue leaf. Now assign any blue node say b_r as a root. Note that it always exists. Now b_r is at level 0 and b_r has degree at least 2. Otherwise, b_r is a leaf with blue color. We put all the adjacent vertices of b_r in level 1. This level consists of only red vertices. In level 2 we put all the adjacent vertices of level 1. So level 2 consists of only blue vertices. This way we traverse all the vertices in T and let that we stop at k^{th}-level. k cannot be even as all the vertices in even level are blue. So k must be odd. Now for each $0 \leqslant i \leqslant \frac{k-1}{2}$, in the vertices of (level $2i$ ∪ level $(2i + 1)$), number of blue vertices is at most the number of red vertices. Which leads to the contradiction that $r < b$. Hence there exists at least one leaf with blue color.   □

Now we describe the algorithm. We first find a spanning tree T in G. If $r = b$ then T itself is a maximum balanced subtree (subgraph also) of G. Without loss of generality assume that $r < b$. So by Lemma 7, T has at least 1 blue leaf. Now we remove that blue leaf from T. Using similar reason, we repetitively remove $(b - r)$ blue vertices from T. Finally, T becomes balanced subgraph of G, with r vertices of each color.

Running Time: Finding a spanning tree in G requires $O(n^2)$ time. To find all the leaves in the tree T requires $O(n)$ time (breadth first search). Hence the total time is needed is $O(n^2)$.

Now, we state the following theorem.

Theorem 7. *Given a bipartite graph G with a proper 2 coloring (r red or b blue vertices), then in $O(n^2)$ time we can find a balanced connected subgraph in G having $\min\{b, r\}$ vertices of each color.*

3.4   Graphs of Diameter 2

In this section, we give a polynomial time algorithm for the *BCS*-problem where the input graph has diameter 2. Let $G(V, E)$ be such a graph which contains b blue vertex set B and r red vertex set R. We find a balanced connected subgraph H of G having $\min\{b, r\}$ vertices of each color. Assume that $b < r$. This can be done in two phases. In phase 1, we generate an induced connected subgraph G' of G such that (i) G' contains all the vertices in B, and (ii) the number of vertices in G' is at most $(2b - 1)$. In phase 2, we find H from G'.

Phase 1: To generate G', we use the following result.

Lemma 8. *Let $G = (V, E)$ be a graph of diameter 2. Then for any pair of non adjacent vertices u and v from G, there always exists a vertex w such that both $(u, w) \in E$ and $(v, w) \in E$.*

We first include B in G'. Now we have the following two cases.
Case 1: The induced subgraph $G[B]$ of B is connected. In this case, G' is $G[B]$.
Case 2: The induced subgraph $G[B]$ of B is not connected. Assume that $G[B]$ has $k(> 1)$ components. Let $B_1, B_2, \ldots, B_k$ be k disjoints sets of vertices such that each induced subgraph $G[B_i]$ of B_i in G is connected. Now using Lemma 8, any two vertices $v_i \in B_i$ and $v_j \in B_j$ are adjacent to a vertex say $u_\ell \in R$. We repetitively apply Lemma 8 to merge all the k subgraphs into a larger graph. We need at most $(k - 1)$ red vertices to merge k subgraph. We take this larger graph as the graph G'.

Phase 2: In this phase, we find the balanced connected subgraph H with b vertices of each color. Note that the graph G' generated in phase 1 contains b blue and at most $(b - 1)$ red vertices. Assume that G' contains b' red vertices. We add $(b - b')$ red vertices from $G \setminus G'$ to G'. This is possible since G in connected.

Running Time: In phase 1, first finding all the blue vertices and it's induced subgraph takes $O(n^2)$ time. Now to merge all the k components into a single component which is G' needs $O(n^2)$ time. In phase 2, adding $(b - b')$ red vertices to G' takes $O(n^2)$ time as well. Hence, total time requirement is $O(n^2)$.

Theorem 8. *Given a graph $G = (V, E)$ of diameter 2, where the vertices in G are colored either red or blue. If G has b blue and r red vertices then, in $O(n^2)$ time we can find a balanced connected subgraph in G having $\min\{b, r\}$ vertices of each color.*

Acknowledgement. We thank Florian Sikora for pointing out the connection with the Graph Motif problem.

References

1. Aichholzer, O., et al.: Balanced islands in two colored point sets in the plane. arXiv preprint arXiv:1510.01819 (2015)
2. Balachandran, N., Mathew, R., Mishra, T.K., Pal, S.P.: System of unbiased representatives for a collection of bicolorings. arXiv preprint arXiv:1704.07716 (2017)
3. Bereg, S., et al.: Balanced partitions of 3-colored geometric sets in the plane. Discret. Appl. Math. **181**, 21–32 (2015)
4. Betzler, N., van Bevern, R., Fellows, M.R., Komusiewicz, C., Niedermeier, R.: Parameterized algorithmics for finding connected motifs in biological networks. IEEE/ACM Trans. Comput. Biol. Bioinform. **8**(5), 1296–1308 (2011)
5. Biniaz, A., Maheshwari, A., Smid, M.H.: Bottleneck bichromatic plane matching of points. In: CCCG (2014)
6. Böcker, S., Rasche, F., Steijger, T.: Annotating fragmentation patterns. In: Salzberg, S.L., Warnow, T. (eds.) WABI 2009. LNCS, vol. 5724, pp. 13–24. Springer, Heidelberg (2009). https://doi.org/10.1007/978-3-642-04241-6_2
7. Bonnet, É., Sikora, F.: The graph motif problem parameterized by the structure of the input graph. Discret. Appl. Math. **231**, 78–94 (2017)
8. Crowston, R., Gutin, G., Jones, M., Muciaccia, G.: Maximum balanced subgraph problem parameterized above lower bound. Theor. Comput. Sci. **513**, 53–64 (2013)
9. Derhy, N., Picouleau, C.: Finding induced trees. Discret. Appl. Math. **157**(17), 3552–3557 (2009)
10. Dumitrescu, A., Kaye, R.: Matching colored points in the plane: some new results. Comput.Geom. **19**(1), 69–85 (2001)
11. Dumitrescu, A., Pach, J.: Partitioning colored point sets into monochromatic parts. Int. J. Comput. Geom. Appl. **12**(05), 401–412 (2002)
12. El-Kebir, M., Klau, G.W.: Solving the maximum-weight connected subgraph problem to optimality. CoRR abs/1409.5308 (2014)
13. Feige, U., Peleg, D., Kortsarz, G.: The dense k-subgraph problem. Algorithmica **29**(3), 410–421 (2001)
14. Fellows, M.R., Fertin, G., Hermelin, D., Vialette, S.: Upper and lower bounds for finding connected motifs in vertex-colored graphs. J. Comput. Syst. Sci. **77**(4), 799–811 (2011)
15. Garey, M.R., Johnson, D.S.: Computers and Intractability: A Guide to the Theory of NP-Completeness. W. H. Freeman, New York (1979)

16. Johnson, D.S.: The NP-completeness column: an ongoing guide. J. Algorithms **6**(1), 145–159 (1985)
17. Kaneko, A., Kano, M.: Discrete geometry on red and blue points in the plane—a survey—. In: Aronov, B., Basu, S., Pach, J., Sharir, M. (eds.) Discrete and Computational Geometry, vol. 25, pp. 551–570. Springer, Heidelberg (2003). https://doi.org/10.1007/978-3-642-55566-4_25
18. Kaneko, A., Kano, M., Watanabe, M.: Balancing colored points on a line by exchanging intervals. J. Inf. Process. **25**, 551–553 (2017)
19. Kierstead, H.A., Trotter, W.T.: Colorful induced subgraphs. Discret. Math. **101**(1–3), 165–169 (1992)
20. Lacroix, V., Fernandes, C.G., Sagot, M.: Motif search in graphs: application to metabolic networks. IEEE/ACM Trans. Comput. Biol. Bioinform. **3**(4), 360–368 (2006)

Covering and Packing of Triangles Intersecting a Straight Line

Supantha Pandit[(✉)]

Stony Brook University, Stony Brook, NY, USA
pantha.pandit@gmail.com

Abstract. We study the four geometric optimization problems: *set cover*, *hitting set*, *piercing set*, and *independent set* with *right-triangles* (a triangle is a right-triangle whose base is parallel to the x-axis, perpendicular is parallel to the y-axis, and the slope of the hypotenuse is -1). The input triangles are constrained to be intersecting a *straight line*. The straight line can either be a *horizontal* or an *inclined* line (a line whose slope is -1). A right-triangle is said to be a λ-*right-triangle*, if the length of both its base and perpendicular is λ. For 1-right-triangles where the triangles intersect an inclined line, we prove that the set cover and hitting set problems are NP-hard, whereas the piercing set and independent set problems are in P. The same results hold for 1-right-triangles where the triangles are intersecting a horizontal line instead of an inclined line. We prove that the piercing set and independent set problems with right-triangles intersecting an inclined line are NP-hard. Finally, we give an $n^{O(\lceil \log c \rceil + 1)}$ time exact algorithm for the independent set problem with λ-right-triangles intersecting a straight line such that λ takes more than one value from $[1, c]$, for some integer c. We also present $O(n^2)$ time dynamic programming algorithms for the independent set problem with 1-right-triangles where the triangles intersect a horizontal line and an inclined line.

Keywords: Set cover · Hitting set · Piercing set · Independent set · Horizontal line · Inclined line · Diagonal line · NP-hard · Right triangles · Dynamic programming.

1 Introduction

The set cover, hitting set, piercing set, and independent set problems are the four well-studied optimization problems in computer science. Various variations and special cases of such problems received attention in the literature due to their numerous applications in different fields. In this paper, we study special cases of these problems where the given objects intersect a *straight line* that is either a *horizontal* or an *inclined* line (a line whose slope is -1). We formally define these problems as follows.

S. Pandit—Partially supported by the Indo-US Science & Technology Forum (IUSSTF) under the SERB Indo-US Postdoctoral Fellowship scheme with grant number 2017/94, Department of Science and Technology, Government of India.

© Springer Nature Switzerland AG 2019
S. P. Pal and A. Vijayakumar (Eds.): CALDAM 2019, LNCS 11394, pp. 216–230, 2019.
https://doi.org/10.1007/978-3-030-11509-8_18

Set cover problem with objects intersecting a straight line ($SCPI$)
Given a set P of points, a set $\mathcal{O}$ of objects, and a straight line L such that all the objects in $\mathcal{O}$ intersect L, the objective is to find a subset $\mathcal{O}' \subseteq \mathcal{O}$ of objects with minimum cardinality that covers all the points in P.

Hitting set problem with objects intersecting a straight line ($HSPI$) Given a set P of points, a set $\mathcal{O}$ of objects, and a straight line L such that all the objects in $\mathcal{O}$ intersect L, the objective is to find a subset $P' \subseteq P$ of points with minimum cardinality that hits all the objects in $\mathcal{O}$.

Piercing set problem with objects intersecting a straight line ($PSPI$) Given a set $\mathcal{O}$ of objects and a straight line L such that all the objects in $\mathcal{O}$ are intersecting the line L, the objective is to find a set P' of points with minimum cardinality that hits all the objects in $\mathcal{O}$.

Independent set problem with objects intersecting a straight line ($ISPI$) Given a set $\mathcal{O}$ of objects and a straight line L such that all the objects in $\mathcal{O}$ are intersecting the line L, the objective is to find a subset $\mathcal{O}' \subseteq \mathcal{O}$ of objects with maximum cardinality such that any pair of objects in $\mathcal{O}'$ do not intersect each other.

A right angled triangle is said to be a *right-triangle* if its base is parallel to the x-axis, perpendicular is parallel to the y-axis, and hypotenuse has a slope of -1. If the length of both base and perpendicular of a right-triangle is λ then we say that this right-triangle is a λ-*right-triangle*. In this paper we consider right-triangles and λ-right-triangles. We assume that the slope of the inclined line is -1. We can apply a *rotation* on the given input configuration such that the inclined line L becomes parallel to the x-axis and all the triangles are tilted at the same angle with respect to the x-axis.

1.1   Previous Work

Chepoi and Felsner [2] first considered the piercing set and independent set problems where the given geometric objects are axis-parallel rectangles intersecting an axis-monotone curve. For both of these problems they gave factor 6 approximation algorithms. Correa et al. [3] considered the same problems, where the rectangles are intersecting an inclined line. They showed that independent set problem with axis-parallel rectangles is NP-complete even when each of the rectangles is touching the inclined line at a corner. Further, they gave a factor two and factor four approximation algorithms for independent set and piercing set problems respectively. They optimally solved independent set problem in quadratic time by improving a cubic time algorithm of Lubiw [10] when the axis-parallel rectangles are on one side of the inclined line and touch the line at a single point. Recently in [11], several variations of the set cover, hitting set, piercing set, and independent set problems with axis-parallel rectangles, squares, unit-height rectangles are considered where the objects intersect an inclined line.

Fraser et al. [5] proved that the minimum hitting set and minimum set cover problems on unit disks are NP-complete, when the set of points and the set of disk centers lie inside a strip of any non-zero height. Recently, Das et al. [4] gave a cubic time algorithm for the maximum independent set problem on disks with diameter 1 such that the disk centers lie inside a unit height strip.

The independent set problem with axis-parallel line segments is NP-complete [8]. This implies that the independent set problem with triangles is NP-complete [1,8]. Further, the independent set problem can be solved optimally in cubic time for right-triangles such that right-angled corners of these triangles are on an inclined line [1].

1.2 Our Contributions

We summarize our contributions in Table 1.

Table 1. Our contributions are shown in bold colored text. ($^*n$ is the number of λ-right-triangles where λ takes more than one value in $[1, c]$. This result also holds for λ-right-triangles intersecting a horizontal line).

Geometric Objects	SCPI	HSPI	PSPI	ISPI
1-right-triangles intersecting an inclined line	NP-hard **Theorem 1**	NP-hard **Theorem 3**	P **Theorem 5**	P **Theorem 10**
1-right-triangles intersecting a horizontal line	NP-hard **Theorem 2**	NP-hard **Theorem 4**	P **Theorem 6**	P **Theorem 11**
right-triangles intersecting an inclined line	NP-hard **Theorem 1**	NP-hard **Theorem 3**	NP-hard **Theorem 7**	NP-hard **Theorem 8**
λ-right-triangles intersecting an inclined line ($\lambda \in [1, c]$, for an integer c)	NP-hard **Theorem 1**	NP-hard **Theorem 3**	**Open question ? ?**	$n^{O(\lceil \log c \rceil + 1)^*}$ time exact algorithm **Theorem 9**

2   Set Cover Problem

2.1   1-Right-Triangles Intersecting an Inclined Line

In this section, we prove that the SCPI problem with 1-right-triangles intersecting an inclined line is NP-hard. We give a reduction from the NP-hard problem *planar vertex cover* such that the degree of each vertex is at most 3 (*PVC-3*) [6]. The reduction is composed with two *phases*. Let G be a *PVC-3* problem instance. In phase one, we generate another *PVC-3* problem instance G_1 from

G by adding *dummy* vertices to G. In phase two, we generate an instance S of the *SCPI* problem for 1-right-triangles from G_1.

Phase One: In this phase a *PVC-3* problem instance G_1 is generated from G by adding *dummy* vertices to G. This phase is similar to the NP-hard reduction of Fraser and López-Ortiz [5] for the hitting set problem with unit disk such that the centers of the disks and the given points are located inside a horizontal strip. Since G is planar, we can get a planar embedding of G where no two vertices have the same x- or y-coordinates. Then if an edge e is incident to a vertex $v \in V$, we say that e can connect to v either from right direction or from left direction. As a result, the edges that are incident to a vertex in either of the directions can be ordered in y-direction. We now add dummy vertices to G in the following four *steps*.

Step 1: If $v \in V$ is a vertex of degree 3 and all the edges that incident to v are connected to it either from left direction or from right direction, then we modify the bottom edge like an *angled* shape and add a dummy vertex at the corner of the angle. See triangle shaped vertex in Fig. 1(b).

Step 2: Draw vertical lines through each vertices and add a dummy vertex at the intersection point between a vertical line and an edge of G. See square shaped vertices in Fig. 1(c).

Step 3: If the difference between the number of vertices in two consecutive vertical lines differ by more than 1 then add a vertical line between these two consecutive vertical lines. Add a dummy vertex at the intersection point between a newly added vertical line and an edge of G. See tilted square shaped vertices in Fig. 1(d).

Step 4: Let us consider an edge $e \in G$ before Step 1. If the number of dummy vertices added to e is odd, add a dummy vertex to this edge. This dummy vertex is added next to a degree 2 vertex of e. See star shaped vertices in Fig. 1(e).

Lemma 1 ([5]). *If we add $2q$ number of dummy vertices to an edge of a planar graph G, then the size of the vertex cover of G is precisely increased by q.*

Repetitively applying Lemma 1 in G, we say that the *PVC-3* problem on G_1 constructed in phase one is NP-hard.

Phase Two: In this phase, an instance S of the *SCPI* problem for 1-right-triangles is created from G_1. For each vertex $v \in G_1$, take a 1-right-triangle t_v and for each edge $e \in G_1$, take a point p_e in S. If e is an edge incident on two vertices u and v in G_1 then both the two 1-right-triangle t_u and t_v cover the point p_e. Now we describe the placement of the triangles and points.

Consider the graph G_{S3} generated after step 3 in phase one. Let L_1 and L_2 be two consecutive vertical lines in G_{S3}. Let $u_1, u_2, \ldots, u_\tau$ be the vertices on L_1 that are ordered from bottom to top. Similarly, let $v_1, v_2, \ldots, v_\kappa$ be the vertices on L_2 that are ordered from bottom to top. The corner points of $t_{u_1}, t_{u_2}, \ldots, t_{u_\tau}$ are on a vertical line ℓ_1 and the corner points of $t_{v_1}, t_{v_2}, \ldots, t_{v_\kappa}$ are another vertical

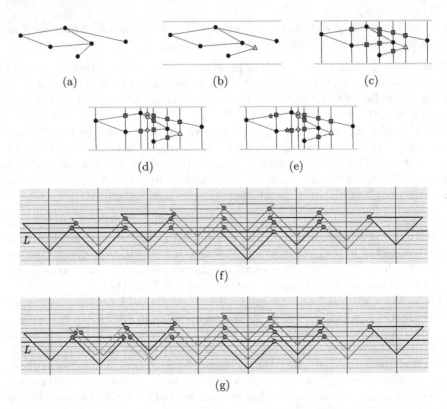

Fig. 1. (a) An instance of the *PVC-3* problem. (b)–(e) The *PVC-3* problem instances are generated from the instance in (a) during phase one. (f) An instance of the *SCPI* problem with 1-right-triangles intersecting an inclined line is generated from the graph G_{S3} (i.e., for (d)). (g) An instance of the *SCPI* problem with 1-right-triangles intersecting an inclined line is generated from the instance in (e).

line ℓ_2. The two vertical lines ℓ_1 and ℓ_2 are consecutive in S (the generated instance) and the distance between them is $2\sqrt{2} - \delta$, where δ is a very small constant. The distance between two consecutive corners on a single vertical line is 2δ. We choose δ very small (so close to zero) such that for each vertical line ℓ all the triangles on ℓ must have a common intersection point and we can find a horizontal line that intersects all the triangles (see line L in Fig. 1(g)). Now we have three cases:

Case (i) $\tau = \kappa$:

Here the difference in the y-coordinates of the corners of t_{u_i} and t_{v_i} is exactly δ, for all $i = 1$ to k.

Case (ii) $\tau < \kappa$:

Let u_i have two edges connecting to v_j and v_{j+1}. Here, the difference in y-coordinates of the corners of both (t_{u_i}, t_{v_j}) and $(t_{v_i}, t_{v_{j+1}})$ are exactly δ each.

This is possible since the difference in the number of vertices of G_{S3} on L_1 and L_2 is at most 1.

Case (iii) $\tau > \kappa$:
This case is similar to Case (ii).

Let there be an edge e between v_i and v'_j in G_1. Then the point p_e is covered by the two 1-right-triangles t_{v_i} and $t_{v'_j}$. This construction is shown in Fig. 1(f). Let v^* be a dummy vertex added on an edge $e = (v_i, v'_j)$ in Step 4 of the construction of G_1 from G. The edge e can be divided into two edges $e_1 = (v_i, v^*)$ and $e_2 = (v^*, v'_j)$. We slightly move the triangle $t_{v'_j}$ to the right and place a triangle t_{v^*} to the left of $t_{v'_j}$. The point p_{e_1} now corresponds to the point p_e. Add one extra point p_{e_1} corresponding to e_1 and place it suitably such that it only covered by the two triangles t_{v_i} and t_{v^*}. The whole construction is shown in Fig. 1(g). We rotate the above configuration such that L makes an angle $135°$ with the x-axis. Notice that, finding a minimum vertex cover in G_1 is equivalent to finding minimum number of 1-right-triangles covering all points in S.

Theorem 1. *The minimum set cover problem with 1-right-triangles intersecting an inclined line is* NP-*hard.*

2.2   1-Right-Triangles Intersecting a Horizontal Line

In this section, we prove that the *SCPI* problem with 1-right-triangles is NP-hard, where the triangles intersect a horizontal line. The reduction is from the *PVC-3* problem. Let G be an instance of the *PVC-3* problem. As in Sect. 2.1, here we describe the reduction in two phases. In phase one, we generate a *PVC-3* problem instance G_1 from G by adding dummy vertices to G and in phase two, we generate an instance S of the *SCPI* problem for 1-right-triangles from G_1.

The phase one is exactly identical to the phase one of Sect. 2.1. The phase two is similar to the phase two of Sect. 2.1. For each vertex $v \in G_1$, take a 1-right-triangle t_v and for each edge $e \in G_1$, take a point p_e in S. If e is an edge incident on two vertices u and v in G_1 then both the two 1-right-triangle t_u and t_v cover the point p_e. The placement of the points and triangles is similar to that of in phase two in Sect. 2.1. In Fig. 2, we give a construction of a *SCPI* problem instance for the *VCP-3* problem instance G in Fig. 1(a). Hence we conclude the following theorem.

Theorem 2. *The minimum set cover problem with 1-right-triangles intersecting a horizontal line is* NP-*hard.*

3   Hitting Set Problem

In this section, we prove that the *HSPI* problem with 1-right-triangles intersecting either an inclined line or a horizontal line is NP-hard. The reductions are similar to the reduction in Sect. 2.1 with some nontrivial modifications.

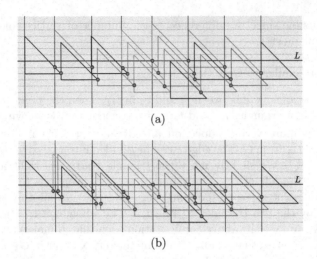

(a)

(b)

Fig. 2. (a) An instance of the *SCPI* problem for 1-right-triangles intersecting a horizontal line is generated from G_{S3} in Sect. 2.1. (b) An instance of the *SCPI* problem for 1-right-triangles intersecting a horizontal line is generated from the instance (b) in Fig. 1(a).

Theorem 3. *The minimum hitting set problem with 1-right-triangles intersecting an inclined line is* NP-*hard.*

Theorem 4. *The minimum hitting set problem with 1-right-triangles intersecting a horizontal line is* NP-*hard.*

4   Piercing Set Problem

4.1   1-Right-Triangles Intersecting an Inclined Line

In this section, we prove that the *PSPI* problem with 1-right-triangles intersecting an inclined line can be solved in polynomial time. To proceed further, we require the following *prerequisites*. For a 1-right-triangle the radius of its *incircle* is $\frac{2-\sqrt{2}}{2}$. A square with side length $\sqrt{2}-1$ is fully contained inside this incircle and hence inside the triangle. Further, the maximum distance from any point inside the triangle to its *incenter* is $\sqrt{2-\sqrt{2}}$.

We now describe the dynamic programming to solve this problem as follows. Let $\mathbb{T}$ be a set of n 1-right-triangles. All the triangles are intersecting an inclined line L. Rotate L such that L becomes horizontal. Let S be a horizontal strip of height $2\sqrt{2-\sqrt{2}}$ such that L divides it into two equal parts. Observe that, the incenters of all the triangles are inside S. We further partition S into rectangular *regions* of length $\sqrt{2}$ and height $2\sqrt{2-\sqrt{2}}$. Let $R_1, \ldots, R_r$ be the regions named according to left to right. Note that r is at most $2n$, since a triangle can intersect at most two regions. We now take a dummy region R_0 before R_1 and a dummy region R_{r+1} after R_r. Let $\mathbb{T}_i$ be the subset of 1-right-triangles that intersect region R_i. Also let $\mathbb{T}_i^{in}$ be the set of 1-right-triangles in $\mathbb{T}_i$ whose incenters are inside R_i and let $\mathbb{T}_i^{out} = \mathbb{T}_i \setminus \mathbb{T}_i^{in}$ be the set of triangles in $\mathbb{T}_i$ whose incenters are outside R_i. We now have the following lemma.

Lemma 2. *The minimum number of points required to pierce the triangles in $\mathbb{T}_i^{in}$ for any region R_i is at most 16.*

Proof. The dimension of R_i is $\sqrt{2} \times 2\sqrt{2 - \sqrt{2}}$. Hence, 16 squares $s_1, s_2, \ldots, s_{16}$ each of length $\sqrt{2} - 1$, 4 in a column and 4 in a row, fully cover R_i. Hence, using the above prerequisites we say that the center c_i of square s_i pierces all the triangles whose incenters are inside s_i. Therefore, we have the desired result. $\square$

Lemma 3. *Any optimal piercing set contains at most 48 points from region R_i.*

Proof. Observe that $\mathbb{T}_i^{out} \subseteq \mathbb{T}_{i-1}^{in} \cup \mathbb{T}_{i+1}^{in}$. Therefore, by Lemma 2 the triangles in $\mathbb{T}_i$ can be pierced by 48 points, 16 points each from regions R_{i-1}, R_i, and R_{i+1}. Since points inside R_i can only pierce triangles in $\mathbb{T}_i$, if an optimal solution OPT contains more than 48 points from region R_i, we can replace them by 48 points in regions R_{i-1}, R_i, R_{i+1} without leaving any triangle to be pierced. This contradicts that OPT was an optimal piercing set. $\square$

Let the triangles in $\mathbb{T}_i$ divide R_i into m_i subregions. Since any two 1-right-triangles can intersect in at most 2 points on their boundary, $m_i = O(|S_i|^2) = O(n^2)$. Let P_i be a set of m_i points, with one point picked from each of the m_i subregions. We can assume that optimal piercing set is a subset of $\bigcup_{i=0}^{r+1} P_i$.

For $0 \leq i \leq r$, let $T(i, U_1, U_2)$ where $U_1 \subseteq P_i$ and $U_2 \subseteq P_{i+1}$ denote the cost of an optimal piercing set H for the triangles that lie completely inside $\bigcup_{j=i}^{r+1} R_j$ such that $H \cap P_i = U_1$ and $H \cap P_{i+1} = U_2$. Note that by Lemma 3, we can assume that both U_1 and U_2 have at most 48 points each. $T(i, U_1, U_2)$ satisfies the following recurrence:

1. If $U_1 \bigcup U_2$ does not pierce all triangles that lie completely inside $R_i \bigcup R_{i+1}$, then $T(i, U_1, U_2) = \infty$.
2. Otherwise, $T(i, U_1, U_2) = \min_{U_3 \subseteq P_{i+2}, |U_3| \leq 48} T(i+1, U_2, U_3) + |U_1|$.

The optimal piercing set is now given by $\min_{U_1 \subseteq P_0, U_2 \subseteq P_1} T(0, U_1, U_2)$ and can be obtained from the above recurrences by dynamic programming.

We now analyze the time required to find the optimal piercing set. As each P_i has $O(n^2)$ points, the number of 48 element subsets of P_i is $O(n^{96})$. Therefore, the number of subproblems $T(i, U_1, U_2)$ defined above are at most $O(n^{193})$. Each subproblem depends on $O(n^{96})$ smaller subproblems. By allowing an extra bookkeeping cost the total time taken to compute the optimal piercing set is $n^{O(1)}$. This leads to the following theorem.

Theorem 5. *The minimum piercing set problem with 1-right-triangles intersecting an inclined line can be solved in polynomial time.*

4.2   1-Right-Triangles Intersecting a Horizontal Line

The algorithm and analysis is similar to the algorithm and its analysis described in Sect. 4.1 with the following differences. Since the triangles are intersecting a horizontal line, the maximum distance from the horizontal line to an *incenter* of a triangle is $\frac{2-\sqrt{2}}{2}$. Hence, we choose the height of S as $2 - \sqrt{2}$. Further, we choose the length and height of each region R_i as 1 and $2 - \sqrt{2}$ respectively. Using the prerequisites and similar analysis of Lemmas 2 and 3 in Sect. 4.1 ensure that, for 1-right-triangles intersecting a horizontal line any optimal piercing set contains at most 18 points from region R_i. The rest of the analysis is similar to the analysis described in Sect. 4.1 with different constants involved in the running time.

Theorem 6. *The minimum piercing set problem with 1-right-triangles intersecting a horizontal line can be solved in polynomial time.*

4.3   Right-Triangles Intersecting an Inclined Line

In this section, we prove that the *PSPI* problem with right-triangles intersecting an inclined line is NP-hard. We give a reduction from the following rectilinear embedding of planar 3 satisfiability *RP-3-SAT* problem. Lichtenstein [9] proved that the planar 3 satisfiability problem is NP-complete. Later on, Knuth and Raghunathan [7] proved that the *RP-3-SAT* is NP-complete.

Definition 1 (*RP-3-SAT* [7]). *Let ϕ be a 3 cnf formula where each clause in ϕ contains exactly 3 literals. The variables are placed on a horizontal line and the three legged clauses connects to these variables either from above or from below by rectilinear line segments such that no two line segments can intersect each other (see Fig. 3(a)). The objective is to find an assignment of truth values to the variables such that the formula ϕ becomes satisfiable.*

Here for our purpose, we modify the *RP-3-SAT* problem in the following way. This modification is similar in [11].

Definition 2 (*Modified-RP-3-SAT* [11]). *Let ϕ be a 3 cnf formula where each clause in ϕ contains exactly 3 literals. The variables are placed on an inclined line and the clauses shaped "⌊ " or "⌉ " connect to the variables either from the above or below of the inclined line such that no two line segments that form the clause shapes can intersect each other (see Fig. 3(b)). The objective is to find an assignment of truth values to the variables such that the formula ϕ becomes satisfiable.*

We assume that α to be the maximum number of clauses that connect to a variable of ϕ either from above or below. Next we assume that $q = 4\alpha + 1$. We now describe the construction in terms of *variable gadgets*, *clause gadgets*, and *variable-clause-interaction* as follows.

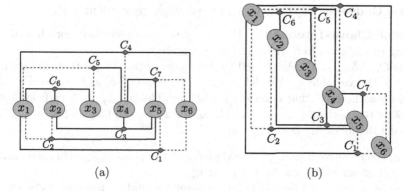

(a)   (b)

Fig. 3. (a) Representation of a *Rectilinear-Planar-3-SAT* formula $\phi = C_1 \wedge C_2 \wedge C_3 \wedge C_4 \wedge C_5 \wedge C_6$, where $C_1 = (x_1 \vee \overline{x}_2 \vee \overline{x}_3)$, $C_2 = (\overline{x}_1 \vee x_3 \vee \overline{x}_5)$, $C_3 = (\overline{x}_3 \vee x_4 \vee \overline{x}_5)$, $C_4 = (\overline{x}_2 \vee \overline{x}_3 \vee x_4)$, $C_5 = (x_1 \vee x_2 \vee \overline{x}_4)$, and $C_6 = (x_1 \vee \overline{x}_4 \vee x_5)$. (b) Representation of the formula ϕ in (a) after modification. In both the figures solid (resp. dotted) lines represents that the variable is positively (resp. negatively) present in that clause.

Variable Gadgets: The variable gadgets are shown in Fig. 4(a)). For a variable x_i, we take $2q$ right-triangles $\{g_1^i, g_2^i, \ldots, g_{2q}^i\}$ such that q right-triangles are placed in both above and below the inclined line that are also intersecting the inclined line. Further, any two consecutive right-triangles intersect at a single point and all the $2q$ right-triangles form a cycle of length $2q$. We named these intersection points $P^i = \{p_1^i, p_2^i, \ldots, p_{2q}^i\}$ as *piercing points*. Now to pierce the $2q$ right-triangles , we can select points from the piercing points. Since the piercing points also form a cycle of length $2q$, there are two optimal piercing sets of points $P_{odd}^i = \{p_1^i, p_3^i, \ldots, p_{2q-1}^i\}$ and $P_{even}^i = \{p_2^i, p_4^i, \ldots, p_{2q}^i\}$ for each variable gadget. These two sets corresponding to the truth values of the variable x_i.

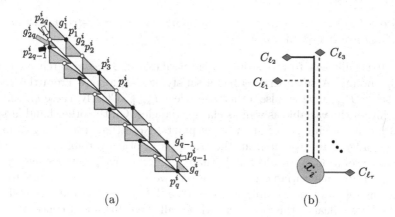

(a)   (b)

Fig. 4. (a) Structure of a gadget for the variable x_i. (b) Naming convention of the clauses that connect to the variable x_i.

Clause Gadgets: For clause C, we take a single right-triangle g^c.

Variable-Clause-Interaction: Here we give the description for clauses connecting to variables from right and left separately.

Let $C_{\ell_1}, C_{\ell_2}, \ldots, C_{\ell_r}$ be the left to right order of the clauses that connect from above to the variable x_i. We define the α-*th* clause of x_i as C_{ℓ_α} (see Fig. 4(b)). Consider a clause C that connects to the variables x_i, x_j, and x_k from above. Let C be the α-th, β-th, and γ-th clause for x_i, x_j, and x_k respectively. Then we make the following (see Fig. 5(a)).

- If x_i is a negative literal, g^c should pierce by the point $p^i_{4\alpha-1}$. If x_i is a positive literal, g^c should pierce by the point $p^i_{4\alpha}$.
- If x_j is a negative literal, modify (possibly extending the corners of the triangles and moving the triangles towards left, right, up, or down so that their lower-left corner remain touching the inclined line) the two triangles $g^j_{4\beta-1}$ and $g^j_{4\beta}$ such that both of them plus g^c have a common intersection point. Also, modify the triangle $g^j_{4\beta-2}$ such that it remain intersect with $g^j_{4\beta-1}$. If x_j is a positive literal, modify the two triangles $g^j_{4\beta}$ and $g^j_{4\beta+1}$ such that both of them plus g^c have a common intersection point. Also, modify the triangle $g^j_{4\beta-1}$ such that it remain intersect with $g^j_{4\beta}$.
- If x_k is a negative literal, modify the two triangles $g^k_{4\gamma-1}$ and $g^k_{4\gamma}$ such that both of them plus g^c have a common intersection point. Also, modify the triangle $g^k_{4\gamma-2}$ such that it remains intersecting with $g^k_{4\gamma-1}$. If x_k is a positive literal, modify the two triangles $g^k_{4\gamma}$ and $g^k_{4\gamma+1}$ such that both of them plus g^c have a common intersection point. Also, modify the triangle $g^k_{4\gamma-1}$ such that it remain intersect with $g^k_{4\gamma}$.

A similar construction can be done for a clause C that connects to the variables x_i, x_j, and x_k from below (see Fig. 5(b)). Note that, here the index of the triangles are different. Clearly, the construction takes polynomial time.

Theorem 7. *The minimum piercing set problem with right-triangles intersecting an inclined line is* NP-*hard.*

Proof. It is sufficient to prove that, ϕ is satisfiable iff S has a piercing set of at most qn points. Assume that, ϕ has a satisfying assignment. From the gadget of x_i, select P^i_{odd} if x_i is false, otherwise select P^i_{even}. Clearly, these qn selected points hit all the variable as well as clause triangles. On the other hand, assume that S has a piercing set of at most qn points. Note that, the triangles in one variable gadget are disjoint from the triangles in other variable gadget. Then, any qn size solution must select q alternate representative points i.e., either P^i_{odd} or P^i_{even} from each variable gadget. Hence, we can set binary values to each variable as follows. Set variable x_i to be true if P^i_{even} is selected, otherwise set x_i to be false. Clearly, this solution satisfies all the clauses and hence ϕ. □

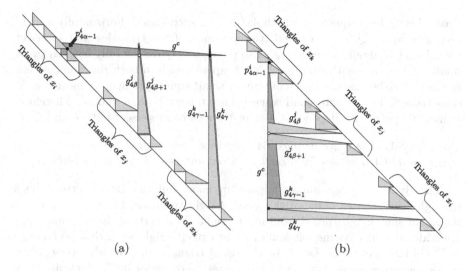

Fig. 5. (a) Interconnection between the gadgets for the variables x_i, x_j, and x_k and the rectangle for the clause C where C connects the variables form the right. (b) Interconnection between the gadgets for the variables x_i, x_j, and x_k and the rectangle for the clause C where C connects the variables form the left.

5   Independent Set Problem

5.1   Right-Triangles Intersecting an Inclined Line

In this section, we prove that the *ISPI* problem with right-triangles intersecting an inclined line is NP-hard. The reduction is along the same line as the reduction of the *ISPI* problem with rectangles such that the rectangles are intersecting an inclined line. Correa et al. [3] proved that this problem is NP-hard (see Section 4, Theorem 5 of [3]).

Theorem 8. *The maximum independent set problem with right-triangles intersecting an inclined line is NP-hard.*

5.2 λ-Right-Triangles Intersecting an Inclined Line

In this section, we present an exact algorithm for the *ISP* problem with λ-right-triangles where the triangles are intersecting an inclined line and λ takes more than one value from $[1, c]$, for some positive integer c. Let us assume that $\mathbb{T}$ be a set of n λ-right-triangles that are intersecting an inclined line L. After a rotation we ensure that L becomes parallel to the x-axis. We first group the triangles in $\mathbb{T}$ into $k = \lceil \log c \rceil + 1$ groups $g_1, g_2, \ldots, g_k$, where the i-th group g_i contains triangles with $\lambda \in [2^{i-1}, 2^i)$. We now consider the following lemma.

Lemma 4. *There are at most 64 pairwise independent λ-right-triangles with $\lambda \in [1, 2)$ such that they intersect both L and a vertical line L'.*

Proof. Let Q be a square of length $4\sqrt{2} - \sqrt{2}$ such that L horizontally and L' vertically bisect Q. Observe that, the incenters of the triangles that intersect both L and L' should be inside Q. Also the square Q can be fully covered by 64 small squares each with length $\sqrt{2} - 1$, 8 squares each in both row and column. A single λ-right-triangle fully contains a small square. Hence, at most one λ-right-triangle for a single small square can intersect both L and L'. Therefore, at most 64 pairwise independent λ-right-triangle can intersect both L and L'. □

Corollary 1. *There are at most $64k$ pairwise independent λ-right-triangles, a maximum of 64 triangles from each g_i, which intersect L and a vertical line.*

The dynamic programming algorithm is as follows. Draw vertical lines through every corner point of the given λ-right-triangles. Let $\ell_1, \ell_2, \ldots, \ell_\tau$ be the set of such lines ordered from left to right. Two vertical lines ℓ_0, and $\ell_{\tau+1}$ are added one at extreme left and other at extreme right such that no triangle in $\mathbb{T}$ will intersect them. Let T_i be the set of triangles in $\mathbb{T}$ which intersect line ℓ_i. Define subproblem $S(T_i', i)$ to be the set of triangles in $\mathbb{T}$ which are fully contained inside the region left to ℓ_i plus an independent set of triangles $T_i' \subseteq T_i$ and let $V(T_i', i)$ be the value of an optimal solution for this subproblem. The value satisfies the following recurrence.

$$V(T_i', i) = \max_{T_{i-1}' \subseteq T_{i-1}, |T_{i-1}'| \leq 64k} V(T_{i-1}', i - 1) + |T_i' \setminus T_{i-1}'|,$$

where T_{i-1}' is an independent set.

Optimal Solution: The optimal solution can be found by evaluating $V(\emptyset, \tau+1)$.

Running Time: Note that τ is $O(n)$, since there are $3n$ corner points. The total number of sunproblems are $O(n^{64k} + 1)$ and each subproblem is depends on $O(n^{64k})$ subproblems. Hence, the total time taken is $n^{O(k)}$), i.e., $n^{O(\lceil \log c \rceil + 1)}$.

Theorem 9. *The maximum independent set problem with λ-right-triangles intersecting an inclined line can be solved in $n^{O(\lceil \log c \rceil + 1)}$ time, where n is the number of λ-right-triangles and λ takes values from $[1, c]$.*

Corollary 2. *The maximum independent set problem with 1-right-triangles intersecting an inclined line can be solved in polynomial time.*

Remark 1. Theorem 9 is true for λ-right-triangles intersecting a horizontal line. As a corollary, we note that the maximum independent set problem with 1-right-triangles intersecting a horizontal line can be solved in polynomial time.

5.3   1-Right-Triangles Intersecting a Straight Line

We design a dynamic programming algorithm for the *ISP* problem with 1-right-triangles intersecting a horizontal line. The algorithm runs in $O(n^2)$ time. We also design a dynamic programming algorithm for the *ISP* problem with 1-right-triangles intersecting an inclined line with the same running time.

Let $T = \{t_1, t_2, \ldots, t_n\}$ be a set of n 1-right-triangles intersecting a horizontal line. For a triangle $t \in T$, let $x(t)$ be the x-coordinate of the perpendicular of t.

Lemma 5. *Let $t_1, t_2, t_3 \in T$ be three 1-right-triangles intersecting a horizontal line L such that $x(t_1) < x(t_2) < x(t_3)$. If t_1, t_2 are non-intersecting and t_2, t_3 are non-intersecting, then t_1, t_3 are non-intersecting.*

Sort the triangles in T using the x-coordinates of their perpendiculars and let $t_1, t_2, \ldots, t_n$ be this order. Define a subproblem $T(i)$ as the set of all triangles $\{t_1, t_2, \ldots t_i\}$. Let $O(i)$ be an optimal set of triangles in $T(i)$ with value $OPT(i)$. We use the following recursion whose correctness follows from Lemma 5.

$$OPT(i) = \max\{OPT(j) + 1, OPT(i-1)\}$$

The optimal solution of the problem can be found by calling the function $OPT(n)$ with the base case as $OPT(1) = 1$. Thus we conclude:

Theorem 10. *The maximum independent set problem with 1-right-triangles intersecting a horizontal line can be solved in $O(n^2)$ time.*

For the 1-right-triangles intersect an inclined line L, we first rotate the given configuration such that L becomes parallel to the x axis. Here, for a triangle $t \in T$, $x(t)$ defines the x-coordinate of the incenter of t. Here also we get a similar result like Lemma 5 and a similar dynamic programming above concludes:

Theorem 11. *The maximum independent set problem with 1-right-triangles intersecting an inclined line can be solved in $O(n^2)$ time.*

References

1. Catanzaro, D., et al.: Max point-tolerance graphs. Discret. Appl. Math. **216**, 84–97 (2017)
2. Chepoi, V., Felsner, S.: Approximating hitting sets of axis-parallel rectangles intersecting a monotone curve. Comput. Geom. **46**(9), 1036–1041 (2013)
3. Correa, J., Feuilloley, L., Pérez-Lantero, P., Soto, J.A.: Independent and hitting sets of rectangles intersecting a diagonal line: algorithms and complexity. Discret. Comput. Geom. **53**(2), 344–365 (2015)
4. Das, G.K., De, M., Kolay, S., Nandy, S.C., Sur-Kolay, S.: Approximation algorithms for maximum independent set of a unit disk graph. Inf. Process. Lett. **115**(3), 439–446 (2015)
5. Fraser, R., López-Ortiz, A.: The within-strip discrete unit disk cover problem. Theor. Comput. Sci. **674**, 99–115 (2017)
6. Garey, M.R., Johnson, D.S.: The rectilinear steiner tree problem is NP-complete. SIAM J. Appl. Math. **32**(4), 826–834 (1977)
7. Knuth, D.E., Raghunathan, A.: The problem of compatible representatives. SIAM J. Discret. Math. **5**(3), 422–427 (1992)

8. Kratochvíl, J., Nešetřil, J.: INDEPENDENT SET and CLIQUE problems in intersection-defined classes of graphs. Commentationes Mathematicae Universitatis Carolinae **031**(1), 85–93 (1990)
9. Lichtenstein, D.: Planar formulae and their uses. SIAM J. Comput. **11**(2), 329–343 (1982)
10. Lubiw, A.: A weighted min-max relation for intervals. J. Comb. Theory Ser. B **53**(2), 151–172 (1991)
11. Mudgal, A., Pandit, S.: Covering, hitting, piercing and packing rectangles intersecting an inclined line. In: Lu, Z., Kim, D., Wu, W., Li, W., Du, D.-Z. (eds.) COCOA 2015. LNCS, vol. 9486, pp. 126–137. Springer, Cham (2015). https://doi.org/10.1007/978-3-319-26626-8_10

H-Free Coloring on Graphs
with Bounded Tree-Width

N. R. Aravind[1], Subrahmanyam Kalyanasundaram[1],
and Anjeneya Swami Kare[2(✉)]

[1] Department of Computer Science and Engineering, IIT Hyderabad,
Hyderabad, India
{aravind,subruk}@iith.ac.in
[2] School of Computer and Information Sciences, University of Hyderabad,
Hyderabad, India
askcs@uohyd.ac.in

Abstract. Let H be a fixed undirected graph. A vertex coloring of an undirected input graph G is said to be an H-FREE COLORING if none of the color classes contain H as an induced subgraph. The H-FREE CHROMATIC NUMBER of G is the minimum number of colors required for an H-FREE COLORING of G. This problem is NP-complete and is expressible in monadic second order logic (MSOL). The MSOL formulation, together with Courcelle's theorem implies linear time solvability on graphs with bounded tree-width. This approach yields an algorithm with running time $f(||\varphi||, t) \cdot n$, where $||\varphi||$ is the length of the MSOL formula, t is the tree-width of the graph and n is the number of vertices of the graph. The dependency of $f(||\varphi||, t)$ on $||\varphi||$ can be as bad as a tower of exponentials.

In this paper, we provide an explicit combinatorial FPT algorithm to compute the H-FREE CHROMATIC NUMBER of a given graph G, parameterized by the tree-width of G. The techniques are also used to provide an FPT algorithm when H is forbidden as a subgraph (not necessarily induced) in the color classes of G.

1 Introduction

Let G be an undirected graph. The classical q-COLORING problem asks to color the vertices of the graph using at most q colors such that no pair of adjacent vertices are of the same color. The CHROMATIC NUMBER of the graph is the minimum number of colors required for q-coloring the graph and is denoted by $\chi(G)$. The graph coloring problem has been extensively studied in various settings.

In this paper we consider a generalization of the graph coloring problem called H-FREE q-COLORING which asks to color the vertices of the graph using at most q colors such that none of the color classes contain H as an induced subgraph. Here, H is any fixed graph, $|V(H)| = r$, for some fixed r. The H-FREE CHROMATIC NUMBER is the minimum number of colors required to H-free color

© Springer Nature Switzerland AG 2019
S. P. Pal and A. Vijayakumar (Eds.): CALDAM 2019, LNCS 11394, pp. 231–244, 2019.
https://doi.org/10.1007/978-3-030-11509-8_19

the graph. Note that when $H = K_2$, the H-FREE q-COLORING problem is same as the classical q-COLORING problem.

For $q \geq 3$, H-FREE q-COLORING problem is NP-complete as the q-COLORING problem is NP-complete. The 2-COLORING problem is polynomial time solvable as it is equivalent to decide whether the graph is bipartite. The H-FREE 2-COLORING problem has been shown to be NP-complete as long as H has 3 or more vertices [1]. A variant of H-FREE COLORING problem which we call H-(SUBGRAPH)FREE q-COLORING which asks to color the vertices of the graph such that none of the color classes contain H as a subgraph (not necessarily induced) is studied in [2,3].

Graph bipartitioning (2-coloring) problems with other constraints have been explored in the past. Many variants of 2-coloring have been shown to be NP-hard. Recently, Karpiński [4] studied a problem which asks to color the vertices of the graph using 2 colors such that there is no monochromatic cycle of a fixed length. The degree bounded bipartitioning problem asks to partition the vertices of G into two sets A and B such that the maximum degree in the induced subgraphs $G[A]$ and $G[B]$ are at most a and b respectively. Xiao and Nagamochi [5] proved that this problem is NP-complete for any non-negative integers a and b except for the case $a = b = 0$, in which case the problem is equivalent to testing whether G is bipartite. Other variants that place constraints on the degree of the vertices within the partitions have also been studied [6,7]. Wu, Yuan and Zhao [8] showed the NP-completeness of the variant that asks to partition the vertices of the graph G into two sets such that both the induced graphs are acyclic. Farrugia [9] showed the NP-completeness of a problem called $(\mathcal{P}, \mathcal{Q})$-coloring problem. Here, $\mathcal{P}$ and $\mathcal{Q}$ are any additive induced-hereditary graph properties. The problem asks to partition the vertices of G into A and B such that $G[A]$ and $G[B]$ have properties $\mathcal{P}$ and $\mathcal{Q}$ respectively.

For a fixed q, the H-FREE q-COLORING problem can be expressed in monadic second order logic (MSOL) [10]. The MSOL formulation together with Courcelle's theorem [11,12] implies linear time solvability on graphs with bounded tree-width. This approach yields an algorithm with running time $f(||\varphi||, t) \cdot n$, where $||\varphi||$ is the length of the MSOL formula, t is the tree-width of the graph and n is the number of vertices of the graph. The dependency of $f(||\varphi||, t)$ on $||\varphi||$ can be as bad as a tower of exponentials.

In this paper we present explicit combinatorial algorithms for the H-FREE q-COLORING problem. We have the following results:

- $O(q^{4t^r} \cdot n)$ time algorithm for the H-FREE q-COLORING problem for any arbitrary fixed graph H on r vertices.
- $O(2^{t+r \log t} \cdot n)$ time algorithm for K_r-Free 2-Coloring problem, where K_r is a complete graph on r vertices.
- $O(2^{3t^2} \cdot n)$ time algorithm for C_4-Free 2-Coloring problem, where C_4 is a cycle on 4 vertices.

From the above we get the explicit FPT algorithm for H-FREE CHROMATIC NUMBER problem. The techniques can also be extended to obtain analogous results for the H-(SUBGRAPH)FREE q-COLORING.

2   Preliminaries

For a vertex set $S \subseteq V$, the subgraph induced by S is denoted by $G[S]$. A graph G is said to be H-free if G does not have H as an induced subgraph. We follow the standard graph theoretic terminology from [13].

A parameterized problem is a language $L \subseteq \Sigma^* \times \mathbb{N}$, where Σ is a fixed and finite alphabet. For $(x, k) \in \Sigma^* \times \mathbb{N}$, k is referred to as the parameter. A parameterized problem L is *fixed parameter tractable (FPT)* if there is an algorithm A, a computable non-decreasing function $f : \mathbb{N} \to \mathbb{N}$ and a constant c such that, given $(x, k) \in \Sigma^* \times \mathbb{N}$ the algorithm A correctly decides whether $(x, k) \in L$ in time bounded by $f(k).|x|^c$. For more details on parameterized algorithms refer to [14]. A *tree decomposition* of G is a pair $(T, \{X_i, i \in I\})$, where for $i \in I$, $X_i \subseteq V$ (usually called bags) and T is a tree with elements of I as the nodes such that:

1. For each vertex $v \in V$, there is an $i \in I$ such that $v \in X_i$.
2. For each edge $\{u, v\} \in E$, there is an $i \in I$ such that $\{u, v\} \subseteq X_i$.
3. For each vertex $v \in V$, $T[\{i \in I | v \in X_i\}]$ is connected.

The width of the tree decomposition is $\max_{i \in I}(|X_i| - 1)$. The tree-width of G is the minimum width taken over all tree decompositions of G and we denote it as t. For more details on tree-width, we refer the reader to [15]. A rooted tree decomposition is called a *nice tree decomposition*, if every node $i \in I$ is one of the following types:

1. Leaf Node: For a leaf node i, $X_i = \emptyset$.
2. Introduce Node: An introduce node i has exactly one child j and there is a vertex $v \in V \backslash X_j$ such that $X_i = X_j \cup \{v\}$.
3. Forget Node: A forget node i has exactly one child j and there is a vertex $v \in V \backslash X_i$ such that $X_j = X_i \cup \{v\}$.
4. Join Node: A join node i has exactly two children j_1 and j_2 such that $X_i = X_{j_1} = X_{j_2}$.

The notion of *nice tree decomposition* was introduced by Kloks [16]. Every graph G has a nice tree decomposition with $|I| = O(n)$ nodes and width equal to the tree-width of G. Moreover, such a decomposition can be found in linear time if the tree-width is bounded.

2.1   Overview of the Techniques Used

In the rest of the paper, we assume that the nice tree decomposition is given. Let i be a node in the nice tree decomposition, X_i is the bag of vertices associated with the node i. Let T_i be the subtree rooted at the node i and $G[T_i]$ denote the graph induced by all the vertices in T_i.

We use dynamic programming on the nice tree decomposition. We process the nodes of the nice tree decomposition according to its post order traversal. We say that a partition (A, B) of G is a *valid* partition if neither $G[A]$ nor $G[B]$ has

H as an induced subgraph. At each node i, we check each bipartition (A_i, B_i) of the bag X_i to see if (A_i, B_i) leads to a valid partition in the graph $G[T_i]$. For each partition, we also keep some extra information that will help us to detect if the partition leads to an invalid partition at some ancestral (parent) node. We have four types of nodes in the tree decomposition – leaf, introduce, forget and join nodes. In the algorithm, we explain the procedure for updating the information at each of these nodes and consequently, to certify whether a partition is valid or not. During the description of the algorithms, we refer to the set $V(T_i) \backslash X_i$, i.e., the vertices in the subtree T_i but not in the bag X_i, as *forgotten vertices* of the subtree T_i.

In Sect. 3, we start the discussion with H-FREE 2-COLORING problems. In Sects. 3.1 and 3.2, we discuss the algorithm for the cases when $H = K_r$ and $H = C_4$ respectively before moving on to the case of general H in Sect. 3.3. In Sect. 4, we give the algorithm for H-FREE q-COLORING problem. In Sect. 5, we give the algorithm for H-(SUBGRAPH)FREE q-COLORING problem. Presenting the algorithms for $H = K_r$ and $H = C_4$ initially will help in the exposition, as they will help to understand the setup before moving to the more involved general case.

3   Algorithms for H-FREE 2-COLORING Problems

3.1 K_r-Free 2-Coloring

In this section, we consider the H-FREE 2-COLORING problem when $H = K_r$, a complete graph on r vertices.

Let $\Psi = (A_i, B_i)$ be a partition of a bag X_i. We set $M_i[\Psi]$ to 1 if there exists a partition (A, B) of $V(T_i)$ such that $A_i \subseteq A$, $B_i \subseteq B$ and both $G[A]$ and $G[B]$ are K_r-free. Otherwise, $M_i[\Psi]$ is set to 0.

Leaf Node: For a leaf node $\Psi = (\emptyset, \emptyset)$ and $M_i[\Psi] = 1$. This step takes constant time.

Introduce Node: Let j be the only child of the node i. Let v be the lone vertex in $X_i \backslash X_j$. Let $\Psi = (A_i, B_i)$ be a partition of X_i. If $G[A_i]$ or $G[B_i]$ has K_r as a subgraph, we set $M_i[\Psi]$ to 0. Otherwise, we use the following cases to compute $M_i[\Psi]$ value. Since v cannot have forgotten neighbors, it can form a K_r only within the bag X_i.

Case 1: $v \in A_i$, $M_i[\Psi] = M_j[\Psi']$, where $\Psi' = (A_i \backslash \{v\}, B_i)$.
Case 2: $v \in B_i$, $M_i[\Psi] = M_j[\Psi']$, where $\Psi' = (A_i, B_i \backslash \{v\})$.

The total number of Ψ's for X_i is 2^{t+1}, for each Ψ checking if $G[A_i]$ or $G[B_i]$ contains K_r as subgraph can be done in $(t + 1)^r r^2$ time. Hence the total time complexity at the introduce node is $O(2^t t^r)$.

Forget Node: Let j be the only child of the node i. Let v be the lone vertex in $X_j \backslash X_i$. Let $\Psi = (A_i, B_i)$ be a partition of X_i. If $G[A_i]$ or $G[B_i]$ has K_r as a subgraph, we set $M_i[\Psi]$ to 0. Otherwise, $M_i[\Psi] = \max\{M_j[\Psi'], M_j[\Psi'']\}$, where, $\Psi' = (A_i \cup \{v\}, B_i)$ and $\Psi'' = (A_i, B_i \cup \{v\})$.

The total number of Ψ's for X_i is 2^t, for each Ψ checking if $G[A_i]$ or $G[B_i]$ contains K_r as subgraph can be done in $t^r r^2$ time. Hence the total time complexity at the forget node is $O(2^t t^r)$.

Join Node: Let j_1 and j_2 be the children of the node i. $X_i = X_{j_1} = X_{j_2}$ and $V(T_{j_1}) \cap V(T_{j_2}) = X_i$. Let $\Psi = (A_i, B_i)$ be a partition of X_i. If $G[A_i]$ or $G[B_i]$ has K_r as a subgraph, we set $M_i[\Psi]$ to 0. Otherwise, we use the following expression to compute $M_i[\Psi]$ value. Since there are no edges between $V(T_{j_1}) \backslash X_i$ and $V(T_{j_2}) \backslash X_i$, a K_r cannot contain forgotten vertices from both T_{j_1} and T_{j_2}.

$$M_i[\Psi] = \begin{cases} 1, & \text{If } M_{j_1}[\Psi] = 1 \text{ and } M_{j_2}[\Psi] = 1. \\ 0, & \text{Otherwise.} \end{cases} \tag{1}$$

The total number of Ψ's for X_i is 2^{t+1}, for each Ψ checking if $G[A_i]$ or $G[B_i]$ contains K_r as subgraph can be done in $(t+1)^r r^2$ time. Hence the total time complexity at the join node is $O(2^t t^r)$.

The correctness of the algorithm is implied from the correctness of $M_i[\Psi]$ values, which can be proved using bottom up induction on the nice tree decomposition. G has a valid bipartitioning if there exists a Ψ such that $M_r[\Psi] = 1$, where r is the root node of the nice tree decomposition. The total time complexity of the algorithm is $O(2^t t^r \cdot n) = O(2^{t+r \log t} \cdot n)$. With this we state the following theorem.

Theorem 1. *There is an $O(2^{t+r \log t} \cdot n)$ time algorithm that solves the H-FREE 2-COLORING problem when $H = K_r$, on graphs with tree-width at most t.*

3.2 C_4-Free 2-Coloring

In this section, we describe the combinatorial algorithm for the H-FREE 2-COLORING problem for the case when $H = C_4$, a cycle of length 4.

Note that an induced cycle of length 4 is formed when a pair of non-adjacent vertices have two non-adjacent neighbors. If a graph has no induced C_4 then any non-adjacent vertex pairs cannot have two or more non-adjacent vertices as neighbors. They can have neighbors which are pairwise adjacent. We keep track of such vertex pairs as they can form an induced C_4 at some ancestral (introduce/join) nodes. Let X_i be a bag at the node i of the nice tree decomposition. We consider partitions (A_i, B_i) of the bag X_i and see if they lead to a valid partition (A, B) of $V(T_i)$. For each non-adjacent pair of vertices from A_i (similarly B_i), we also guess if the pair has a common forgotten neighbor in part A (similarly B) of the partition. We check if the above guesses lead to a valid partitioning in the subgraph $G[T_i]$, which is the graph induced by the vertices in the node i and all its descendant nodes. In this section, we use the standard notation of $\binom{S}{2}$ to denote the set of all 2-subsets of a set S.

Let $\Psi = (A_i, B_i, P_i, Q_i)$ be a 4-tuple defined as follows: (A_i, B_i) is a partition of X_i, $P_i \subseteq \binom{A_i}{2}$ and $Q_i \subseteq \binom{B_i}{2}$. Intuitively, P_i and Q_i are the set of those non-adjacent pairs that have common forgotten neighbor.

We define $M_i[\Psi]$ to be 1 if there is a partition (A, B) of $V(T_i)$ such that:

1. $A_i \subseteq A$ and $B_i \subseteq B$.
2. Every pair in P_i has a common neighbor in $A \backslash A_i$.
3. Every pair in $\binom{A_i}{2} \backslash P_i$ does not have a common neighbor in $A \backslash A_i$.
4. Every pair in Q_i has a common neighbor in $B \backslash B_i$.
5. Every pair in $\binom{B_i}{2} \backslash Q_i$ does not have a common neighbor in $B \backslash B_i$.
6. $G[A]$ and $G[B]$ are C_4-free.

Otherwise, $M_i[\Psi]$ is set to 0. Suppose there exists a 4-tuple Ψ such that $M_r[\Psi] = 1$, where r is the root of the nice tree decomposition. Then the above conditions 1 and 6 ensure that G can be partitioned in the required manner.

When one of the following occurs, it is easy to see that the 4-tuple does not lead to a required partition. We say that the 4-tuple Ψ is *invalid* if one of the below cases occur:

(i) $G[A_i]$ or $G[B_i]$ contains an induced C_4.
(ii) There exists a pair $\{x, y\} \in P_i$ such that $\{x, y\} \in E$.
(iii) There exists a pair $\{x, y\} \in Q_i$ such that $\{x, y\} \in E$.

Note that it takes $O(t^4)$ time to check if a given Ψ is invalid. Below we explain how to compute $M_i[\Psi]$ value at each node i.

Leaf Node: For a leaf node i, $\Psi = (\emptyset, \emptyset, \emptyset, \emptyset)$ and $M_i[\Psi] = 1$. This step takes constant time.

Introduce Node: Let j be the only child of the node i. Suppose $v \in X_i$ is the new vertex present in X_i, $v \notin X_j$. Let $\Psi = (A_i, B_i, P_i, Q_i)$ be a 4-tuple of X_i, If Ψ is invalid, we set $M_i[\Psi]$ to 0. Otherwise, we use the following cases to compute the $M_i[\Psi]$ value.

Case 1, $v \in A_i$: If $\exists \{v, x\} \in P_i$ for some $x \in A_i$ or if $\exists \{x, y\} \in P_i$ such that $\{x, y\} \subseteq N(v) \cap A_i$, then $M_i[\Psi] = 0$. Otherwise, $M_i[\Psi] = M_j[\Psi']$, where $\Psi' = (A_i \backslash \{v\}, B_i, P_i, Q_i)$.

As v is a newly introduced vertex, it cannot have any forgotten neighbors. Hence, $\{v, x\} \in P_i \implies M_i[\Psi] = 0$. If x and y have a common forgotten neighbor, they all form an induced C_4, together with v. Hence $\{x, y\} \in P_i \implies M_i[\Psi] = 0$.

Case 2, $v \in B_i$: If $\exists \{v, x\} \in Q_i$ for some $x \in B_i$ or if $\exists \{x, y\} \in Q_i$ such that $\{x, y\} \subseteq N(v) \cap B_i$, then $M_i[\Psi] = 0$. Otherwise, $M_i[\Psi] = M_j[\Psi']$, where $\Psi' = (A_i, B_i \backslash \{v\}, P_i, Q_i)$.

The total number of Ψ's for X_i is $2^{t+1} 2^{(t+1)^2}$. It takes $O(t^4)$ time to check if Ψ is invalid. Hence total time complexity at the introduce node is $O(2^{t^2 + 3t} t^4)$.

Forget Node: Let j be the only child of the node i. Suppose $v \in X_j$ is the vertex missing in X_i, $v \notin X_i$. Let $\Psi = (A_i, B_i, P_i, Q_i)$ be a 4-tuple of X_i, If Ψ is invalid, we set $M_i[\Psi]$ to 0. Otherwise, $M_i[\Psi]$ is computed as follows:

Case 1, $v \in A_j$: If $\exists x, y \in A_i$ such that $xy \notin E$ and $xv, yv \in E$, then v is a common forgotten neighbor for x and y. Hence we set $M_i[\Psi] = 0$ whenever $\{x, y\} \notin P_i$. Otherwise, let $R = \{\{x, y\} | x, y \in A_i \cap N(v)\}$. Some of the vertex pairs in R can still have a common forgotten neighbor (other than v) at node j which is adjacent to v. Also there can be new pairs formed with v at the node j. Let $S = \{\{v, x\} | x \in A_i\}$. We have the following equation.

$$\delta_1 = \max_{X \subseteq S, Y \subseteq R} \{M_j[A_i \cup \{v\}, B_i, (P_i \backslash R) \cup (X \cup Y), Q_i]\}. \tag{2}$$

Case 2, $v \in B_j$: This is analogous to Case 1. We set $M_i[\Psi] = 0$, whenever $\{x, y\} \notin Q_i$. Otherwise, let $R = \{\{x, y\} | x, y \in B_i \cap N(v)\}$ and $S = \{\{v, x\} | x \in B_i\}$.

$$\delta_2 = \max_{X \subseteq S, Y \subseteq R} \{M_j[A_i, B_i \cup \{v\}, P_i, (Q_i \backslash R) \cup (X \cup Y)]\}. \tag{3}$$

If $M_i[\Psi]$ is not set to 0 already, we set $M_i[\Psi] = \max\{\delta_1, \delta_2\}$.

The total number of Ψ's for X_i is $2^t 2^{t^2}$. It takes $O(t^4)$ time to check if Ψ is invalid. The computations of δ_1 and δ_2 requires us to iterate over every subset of S which is of size at most t and every subset of R which is of size at most t^2. Hence, we get a factor of 2^{t+t^2} in the overall time complexity. Thus the total time complexity at the forget node is $O(2^{2t^2 + 2t} t^4)$.

Join Node: Let j_1 and j_2 be the children of the node i. By the property of nice tree decomposition, we have $X_i = X_{j_1} = X_{j_2}$ and $V(T_{j_1}) \cap V(T_{j_2}) = X_i$. There are no edges between $V(T_{j_1}) \backslash X_i$ and $V(T_{j_2}) \backslash X_i$. Let $\Psi = (A_i, B_i, P_i, Q_i)$ be a 4-tuple of X_i. If Ψ is invalid, we set $M_i[\Psi]$ to 0. Otherwise, we use the following expression to compute the value of $M_i[\Psi]$.

A pair $\{x, y\} \in P_i$ can come either from the left subtree or from the right subtree but not from both, for that would imply two distinct non-adjacent common neighbors for x and y and hence an induced C_4. For $X \subseteq P_i$ and $Y \subseteq Q_i$, $\Psi_1 = (A_i, B_i, X, Y)$ and $\Psi_2 = (A_i, B_i, P_i \backslash X, Q_i \backslash Y)$.

$$M_i[\Psi] = \begin{cases} 1, & \exists X \subseteq P_i, Y \subseteq Q_i \text{ such that } M_{j_1}[\Psi_1] = M_{j_2}[\Psi_2] = 1. \\ 0, & \text{Otherwise.} \end{cases} \tag{4}$$

The total number of Ψ's for X_i is $2^{t+1} 2^{(t+1)^2}$. It takes $O(t^4)$ time to check if Ψ is invalid. As we solve the Eq. 4, a factor of $2^{(t+1)^2}$ comes in the overall time complexity. Hence total time complexity at the join node is $O(2^{2t^2 + 5t} t^4)$.

The correctness of the algorithm is implied by the correctness of $M_i[\Psi]$ values, which follows by a bottom-up induction on the nice tree decomposition. G has a valid bipartitioning if there exists a 4-tuple Ψ such that $M_r[\Psi] = 1$, where r is the root of the nice tree decomposition. We have the following theorem.

Theorem 2. *There is an $O(2^{3t^2} \cdot n)$ time algorithm that solves the H-FREE 2-COLORING problem when $H = C_4$ on graphs with tree-width at most t.*

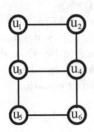

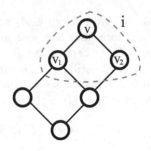

Fig. 1. An example graph H.

Fig. 2. Forming H at an introduce node. Sequence $s = (v, v_2, v_1, \mathrm{fg}, \mathrm{fg}, \mathrm{fg})$.

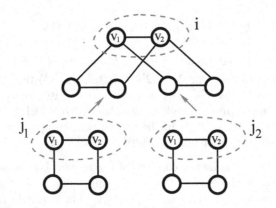

Fig. 3. Forming H at join node. Sequences at node j_1, $s' = (\mathrm{dc}, \mathrm{dc}, v_1, v_2, \mathrm{fg}, \mathrm{fg})$, at node j_2, $s'' = (\mathrm{fg}, \mathrm{fg}, v_1, v_2, \mathrm{dc}, \mathrm{dc})$ gives a sequence $s = (\mathrm{fg}, \mathrm{fg}, v_1, v_2, \mathrm{fg}, \mathrm{fg})$ at node i. The vertices outside the dashed lines are forgotten vertices.

3.3 H-FREE 2-COLORING **Problem**

Let X_i be a bag at node i of the nice tree decomposition. Let (A_i, B_i) be a partition of X_i. We can easily check if $G[A_i]$ or $G[B_i]$ has H as an induced subgraph. Otherwise, we need to see if there is a partition (A, B) of $V(T_i)$ such that $A_i \subseteq A$, $B_i \subseteq B$ and both $G[A]$ and $G[B]$ are H-free. If there is such a partition (A, B), then $G[A]$ and $G[B]$ may have subgraph H', an induced subgraph of H which can lead to H at some ancestral node (introduce node or join node) of the nice tree decomposition (see Figs. 1, 2 and 3).

We perform dynamic programming over the nice tree decomposition. At each node i we guess a partition (A_i, B_i) of X_i and possible induced subgraphs of H that are part of A and B respectively. We check if such a partition is possible. Below we explain the algorithm in detail.

Let the vertices of the graph H be labeled as $u_1, u_2, u_3, \ldots, u_r$. Let (A_i, B_i) be a partition of vertices in the bag X_i. Let (A, B) be a partition of $V(T_i)$ such that $A \supseteq A_i$ and $B \supseteq B_i$. We define Γ_{A_i} as follows:

$$S_{A_i} = \{(w_1, w_2, w_3, \ldots, w_r) | w_\ell \in \{A_i \cup \{\text{fg}, \text{dc}\}\},$$
$$\forall \ell_1 \neq \ell_2, w_{\ell_1} = w_{\ell_2} \implies w_{\ell_1} \in \{\text{fg}, \text{dc}\}\}.$$
$$I_{A_i} = \{s = (w_1, w_2, w_3, \ldots, w_r) \in S_{A_i} | \text{ there exists } \ell_1 \neq \ell_2$$
$$\text{such that } w_{\ell_1} = \text{fg}, w_{\ell_2} = \text{dc and } \{u_{\ell_1}, u_{\ell_2}\} \in E(H)\}.$$
$$\Gamma_{A_i} = S_{A_i} \backslash I_{A_i}.$$

Here 'fg' represents a vertex in $A \backslash A_i$, i.e. the forgotten vertices in A. The label 'dc' (can be thought of as "don't care") represents the vertices that are not part of the subgraph right now, and can potentially be added at some ancestral nodes to form a larger induced subgraph of H.

Similarly, we can define Γ_{B_i} with respect to the sets B_i and B.

A sequence in S_{A_i} corresponds to an induced subgraph H' of H in A as follows:

1. If $w_\ell = \text{fg}$ then u_ℓ is part of $A \backslash A_i$, the forgotten vertices in A.
2. If $w_\ell = \text{dc}$ then u_ℓ is not be part of the subgraph H'.
3. If $w_\ell \in A_i$ then the vertex w_ℓ corresponds to the vertex u_ℓ of H'.

Γ_{A_i} is the set of sequences that can become H in future at some ancestral (introduce/join) node of the tree decomposition. Note that the sequences I_{A_i} are excluded from Γ_{A_i} because a forgotten vertex cannot have an edge to a vertex which will come in future at some ancestral node (introduce or join nodes).

Definition 1 (Induced Subgraph Legal Sequence in Γ_{A_i} with respect to A). *A sequence* $s = (w_1, w_2, w_3, \ldots, w_r) \in \Gamma_{A_i}$ *is legal if the sequence* s *corresponds to an induced subgraph* H' *of* H *within* A *as follows.*

Let $FG(s) = \{\ell | w_\ell = fg\}$, $DC(s) = \{\ell | w_\ell = dc\}$ *and* $VI(s) = [r] \backslash \{FG(s) \cup DC(s)\}$. *Let* H' *be the induced subgraph of* H *formed by* u_ℓ, $\ell \in \{VI(s) \cup FG(s)\}$. *That is* $H' = H[\{u_\ell | \ell \in VI(s) \cup FG(s)\}]$.

If there exist $|FG(s)|$ *distinct vertices* $z_\ell \in A \backslash A_i$ *corresponding to each index in* $FG(s)$ *such that* H' *is isomorphic to* $G[\{w_\ell | \ell \in VI(s)\} \cup \{z_\ell | \ell \in FG(s)\}]$, *then* s *is legal. Otherwise, the sequence is illegal.*

Analogously, we define legal/illegal sequences in Γ_{B_i} with respect to B.

Let $\Psi = (A_i, B_i, P_i, Q_i)$ be a 4-tuple. Here, (A_i, B_i) is a partition of X_i, $P_i \subseteq \Gamma_{A_i}$ and $Q_i \subseteq \Gamma_{B_i}$.

We define $M_i[\Psi]$ to be 1 if there is a partition (A, B) of $V(T_i)$ such that:

1. $A_i \subseteq A$ and $B_i \subseteq B$.
2. Every sequence in P_i is legal with respect to A.
3. Every sequence in Q_i is legal with respect to B.
4. Every sequence in $\Gamma_{A_i} \backslash P_i$ is illegal with respect to A.
5. Every sequence in $\Gamma_{B_i} \backslash Q_i$ is illegal with respect to B.
6. Neither $G[A]$ nor $G[B]$ contains H as an induced subgraph.

Otherwise $M_i[\Psi]$ is set to 0.

We call a 4-tuple Ψ as invalid if one of the following conditions occur. If Ψ is invalid we set $M_i[\Psi]$ to 0.

1. There exists a sequence $s \in P_i$ such that s does not contain dc.
2. There exists a sequence $s \in Q_i$ such that s does not contain dc.

As $|P_i| + |Q_i| \leq (t+5)^r$, it takes $(t+5)^r r$ time to check if Ψ is invalid.

Now we explain how to compute $M_i[\Psi]$ values at the leaf, introduce, forget and join nodes of the nice tree decomposition.

Leaf Node: Let i be a leaf node, $X_i = \emptyset$, for $\Psi = (A_i, B_i, P_i, Q_i)$, we have $M_i[\Psi] = 1$. Here $A_i = B_i = \emptyset$, $P_i \subseteq \{([dc]^r)\}$ and $Q_i \subseteq \{([dc]^r)\}$. This step takes constant time.

Introduce Node: Let i be an introduce node and j be the child node of i. Let $\{v\} = X_i \backslash X_j$. Let $\Psi = (A_i, B_i, P_i, Q_i)$ be a 4-tuple at node i. If Ψ is invalid we set $M_i[\Psi] = 0$. Otherwise depending on whether $v \in A_i$ or $v \in B_i$ we have two cases. We discuss only the case $v \in A_i$, the case $v \in B_i$ can be analogously defined.

$v \in A_i$: We set $M_i[\Psi] = 0$, if there exists an illegal sequence s (in P_i) containing v or if there exists a trivial legal sequence s containing v but s is not in P_i. That is, we set $M_i[\Psi] = 0$ if one of the following $(\star)$ conditions occurs:

[$\star$ Conditions]

1. $\exists \ell_1 \neq \ell_2$, such that $w_{\ell_1} = v$, $w_{\ell_2} \in A_i$, $\{u_{\ell_1}, u_{\ell_2}\} \in E(H)$ but $\{v, w_{\ell_2}\} \notin E(G)$.
2. $\exists \ell_1 \neq \ell_2$, such that $w_{\ell_1} = v$, $w_{\ell_2} \in A_i$, $\{u_{\ell_1}, u_{\ell_2}\} \notin E(H)$ but $\{v, w_{\ell_2}\} \in E(G)$.
3. $\exists \ell_1 \neq \ell_2$, such that $w_{\ell_1} = v$, $w_{\ell_2} = $ fg, $\{u_{\ell_1}, u_{\ell_2}\} \in E(H)$.
4. Let $s = (w_1, w_2, w_3, \ldots, w_r) \in \Gamma_{A_i} \backslash P_i$. There exists ℓ_1 such that $w_{\ell_1} = v$ and for all $\ell_2 \neq \ell_1$, $w_{\ell_2} \in A_i \cup \{dc\}$. For all $\ell_1 \neq \ell_2$, $w_{\ell_1}, w_{\ell_2} \in A_i$, $\{u_{\ell_1}, u_{\ell_2}\} \in E(H) \Longleftrightarrow \{w_{\ell_1}, w_{\ell_2}\} \in E(G)$.

The conditions 1–3 are to check if a sequence $s \in P_i$ containing the vertex v is an illegal sequence. The condition 4 is to check if a sequence $s \notin P_i$ containing the vertex v is a trivial legal sequence. Otherwise we set $M_i[\Psi] = M_j[\Psi']$, where $\Psi' = (A_i \backslash \{v\}, B_i, P_j, Q_i)$. Here P_j is computed as $P_j = \cup_{s \in P_i} \{\text{Rep}_{dc}(s, v)\}$, where Rep_{dc} is defined as follows:

Definition 2. $Rep_{dc}(s, v) = s'$, sequence s' obtained by replacing v (if present) with dc in s.

Note that, $\text{Rep}_{dc}(s, v) = s$, if v not present in s.

The total number of Ψ's for X_i is $2^{(t+1)}2^{(t+5)^r}$. Checking if Ψ is invalid takes $(t+5)^r r$ time. Checking for illegal sequences containing v (steps 1 to 3 in $\star$ Conditions) takes $(t+5)^r r$ time. Checking for legal sequences containing v not part of P_i/Q_i (steps 4 in $\star$ Conditions) takes $(t+5)^r r^2$. Computing Ψ' takes $(t+5)^r r$. Hence total time complexity is $O(2^{(t+1)}2^{(t+5)^r}(t+5)^{2r}r^2) = O(2^{2t^r})$.

Forget Node: Let i be a forget node and j be the only child of node i. Let $\{v\} = X_j \backslash X_i$. Let $\Psi = (A_i, B_i, P_i, Q_i)$ be a 4-tuple at node i. If Ψ is invalid we set $M_i[\Psi] = 0$. Otherwise, we set $M_i[\Psi] = \max\{\delta_1, \delta_2\}$ where δ_1 and δ_2 are computed as follows:

Computing δ_1: Set $A_j = A_i \cup \{v\}$. As v is the extra vertex in A_j, there could be many possible P_j at node j.

Definition 3. $Rep_{fg}(s, v) = s'$, *sequence s' obtained by replacing v (if present) with fg in s.*

Note that, if s does not contain the vertex v then $\text{Rep}_{fg}(s, v) = s$.

We also extend the definition of Rep_{fg} to a set of sequences as follows:

$$\text{Rep}_{fg}(S, v) = \cup_{s \in S}\{\text{Rep}_{fg}(s, v)\}.$$

Note that, if s is a legal sequence at the node j with respect to A, then $\text{Rep}_{fg}(s, v)$ is also a legal sequence at node i with respect to A.

$$\delta_1 = \max_{\substack{P_j \subseteq \Gamma_{A_j} \\ \text{Rep}_{fg}(P_j, v) = P_i}} \{M_j[(A_j, B_i, P_j, Q_i)]\}$$

Computing δ_2: $B_j = B_i \cup \{v\}$. It is analogous to computing δ_1 but we process on B.

The total number of Ψ's for X_i is $2^t(t+4)^r$. Checking for invalid case takes $(t+4)^r r$ time. computing δ_1 and δ_2 takes $2^{(t+4)^r}(t+4)^r r$ time. Hence the total time complexity is $O(2^t 2^{2(t+4)^r}(t+4)^{2r}r^2) = O(2^{3t^r})$.

Join Node: Let i be a join node, j_1, j_2 be the left and right children of the node i respectively. $X_i = X_{j_1} = X_{j_2}$ and there are no edges between $V(T_{j_1}) \backslash X_i$ and $V(T_{j_2}) \backslash X_i$. Let $\Psi = (A_i, B_i, P_i, Q_i)$ be a 4-tuple at node i. If Ψ is invalid we set $M_i[\Psi] = 0$. Otherwise, we compute $M_i[\Psi]$ value as follows:

Definition 4. *Let $s = (w_1, w_2, w_3, \ldots, w_r)$, $s' = (w'_1, w'_2, w'_3, \ldots, w'_r)$ and $s'' = (w''_1, w''_2, w''_3, \ldots, w''_r)$ be three sequences. We say that $s = Merge(s', s'')$ if the following conditions are satisfied.*

1. $\forall \ell \; w_\ell \in X_i \implies w'_\ell = w''_\ell = w_\ell$.
2. $\forall \ell \; w_\ell = fg \implies$ either $(w'_\ell = fg$ and $w''_\ell = dc)$ or $(w'_\ell = dc$ and $w''_\ell = fg)$.
3. $\forall \ell \; w_\ell = dc \implies w'_\ell = w''_\ell = dc$.

Note that, if $s' \in \Gamma_{A_{j_1}}$ and $s'' \in \Gamma_{A_{j_2}}$ are legal sequences at node j_1 and j_2 respectively then s is a legal sequence at node i with respect to A. We extend the Merge operation to sets of sequences as follows:

$$\text{Merge}(S_1, S_2) = \{s | \exists s' \in S_1, s'' \in S_2 \text{ such that } s = \text{Merge}(s', s'')\}.$$

We set $M_i[\Psi] = 1$ if there exists $P_{j_1}, Q_{j_1}, P_{j_2}$ and Q_{j_2} such that the following conditions are satisfied:

(i) $P_i = \text{Merge}(P_{j_1}, P_{j_2})$,   (ii) $Q_i = \text{Merge}(Q_{j_1}, Q_{j_2})$,

(iii) $M_{j_1}[A_i, B_i, P_{j_1}, Q_{j_1}] = 1$, and   (iv) $M_{j_2}[A_i, B_i, P_{j_2}, Q_{j_2}] = 1$.

The total number of Ψ's for X_i is $2^{(t+1)}2^{(t+5)^r}$. Checking if Ψ is invalid takes $(t+5)^r r$. A factor of $4^{(t+5)^r}(t+5)^r r$ comes as we try all possible $P_{j_1}, Q_{j_1}, P_{j_2}, Q_{j_2}$. Hence the total time complexity at join node is $O(2^{(t+1)}2^{3(t+5)^r}(t+5)^r r) = O(2^{4t^r})$.

The graph has a valid bipartitioning if there exists a Ψ such that $M_r[\Psi] = 1$, where r is the root node of the nice tree decomposition. The correctness of the algorithm is implied by the correctness of $M_i[\Psi]$ values, which can be proved using a bottom up induction on the nice tree decomposition. Thus we get the following:

Theorem 3. *There is an $O(2^{4t^r} \cdot n)$ time algorithm that solves the H-FREE 2-COLORING problem for any arbitrary fixed H, on graphs with tree-width at most t.*

4   Algorithm for H-FREE q-COLORING Problem

We note that our techniques extend in a straightforward manner to solve the H-FREE q-COLORING problem. In this case, we have to consider tuples Ψ that have $2q$ sets. That is $\Psi = (A_i^1, A_i^2, \ldots, A_i^q, P_i^1, P_i^2, \ldots, P_i^q)$. Here $A_i^j \subseteq X_i$ and $P_i^j \subseteq \Gamma_{A_i^j}$. The operations at the leaf, introduce and forget nodes are very similar to the case of 2-coloring problem. At introduce and forget nodes we will have q cases instead of 2 cases. At the join node we need to define the Merge operation on q sets instead of 2 sets. Below is the modified definition of Merge.

Definition 5. *Let $s = (w_1, w_2, w_3, \ldots, w_r)$, $s^1 = (w_1^1, w_2^1, w_3^1, \ldots, w_r^1)$, $s^2 = (w_1^2, w_2^2, w_3^2, \ldots, w_r^2), \ldots, s^q = (w_1^q, w_2^q, w_3^q, \ldots, w_r^q)$ be $q+1$ sequences. We say that $s = \text{Merge}(s^1, s^2, s^3, \ldots, s^q)$ if the following conditions are satisfied.*

1. *$\forall \ell \; w_\ell \in X_i \Longrightarrow w_\ell^1 = w_\ell^2 = \cdots = w_\ell^q = w_\ell$.*
2. *$\forall \ell \; w_\ell = fg \Longrightarrow \exists i$ such that $w_\ell^i = fg$ and $\forall j \neq i, \; w_\ell^j = dc$.*
3. *$\forall \ell \; w_\ell = dc \Longrightarrow w_\ell^1 = w_\ell^2 = \cdots = w_\ell^q = dc$.*

Thus we state the following theorem.

Theorem 4. *There is an $O(q^{4t^r} \cdot n)$ time algorithm that solves the H-FREE q-COLORING problem for any arbitrary fixed H, on graphs with tree-width at most t.*

The H-FREE CHROMATIC NUMBER is at most the chromatic number $\chi(G)$. For graphs with tree-width t, we have $\chi(G) \leq t+1$. Our techniques can also be used to compute the H-FREE CHROMATIC NUMBER of the graph by searching for the smallest q for which there is an H-free q-coloring. We have the following theorem.

Theorem 5. *There is an $O(t^{4t^r} \cdot n \log t)$ time algorithm to compute H-FREE CHROMATIC NUMBER of the graph whose tree-width is at most t.*

5   Algorithm for *H*-(SUBGRAPH)FREE *q*-COLORING Problem

We can solve the H-(SUBGRAPH)FREE 2-COLORING problem using the techniques described in Sect. 3.3. As we are looking for bipartitioning without H as a subgraph, we need to modify the Definition 1 and $(\star)$ conditions.

Instead of Definition 1 we have Definition 6.

Definition 6. (Subgraph Legal Sequence in Γ_{A_i} with respect to A). *A sequence $s = (w_1, w_2, w_3, \ldots, w_r) \in \Gamma_{A_i}$ is legal if the sequence s corresponds to a subgraph H' of H within A as follows.*

Let $FG(s) = \{\ell | w_\ell = fg\}$, $DC(s) = \{\ell | w_\ell = dc\}$ and $VI(s) = [r] \backslash \{FG(s) \cup DC(s)\}$. Let H' be the induced subgraph of H formed by u_ℓ, $\ell \in \{VI(s) \cup FG(s)\}$. That is $H' = H[\{u_\ell | \ell \in VI(s) \cup FG(s)\}]$.

If there exist $|FG(s)|$ distinct vertices $z_\ell \in A \backslash A_i$ corresponding to each index in $FG(s)$ such that H' is a subgraph of $G[\{w_\ell | \ell \in VI(s)\} \cup \{z_\ell | \ell \in FG(s)\}]$, then s is legal. Otherwise, the sequence is illegal.

At the introduce node, instead of $(\star)$ conditions we have to check the following $(\star\star)$ conditions:

[$\star\star$ Conditions]

1. $\exists \ell_1 \neq \ell_2$, such that $w_{\ell_1} = v$, $w_{\ell_2} \in A_i$, $\{u_{\ell_1}, u_{\ell_2}\} \in E(H)$ but $\{v, w_{\ell_2}\} \notin E(G)$.
2. $\exists \ell_1 \neq \ell_2$, such that $w_{\ell_1} = v$, $w_{\ell_2} = \text{fg}$, $\{u_{\ell_1}, u_{\ell_2}\} \in E(H)$.
3. Let $s = (w_1, w_2, w_3, \ldots, w_r) \in \Gamma_{A_i} \backslash P_i$. There exists ℓ_1 such that $w_{\ell_1} = v$ and for all $\ell_2 \neq \ell_1$, $w_{\ell_2} \in A_i \cup \{dc\}$. For all $\ell_1 \neq \ell_2$, $w_{\ell_1}, w_{\ell_2} \in A_i$, $\{u_{\ell_1}, u_{\ell_2}\} \in E(H) \implies \{w_{\ell_1}, w_{\ell_2}\} \in E(G)$.

Thus we get the following:

Theorem 6. *There is an $O(q^{4t^r} \cdot n)$ time algorithm that solves the H-(SUBGRAPH)FREE q-COLORING problem for any arbitrary fixed H, on graphs with tree-width at most t.*

Theorem 7. *There is an $O(t^{4t^r} \cdot n \log t)$ time algorithm to compute H-(SUBGRAPH)FREE CHROMATIC NUMBER of the graph whose tree-width is at most t.*

References

1. Achlioptas, D.: The complexity of G-free colourability. Discret. Math. **165–166**(Supplement C), 21–30 (1997)
2. Kubicka, E., Kubicki, G., McKeon, K.A.: Chromatic sums for colorings avoiding monochromatic subgraphs. Electron. Notes Discret. Math. **43**, 247–254 (2013)
3. Kubicka, E., Kubicki, G., McKeon, K.A.: Chromatic sums for colorings avoiding monochromatic subgraphs. Discuss. Math. Graph Theory **43**, 541–555 (2015)
4. Karpiński, M.: Vertex 2-coloring without monochromatic cycles of fixed size is NP-complete. Theor. Comput. Sci. **659**(Supplement C), 88–94 (2017)
5. Xiao, M., Nagamochi, H.: Complexity and kernels for bipartition into degree-bounded induced graphs. Theor. Comput. Sci. **659**, 72–82 (2017)
6. Cowen, L.J., Cowen, R.H., Woodall, D.R.: Defective colorings of graphs in surfaces: partitions into subgraphs of bounded valency. J. Graph Theory **10**(2), 187–195 (1986)
7. Bazgan, C., Tuza, Z., Vanderpooten, D.: Degree-constrained decompositions of graphs: bounded treewidth and planarity. Theor. Comput. Sci. **355**(3), 389–395 (2006)
8. Wu, Y., Yuan, J., Zhao, Y.: Partition a graph into two induced forests. J. Math. Study **1**, 1–6 (1996)
9. Farrugia, A.: Vertex-partitioning into fixed additive induced-hereditary properties is NP-hard. Electron. J. Comb. **11**, 46 (2004)
10. Rao, M.: MSOL partitioning problems on graphs of bounded treewidth and clique-width. Theor. Comput. Sci. **377**(1), 260–267 (2007)
11. Courcelle, B.: The monadic second-order logic of graphs. I. Recognizable sets of finite graphs. Inf. Comput. **85**(1), 12–75 (1990)
12. Courcelle, B.: The monadic second-order logic of graphs III: tree-decompositions, minor and complexity issues. Theor. Inform. Appl. **26**, 257–286 (1992)
13. Diestel, R.: Graph Theory. Springer, Heidelberg (2005)
14. Cygan, M., et al.: Parameterized Algorithms. Springer, Heidelberg (2015). https://doi.org/10.1007/978-3-319-21275-3
15. Robertson, N., Seymour, P.: Graph minors. X. Obstructions to tree-decomposition. J. Comb. Theory Ser. B **52**(2), 153–190 (1991)
16. Kloks, T. (ed.): Treewidth: Computations and Approximations. LNCS. Springer, Heidelberg (1994). https://doi.org/10.1007/BFb0045375

The Relative Signed Clique Number
of Planar Graphs is 8

Sandip Das[1], Soumen Nandi[2(✉)], Sagnik Sen[3], and Ritesh Seth[3]

[1] Indian Statistical Institute, Kolkata, India
[2] Birla Institute of Technology and Science Pilani, Hyderabad Campus, Pilani, India
soumen2004@gmail.com
[3] Ramakrishna Mission Vivekananda Educational and Research Institute,
Kolkata, India

Abstract. A simple signed graph (G, Σ) is a simple graph with a +ve or a −ve sign assigned to each of its edges where Σ denotes the set of −ve edges. A cycle is unbalanced if it has an odd number of −ve edges. A vertex subset R of (G, Σ) is a relative signed clique if each pair of non-adjacent vertices of R is part of an unbalanced 4-cycle. The relative signed clique number $\omega_{rs}((G, \Sigma))$ of (G, Σ) is the maximum value of $|R|$ where R is a relative signed clique of (G, Σ). Given a family $\mathcal{F}$ of signed graphs, the relative signed clique number is $\omega_{rs}(\mathcal{F}) = \max\{\omega_{rs}((G, \Sigma))|(G, \Sigma) \in \mathcal{F}\}$. For the family $\mathcal{P}_3$ of signed planar graphs, the problem of finding the value of $\omega_{rs}(\mathcal{P}_3)$ is an open problem. In this article, we close it by proving $\omega_{rs}(\mathcal{P}_3) = 8$.

Keywords: Signed graphs · Relative clique number · Planar graphs

1 Introduction and Main Results

Signed graphs are studied since decades [4,9,10]. However, recently, Naserasr, Rollova, and Sopena [7] introduced and studied homomorphisms of signed graphs and the topic has gained a lot of popularity since [1–3,5,6,8].

In this article, we determine the relative signed clique number of planar graphs, a notion introduced in the seminal paper [7]. We are closing an open problem here by proving that the relative signed clique number of planar graphs is 8. That is why we will keep this article to the point and direct focusing only on the singular result and its proof.

A *signed graph* (G, Σ) is a graph with a +ve or a −ve sign assigned to each of its edges where its *signature* Σ is the set of all −ve edges. Moreover, the set of vertices and edges of (G, Σ) is denoted by $V(G)$ and $E(G)$, respectively. A cycle of a signed graph is *unbalanced* if it has odd number of −ve edges, and is *balanced* otherwise. A *relative signed clique* $R \subseteq V(G)$ is a vertex subset of (G, Σ) such that any two non-adjacent vertices of R is part of an unbalanced 4-cycle. The *relative signed clique number* of (G, Σ) is given by

$$\omega_{rs}((G, \Sigma)) = \max\{|R| \text{ where } R \text{ is a relative signed clique of } (G, \Sigma)\}.$$

© Springer Nature Switzerland AG 2019
S. P. Pal and A. Vijayakumar (Eds.): CALDAM 2019, LNCS 11394, pp. 245–253, 2019.
https://doi.org/10.1007/978-3-030-11509-8_20

The *relative signed clique number* of a family $\mathcal{F}$ of graphs is given by

$$\omega_{rs}(\mathcal{F}) = \max\{\omega_{rs}((G, \Sigma)) \ where \ G \in \mathcal{F} \ and \ \Sigma \ is \ a \ signature\}.$$

Let $\mathcal{P}$ be the family of planar graphs. It is known [7] that $8 \leq \omega_{rs}(\mathcal{P}) \leq 15$. The tightness of this bound was left as an open problem [7]. We close it by proving $\omega_{rs}(\mathcal{P}) = 8$.

Theorem 1. *For the family $\mathcal{P}$ of planar graphs, $\omega_{rs}(\mathcal{P}) = 8$.*

We prove this theorem in the next section.

2   Proof of Theorem 1

The proof is contained in several observations and lemmas provided in this section.

Let (H, Π) be a minimal, with respect to the lexicographic ordering of $(|V(H)|, |E(H)|)$, counter example of Theorem 1. Let R be a maximum relative clique of (H, Σ). Thus $|R| \geq 9$. Moreover, assume a particular planar embedding (on a sphere) of (H, Π) for the rest of this section unless otherwise stated. The vertices of R are called *good* vertices while the other vertices are called *helpers*. Also let $S = V(H) \setminus R$.

Observation 2. *The set S of helpers are independent.*

Proof. If not, we may delete the edges between the vertices of S keeping R a relative signed clique. This contradicts the minimality of (H, Π).   □

If two vertices u, v of a signed graph (G, Σ) are either adjacent or they belong to an unbalanced 4-cycle, then we say that u *reaches* v. Moreover, if $uxvyu$ is an unbalanced 4-cycle, then we say that u *reaches* v *through* x *and* y.

The set of all neighbors of a vertex v of (G, Σ) is denoted by $N(v)$. The degree of v is $deg(v) = |N(v)|$. The set of vertices that are adjacent to v by a +ve edge is the set of +-*neighbors* of v, denoted by $N^+(v)$. Similarly, the set of vertices that are adjacent to v by a −ve edge is the set of −-*neighbors* of v, denoted by $N^-(v)$. Moreover, the +ve and −ve degree of v is $deg^+(v) = |N^+(v)|$ and $deg^-(v) = |N^-(v)|$, respectively.

We can provide a lower bound of the degree of a vertex in S.

Observation 3. *Any $u \in S$ has $deg(u) \geq 4$.*

Proof. Note that a vertex $u \in S$ is useful only at least two of its neighbors reach each other through u. Thus $deg(u) \geq 2$.

If $deg(u) \leq 3$, then we add some edges (if required) between its neighbors to create a clique and delete u. In this so obtained graph, R is still a relative signed clique, a contradiction to the minimality of (H, Π).   □

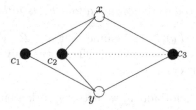

Fig. 1. The configuration F_k.

Now we are going to show that H cannot have subgraphs with certain structures and properties. These subgraphs are called the *forbidden configurations*. Note that, these forbidden configurations are simple graphs, which means that the underlying graph H cannot contain those forbidden configurations as subgraphs.

Drawing Convention: We depict the forbidden configurations in Fig. 1. The black filled circles denote a good vertex and the not filled circles may be good or a helper. When we study a particular configuration, we will use the names given to its vertices in Fig. 1 as reference in the proofs. Sometimes we will describe a new configuration by extending one depicted in Fig. 1. In these cases, the vertices with a name given to it in the figure retains it in the extension.

Let F_k denote a simple graph $K_{2,k}$, where the vertices from the partite set containing k vertices are all good vertices (see Fig. 1). Also suppose that the drawing of F_k in Fig. 1 is the restriction of our assumed embedding of H. This assumption is valid as the graph $K_{2,k}$ has a unique planar embedding on a sphere up to isomorphism and continuous deformation of the sphere. Furthermore, suppose that the region bounded by the cycle $xc_iyc_{i+1}x$ of F_k is called R_i for all $i \in \{1, 2, \cdots, k-1\}$ and the region corresponding to the outerface is called R_0 (considering F_k to be embedded on the plane as in Fig. 1).

We first show that F_k for $k \geq 7$ is a forbidden configuration.

Before going to the proof we need to recall another notion [7]. To *resign* a vertex v of a signed graph (G, Σ) is to switch the signs of all the edges incident to v. Suppose we obtain the graph (G, Σ^*) by resigning a subset of vertices of (G, Σ). It is known [7] that $R \subseteq V(G)$ is a relative signed clique of (G, Σ) if an only if it is a relative signed clique of (G, Σ^*).

Given a signed graph (G, Σ), two vertices u and v *agree* on a third vertex w if $w \in N^\alpha(u) \cap N^\alpha(v)$ for some $\alpha \in \{+, -\}$. On the other hand, two vertices u and v *disagree* on a third vertex w if $w \in N^\alpha(u) \cap N^\beta(v)$ for some $\{\alpha, \beta\} = \{+, -\}$.

Lemma 1. *For all $k \geq 7$, the configuration F_k is forbidden.*

Proof. First switch the $-$ve neighbors of x to obtain the signed graph (H, Π^*). Then at least 4 good vertices from $\{c_1, c_2, \cdots, c_k\}$ agree on both x and y. Observe that it is not possible for these vertices to reach each other as H is planar. □

Now we turn our focus to the cases where $k = 5, 6$.

Lemma 2. *For $k = 5$ and 6, the configuration F_k is forbidden.*

Proof. Note that $|R \setminus V(F_k)| \geq 1$. Assume that $w \in R \setminus V(F_k)$. Without loss of generality let w belongs to the region R_0. Then the only way for w to reach c_3 is through x and y. This creates F_7. □

For proving that F_4 is forbidden, we need to work a bit more. Before proving F_4 is forbidden, we will handle a few other cases.

Start with the configuration F_4 and place a good vertex w_1 in the region R_2 and another good vertex w_2 in the region R_0 to obtain a new configuration called $F_{4,1}$.

Lemma 3. *The configuration $F_{4,1}$ is forbidden.*

Proof. Observe that the only way for w_1 to reach w_2 is through x and y. In that case, F_5 is created. □

The configuration $F_{4,2}$ is obtained by adding three good vertices w_1, w_2 and w_3 to F_4 and placing all them in the region R_2.

Lemma 4. *The configuration $F_{4,2}$ is forbidden.*

Proof. Note that w_1 cannot be adjacent to both x and y, as otherwise F_5 will be created. Thus without loss of generality, suppose w_1 is not adjacent to y. Hence w_1 must reach c_1 through x and c_2. Also w_1 must reach c_4 through x and c_3.

If w_2 is adjacent to x, then w_2 must be adjacent to c_2 and c_3 as well for the same reasons like w_1. However, that contradicts the planarity of H. Therefore, w_2 is not adjacent to x. In that case, w_2 must reach c_1 through y and c_2. Also w_2 must reach c_4 through y and c_3.

Note that w_3 should be adjacent to both c_2 and c_3 as well as one of x or y following arguments similar to those given for w_1 and w_2. However, it is not possible to do so keeping H planar. □

The configuration $F_{4,3}$ is obtained from F_4 by adding two good vertices w_1, w_2 in the region R_2 and one good vertex w_3 in the region R_1.

Lemma 5. *The configuration $F_{4,3}$ is forbidden.*

Proof. The proof of Lemma 4 enables us to assume without loss of generality that w_1 is adjacent to x, c_2 and c_3 while w_2 is adjacent to y, c_2 and c_3.

Now the only way for w_3 to reach w_1 is through x, c_2 and w_2 through y, c_2. This creates F_5. □

Finally, we are ready to prove that F_4 is forbidden.

Lemma 6. *The configuration F_4 is forbidden.*

Proof. Note that $|R \setminus V(F_4)| \geq 3$. Assume that $w_1, w_2, w_3 \in R \setminus V(F_4)$. Without loss of generality let w_1 belongs to the region R_2.

The vertices w_2 and w_3 cannot be in R_0 due to Lemma 3. Moreover, if the three vertices w_1, w_2, w_3 are in three different regions, then a configuration similar to $F_{4,1}$ is created. Thus they must be contained in at most two different regions. Without loss of generality assume that the two regions are R_1 and R_2.

If both w_2 and w_3 are in R_2, then $F_{4,2}$ is created and we are done due to Lemma 4.

Therefore, the only other case left is, without loss of generality, the following: w_2 is in R_2 and w_3 is in R_1. This is precisely the configuration $F_{4,3}$.   □

To prove that the configuration F_3 is forbidden, we need to prove several forbidden sub-configuration of it in the following.

However, before proving the next lemma we would like to establish some additional nomenclatures specific to the configurations $F_{3,i}$ (see descriptions of these in the following). A face f of a graph is expressed by its boundary cycle. The region corresponding to a face f of $F_{3,i}$ is denoted by R_f. This definition of region is well defined as none of the $F_{3,i}$'s are simple cycles.

Let $F_{3,1}$ be the configuration obtained by adding the edges c_1c_2, c_2c_3, xc_2 and yc_2 to the configuration F_3.

Lemma 7. *It is not possible to have a good vertex in the region R of $F_{3,1}$, where $R \in \{R_{xc_1c_2x}, R_{xc_2c_3x}, R_{yc_1c_2y}, R_{yc_2c_3y}\}$.*

Proof. Note that $|R \setminus V(F_{3,1})| \geq 4$. Therefore, we may suppose that $w_1, w_2, w_3, w_4 \in R \setminus V(F_{3,1})$.

Assuming that the statement of the lemma is false, without loss of generality we suppose that w_1 is a vertex in the region $R_{xc_1c_2x}$.

Note that the only way for w_1 to reach c_3 is through x and c_2. If w_i $(i \neq 1)$ is also in $R_{xc_1c_2x}$, then w_i must also reach c_3 is through x and c_2 creating a F_4. Similarly, if w_i is in $R_{xc_2c_3x}$, then w_i must reach c_1 is through x and c_2 also creating a F_4. If w_i is in $R_{yc_2c_3y}$ then w_i cannot reach w_1 at all. Thus w_2, w_3, w_4 must be in $R_{xc_1yc_3x}$ and $R_{yc_1c_2y}$.

Any w_i in $R_{xc_1yc_3x}$ must reach w_1 through x and c_1. Also w_i in $R_{yc_1c_2y}$ must reach w_1 through c_1 and c_2. As we know that at least one of the two regions must have at least 2 vertices, an F_4 is forced.   □

Let $F_{3,1,1}$ be the configuration obtained from $F_{3,1}$ by adding the edge c_1c_3.

Lemma 8. *The configuration $F_{3,1,1}$ is forbidden.*

Proof. Suppose $R \setminus V(F_{3,1}) = \{w_1, w_2, \cdots, w_t\} = W$. We know that $t \geq 4$ as $|R \setminus V(F_{3,1})| \geq 4$.

Moreover, by Lemma 7 we know that all the vertices of W are inside the regions $R_{xc_1c_3x}$ and $R_{yc_1c_3y}$. If some w_i is inside $R_{xc_1c_3x}$ and some w_j $(j \neq i)$ is inside $R_{yc_1c_3y}$, then each element of W must be adjacent to c_1 and c_3 in order to reach w_i or w_j. This creates a F_4.

Thus we can assume that all the vertices from W are inside the region $R_{xc_1c_3x}$. In that case, each $w_i \in W$ must reach c_2 through x, c_1 or x, c_3. As $t \geq 4$, a F_4 is forced.   □

Now we are ready to show that the configuration $F_{3,1}$ is forbidden.

Lemma 9. *The configuration $F_{3,1}$ is forbidden.*

Proof. Suppose, similarly as the previous proof, $R \setminus V(F_{3,1}) = \{w_1, w_2, \cdots, w_t\} = W$ and note that $t \geq 4$ as $|R \setminus V(F_{3,1})| \geq 4$.

Moreover, by Lemma 7 we know that all the vertices of W are inside the region $R_{xc_1yc_3x}$. Any $w_i \in w$ must reach c_2 through x, c_1 or x, c_3 or y, c_1 or y, c_3. Thus without loss of generality we may suppose that w_1 reaches c_2 through x, c_1.

As c_1 and c_3 are non-adjacent due to Lemma 8, w_1 must reach c_3 either by the edge w_1c_3 or through two other vertices. One of these two vertices maybe x. We assume that the other one is h_1. However, we handle these two scenarios separately.

Case 1 - w_1 is adjacent to c_3: In this case, both x, y are good will imply a F_4. Therefore, we may assume that at least one of x or y is a helper. So we have $t \geq 5$.

If we have a good vertex w_2 (say) is inside the face xw_1c_1x, then w_2 must reach c_2 through x, c_1 and c_3 through x, w_1. That means, w_2 must be adjacent to each of x, w_1 and c_1. Similarly, if a good vertex w_3 (say) is inside the face xw_1c_3x, then w_3 must be adjacent to x, w_1 and Thus there can be at most one good vertex inside the face xw_1c_1x or inside the face xw_1c_3x.

If both such w_2 and w_3 exist, then each w_j ($j \geq 4$) must be adjacent to w_1, c_1 and c_3 in order to reach w_1 and w_2. This will create a F_4 as $t \geq 5$.

If such w_1 exists while there is no good vertex inside the face xw_1c_3x, then each w_j ($j \geq 4$) must be adjacent to w_1 and c_1 in order to reach w_2. This will create a F_4 as $t \geq 5$.

Thus we are done with this case.

Case 2 - w_1 reaches c_3 through $h_1 \notin V(F_{3,1})$: We know that c_1 is not adjacent to c_3 due to Lemma 8 and w_1 is not adjacent to c_3 due to the previous case. Thus a good vertex is inside the face xw_1c_1x cannot reach c_3. Hence there is not good vertex inside the face xw_1c_1x.

Furthermore, if h_1 is also a good vertex, then it must be adjacent to x, y or c_1 to reach c_2. If h_1 is adjacent to c_1, then the case can be handled similar to Case 1 of this proof. If h_1 is adjacent to x, then let us try to place at least two more good vertices. Note that if we place any vertex in the faces xh_1c_1x or xh_1c_3x, then that vertex cannot reach c_2 or c_1, respectively. If one good vertex, placed in the face $yc_1w_1h_1c_3y$, reaches c_2 through c_1, c_3, then we cannot place another good vertex anywhere which is able to reach c_1, c_2 and as well as c_3. Thus there can be exactly two more good vertices, say w_3 and w_4, placed in the face $yc_1w_1h_1c_3y$ and w_3, w_4 reaches c_2 through y, c_1 and y, c_3, respectively. Note that it is not possible to place anymore good vertices anywhere which is able to reach c_1, c_2 and as well as c_3. Thus, as (H, Π) is a minimal counter example, x

and y must also be good vertices. However, if we try to make every good vertex reach each other, then we are forced to obtain a triangulation of this graph and are not able to use any helpers. One the other hand, we know due to Naserasr, Rollova and Sopena [7] that our minimal counter example must have at least one helper. Hence, h_1 must be a helper.

With a similar argument one can show that it is not possible to have no good vertex inside the face $xw_1h_1c_3x$.

Thus we must have a good vertex w_2 (say) inside the face $xw_1h_1c_3x$. Note that w_2 must be adjacent to x, c_3 to reach c_2. Moreover, w_2 must be adjacent to either h_1 or w_1 in order to reach c_1. The edge c_1h_1 will reduce our case to something similar to Case 1 of this proof and enable us to argue that w_2 cannot reach c_1 through h_1, or in particular, the edge c_1h_1 cannot be present. As the argument is really similar, we skip the details here. Therefore, the edge w_1w_2 is forced. Moreover, we are forced to place all the other vertices w_j ($j \geq 3$) inside the face $yc_1w_1h_1c_3y$.

Observe that to reach c_2 and w_2, each w_j must be adjacent to two vertices among c_1, y, c_3 and two vertices among w_1, h_1, c_3.

If a good vertex w_3 (say) reaches c_2 through c_1 and c_3, then it is not possible to place w_4 anywhere in a way such that w_4 can reach both c_2, w_2. Therefore, a w_j reaches c_2 only through y, c_1 or y, c_3. Note that if two vertices w_3 and w_4 (say) reaches through the same pair of vertices, then one them will not be able to reach w_2. Thus exactly one vertex w_3 (say) reaches c_2 through y, c_1 and exactly one vertex w_4 (say) reaches c_2 through y, c_3. Note that it is not possible to place anymore good vertices anywhere which is able to reach c_1, c_2, c_3 and as well as w_2. Thus x, y must be a good vertices. But y cannot reach w_2, a contradiction. Thus we are done with this case. $\qquad\square$

Note that the vertices c_1, c_2, c_3 of F_3 cannot induce a path due to Lemma 9. Let $F_{3,2}$ be the configuration obtained from F_3 by adding the edge c_1c_3.

Lemma 10. *The configuration $F_{3,2}$ is forbidden.*

Proof. Suppose $R \setminus V(F_{3,2}) = \{w_1, w_2, \cdots, w_t\} = W$. We know that $t \geq 4$ as $|R \setminus V(F_{3,1})| \geq 4$.

Observe that a good vertex placed in $R_{xc_1c_3x}$ or $R_{yc_1c_3y}$ cannot reach c_2 without creating $F_{3,1}$. Therefore, all the good vertices are inside $R_{xc_1yc_2x}$ and $R_{xc_2yc_3x}$.

Suppose w_1 (say) is inside $R_{xc_1yc_2x}$ and w_2 (say) is inside $R_{xc_2yc_3x}$. If w_1 (or w_2) is adjacent to both x and y, then F_4 is created. Thus, without loss of generality let us assume w_1 is not adjacent to y. In that case, w_1 must reach w_2 and c_3 through x, c_2 and x, c_1, respectively. Similarly, w_2 must reach w_1 and c_1 through x, c_2 and x, c_3, respectively. However, now it will not be possible to place any other good vertex anywhere that can reach each of c_1, c_3, w_1, w_2. Thus without loss of generality every vertex of W must be inside $R_{xc_1yc_2x}$.

Recall that $|W| \geq 4$. Observe that a vertex of W must reach c_3 through x, c_1 or y, c_1. If all of them reaches c_3 through x, c_1 (or y, c_1), then a F_4 is created. Otherwise, the vertices of W cannot reach each other. $\qquad\square$

Now we are ready to prove that the configuration F_3 is forbidden.

Lemma 11. *The configuration F_3 is forbidden.*

Proof. We just need to handle the case when c_1, c_2, c_3 is an independent set due to Lemmas 9 and 10.

However, in this case we cannot place any other good vertex in any region R which can reach the c_i placed outside R without creating F_4.   □

Now we have proved all the forbidden configurations needed for proving our theorem. Recall that we know due to Naserasr, Rollova and Sopena [7] that our minimal counter example must have at least one helper. Let h be such a helper. Also we know due to Observations 2 and 3 that h has at least four good neighbors. Assume that $N(h) = \{v_1, v_2, \cdots, v_t\}$. Also suppose that the vertices $v_1, v_2, \cdots, v_t$ are arranged in a clockwise manner around h in the embedding of H.

Now let us focus on the graph $H[N(h)]$ induced by $N(h)$. This graph cannot have a vertex with degree greater than or equal to 3, as otherwise F_3 will be created in H. Therefore, $H[N(h)]$ is a disjoint union of cycles and paths.

Lemma 12. *The graph $H[N(h)]$ is connected.*

Proof. First let us assume that $H[N(h)]$ is disconnected and each of its components have at least two vertices. Thus we can find four vertices v_1, v_2, v_3, v_4 in $H[N(h)]$ such that the edges v_1, v_2 are in a different component than v_3, v_4. Therefore, v_2 reaches v_4 through h and another vertex h_2 of (H, Π). Now v_1 cannot reach v_3 in (H, Π) without creating a F_3. Thus if $H[N(h)]$ is not connected, then it must have an isolated vertex.

Now without loss of generality let us assume that v_4 is an isolated vertex in $H[N(h)]$. Assume that v_2 reaches v_4 through h and another vertex w of (H, Π). Note that if v_3 reaches v_4 through h_1 as well, then it will create a F_3. Thus v_3 must reach v_4 through some other h_2. If v_3 is not adjacent to v_2, then it cannot reach v_1. Thus we have the edge v_2v_3. Note that h_2 cannot reach v_1 and is thus a helper. Observe that if the faces $v_2v_3v_4h_1v_2$ and $hv_2h_2v_4h$ does not contain any good vertex, then $deg(h) = 2$ which is a contradiction to Observation 3. However, any good vertex w (say) in either of those two faces are not in $N(h)$ due to the assumed ordering of the vertices v_1, v_2, v_3, v_4 around h in our given embedding. Thus w cannot reach v_1.   □

Finally we are ready to prove Theorem 1.

Proof of Theorem 1. So we know that $h[N(h)]$ is connected. Thus assume that either we have a cycle $v_1v_2 \cdots v_tv_1$ or we have a path $v_1v_2 \cdots v_t$. In the later case, v_1 must reach v_t through h and a different vertex h_1.

In either case, it is not possible to place enough good vertices, avoiding the forbidden configurations, and create a counter example. We will show that now. The $+$ operation of subscripts of v_i's are taken modulo t for the rest of this proof. Also let $W = R \setminus N(h)$.

Observe that if v_1 is not adjacent to v_t and $t \geq 5$, then v_1 must reach v_4 through some h_2 (maybe the same as h_1). However, then the only option for v_2 to reach v_5 is through h_2 creating a F_3. Hence $t = 4$. Also if v_1 is adjacent to v_4 and $t \geq 5$, then some two incident edges, say v_1v_t and v_tv_{t-1} of the cycle induced by $N(h)$ have the same sign. Thus v_1 must reach v_{t-1} through some h_2. In that case, v_2 cannot reach v_t. Thus $t = 4$.

Observe that any good vertex w_1 inside the face $hv_iv_{i+1}h$ must be adjacent to h, v_i, v_{i+1} in order to be adjacent to v_{i-1} and v_{i+2}. Thus a face $hv_iv_{i+1}h$ can have at most one good vertex.

Moreover, a good vertex w_1 in $hv_iv_{i+1}h$ can reach a good vertex w_2 in $hv_jv_{j+1}h$ if and only if $|i - j| = 1$. Thus all but at most two good vertices of $W = \{w_1, w_2, \cdots, w_r\}$ must be in the face $v_1v_2 \cdots v_t(h_1)v_1$ (read h_1 depending on the case).

Suppose there is a good vertex w_1 in $hv_iv_{i+1}h$ and a good vertex w_2 in $hv_{i+1}v_{i+2}h$. Then w_3 must reach w_1 and w_2 through v_i, v_{i+1}, v_{i+2} creating a F_3.

Thus there can be at most one vertex of W outside the face $v_1v_2 \cdots v_t(h_1)v_1$. Suppose there is a good vertex w_1 in hv_1v_2h. Each w_i ($i \neq 1$) must reach w_1 through v_1, v_2, creating a F_3.

Therefore, every vertex of W is inside the face $v_1v_2 \cdots v_t(h_1)v_1$. Note that no vertex (other than h) can be adjacent to both v_1, v_3 or both v_2, v_4.

Note that $|W| \geq |R| - |N(h)| \geq 5$ (it is possible to have $h_1 \in W$). Let $S = \{v_1, v_2, v_3, v_4, w_1, w_2, w_3, w_4, w_5\}$. Observe that it is not possible for every vertex of S to reach each other keeping the graph planar.   □

References

1. Beaudou, L., Foucaud, F., Naserasr, R.: Homomorphism bounds and edge-colourings of k_4-minor-free graphs. J. Comb. Theor. Ser. B **124**, 128–164 (2017)
2. Brewster, R.C., Foucaud, F., Hell, P., Naserasr, R.: The complexity of signed graph and edge-coloured graph homomorphisms. Discret. Math. **340**(2), 223–235 (2017)
3. Das, S., Ghosh, P., Prabhu, S., Sen, S.: Relative clique number of planar signed graphs. Discret. Appl. Math. (accepted)
4. Harary, F.: On the notion of balance of a signed graph. Mich. Math. J. **2**(2), 143–146 (1953)
5. Naserasr, R., Sen, S., Sun, Q.: Walk-powers and homomorphism bounds of planar signed graphs. Graphs Comb. **32**(4), 1505–1519 (2016)
6. Naserasr, R., Rollová, E., Sopena, É.: Homomorphisms of planar signed graphs to signed projective cubes. Discret. Math. Theor. Comput. Sci. **15**(3), 1–12 (2013)
7. Naserasr, R., Rollová, E., Sopena, É.: Homomorphisms of signed graphs. J. Graph Theor. **79**(3), 178–212 (2015)
8. Ochem, P., Pinlou, A., Sen, S.: Homomorphisms of 2-edge-colored triangle-free planar graphs. J. Graph Theor. **85**(1), 258–277 (2017)
9. Zaslavsky, T.: Characterizations of signed graphs. J. Graph Theor. **25**(5), 401–406 (1981)
10. Zaslavsky, T.: Signed graphs. Discret. Appl. Math. **4**(1), 47–74 (1982)

Bumblebee Visitation Problem

Sandip Das and Harmender Gahlawat[(⊠)]

Indian Statistical Institute, Kolkata, India
harmendergahlawat@gmail.com

Abstract. Bumblebee visitation problem is defined on connected graphs where a mobile agent, called Bumblebee, moves along the edges under some rules to achieve some optimization function. We prove this problem to be NP-hard for general graphs. We present a linear time algorithm for this problem on trees.

1   Introduction

We study a graph traversal problem where a player Bumblebee, denoted B, visits vertices of a connected undirected labelled graph $G = (V, E, c, m)$. Here V and E denote the vertex set and edge set of the graph respectively. Each node $v \in V$ has been assigned two labels: *capacity $c(v)$* and *marker $m(v)$*. Capacity $c(v)$ is a positive integer whose value decreases dynamically depending on B's moves as described below. Marker $m(v)$, for every vertex $v \in V$, is initially 0 and changes to 1 once B visits v.

Bumblebee B moves along the edges of the graph under the following rules. Initially, B arrives on a vertex. B can move along an edge (u, v) if both $c(u) > 0$ and $c(v) > 0$ at that instant. When B traverses an edge (u, v), both $c(u)$ and $c(v)$ are decremented by 1. Graph traversal based games are popular, have a lot of applications in applied computer science and are widely studied [1,3,4].

Markou introduced this problem in the open problems session in GRAph Searching, Theory and Applications (GRASTA), [2]. He considered two optimization problems on this game. In the first variant, whenever B traverses an edge (u, v), both $c(u)$ and $c(v)$ are decremented by 1, and these two values are collected by player B. In this version, B wants to collect as much as possible. Markou proved this problem to be NP-hard and suggested the following variant.

In the second variant, we want to maximize the optimization function $F(G) = \sum_{v \in V} m(v) \cdot c(v)$ i.e. we want to place and move player B in G, such that the total capacity *remaining at visited vertices* is maximum. We call this problem as BUMBLEBEE VISITATION. As a trivial observation, we note that in BUMBLEBEE VISITATION if all the vertices in the neighbourhood of the present location of B have capacity at most 1, then B will not move anymore.

BUMBLEBEE VISITATION can have applications like the following one as suggested by Markou [2]. "Consider a network. The nodes of the network can host an application but each node has to spend some energy in order to execute the application. Initially each node has energy $c(u)$. The application can migrate in

© Springer Nature Switzerland AG 2019
S. P. Pal and A. Vijayakumar (Eds.): CALDAM 2019, LNCS 11394, pp. 254–262, 2019.
https://doi.org/10.1007/978-3-030-11509-8_21

network from a node u to node v, if nodes u, v are adjacent. For the migration procedure, each one of the nodes u, v spends an energy 2. Where to start the execution of the application and what is the migration tour on the network so that the application can use a maximum energy?"

In this article, we prove this problem to be NP-hard. We extend this NP-hardness result to graphs with Hamiltonian paths, denoted by $\mathcal{G}_{HAM}$. In particular, we show that if vertices with capacity 1 are allowed, then BUMBLEBEE VISITATION is NP-hard even for Hamiltonian graphs.

Theorem 1. *Solving* BUMBLEBEE VISITATION *in general graphs is NP-hard.*

Corollary 1. BUMBLEBEE VISITATION *is NP-hard even when restricted to* $\mathcal{G}_{HAM}$.

Our main result is a linear time algorithm to solve the BUMBLEBEE VISITATION on trees.

Theorem 2. BUMBLEBEE VISITATION *can be computed on trees in linear time.*

Organization: In the rest of this section, we present some definitions and notations required in this article. In Sects. 2 and 3, we prove Theorems 1 and 2 respectively.

1.1 Definitions

We follow standard graph notations from West [6]. All graphs considered here are simple, connected and finite. A tree is a connected acyclic graph. Let $[k] = \{1, 2, \ldots, k\}$.

The Bumblebee Visitation problem is formally defined as follows.

BUMBLEBEE VISITATION
Instance: A labelled Graph $G(V, E, c, m)$ (as defined earlier).
Question: Find the maximum value of $F(G) = \sum_{v \in V} m(v) \cdot c(v)$.

A *Hamiltonian path* in a graph G is a path that contains all vertices of G. We denote $\mathcal{G}_{HAM}$ as the class of graphs containing a Hamiltonian path.

HAMILTONIAN PATH
Instance: A Graph $G(V, E)$.
Question: Does G have a Hamiltonian path?

HAMILTONIAN PATH is known to be NP-complete [5].

2   Proof of Theorem 1: NP-hardness of Bumblebee Visitation

We will give a polynomial time reduction from HAMILTONIAN PATH to BUM-BLEBEE VISITATION.

Consider a graph G on n vertices; and for all $v \in V$, assign $c(v) = 3$. We have the following claim.

Claim. There exists a Hamiltonian path in G if and only if BUMBLEBEE VISI-TATION on G gives $F(G) = 2 + n$.

Proof (of the Claim). Necessity: Suppose G has a Hamiltonian path $v_1 \ldots v_n$. The Bumblebee B moves from v_1 to v_n on this Hamiltonian path. At the end we have $c(v_i) = 2$, for $i = 1, n$; and $c(v_i) = 1$, otherwise. Also for all $v \in V$, $m(v) = 1$. So $F(G) = n + 2$.
Sufficiency: Suppose $F(G) = n + 2$. If B visits a vertex v, then $c(v) \in \{0, 1, 2\}$. Suppose B moves in a *walk* $P = v_1 \ldots v_k$ on $k(\leq n)$ distinct vertices. Every *internal* vertex v_i, for $1 < i < k$, in P contributes at most 1 to $F(G)$. The extreme vertices v_1 and v_k contribute at most 2 to $F(G)$. So $(k - 2) \cdot 1 + 2 \cdot 2 \geq n + 2$. This is possible only when $k = n$; every internal vertex contributes exactly one to $F(G)$; and every extreme vertex contributes exactly two to $F(G)$. This means P starts with v_1 and ends with v_n while entering and leaving every internal vertex exactly once. So P is a Hamiltonian path. This completes the proof of the claim.

Proof of Theorem 1 follows from the claim.

Recall that $\mathcal{G}_{HAM}$ denotes the class of graphs containing a Hamiltonian path. Let $G \in \mathcal{G}_{HAM}$. If we are given G along with its Hamiltonian path and if every vertex $v \in V$ has $c(v) \geq 2$, then we can solve BUMBLEBEE VISITATION by travers-ing every vertex along the Hamiltonian path. However if we allow vertices with capacity 1, then using Theorem 1, we prove that BUMBLEBEE VISITATION is NP-hard for $\mathcal{G}_{HAM}$ even if the Hamiltonian path is given.

Corollary 1. BUMBLEBEE VISITATION is NP-hard even when restricted to $\mathcal{G}_{HAM}$.

Proof. We use the following contradiction argument. If we can solve BUMBLE-BEE VISITATION on $\mathcal{G}_{HAM}$ in polynomial time, then we can solve BUMBLEBEE VISITATION on general graphs in polynomial time. To see this, given a general graph G on n vertices, we fix an ordering of its vertices $v_1, \ldots, v_n$. Now con-struct a graph G' on $2n - 1$ vertices by taking $v_1, \ldots, v_n$ and $n - 1$ new vertices $u_1, \ldots, u_{n-1}$ such that u_i is adjacent to v_i and v_{i+1}, for $i \in [n - 1]$. Clearly $G' \in \mathcal{G}_{HAM}$, as $v_1 u_1 v_2 u_2 \ldots v_{n-1} u_{n-1} v_n$ is a Hamiltonian path. Set $c(u_i) = 1$, for $i \in [n - 1]$. If we can solve BUMBLEBEE VISITATION for G' in polynomial time, then B does not use any of the new vertices u_i, for $i \in [n - 1]$. So the same solution also holds for G, which contradicts Theorem 1.   □

3 Proof of Theorem 2: Linear Time Algorithm for Bumblebee Visitation on Trees

In this section, we present a linear time algorithm to solve BUMBLEBEE VISITA-TION on trees. We need the following definitions.

Consider the BUMBLEBEE VISITATION on a rooted tree T. Let $F(T) = \sum_{v \in V} m(v) \cdot c(v)$. Let $F^*(T)$ be the solution of BUMBLEBEE VISITATION on T i.e. $F^*(T)$ is the maximum possible value of $F(T)$. For every vertex $v \in V(T)$, let r_v (read *return* v) denote the maximum value of $F(T)$ if the Bumblebee B has to start from v and finish at v. Also let s_v (read *stay* v) denote the maximum value of $F(T)$ if B has to start from v (it may/may not finish at v). Clearly $s_v \geq r_v$. Now fix a vertex v in T. Let the subtree (of T) rooted at v be denoted as T_v. Let tr_v denote the maximum value of $F(T_v)$ if B has to start from v and finish at v; and $c(v) \geq 2$ after finishing at v. For a vertex v, if $c(v) = 1$, then we fix $tr_v = 0$ as B cannot return to v. The condition $c(v) \geq 2$ in the definition of tr_v ensures that B can enter v and return from v to its *parent*. Also let ts_v denote the maximum value of $F(T_v)$ if B has to start from v; and $c(v) \geq 1$ after visiting the required vertices in T_v. The condition $c(v) \geq 1$ ensures that B can enter v from its parent.

Let vertex v in a tree T have *children* $u_1, \ldots, u_k$. Let $r_v(u_i)$, for $i \in [k]$, denote the maximum value of $F(T)$ if B starts at v and finishes at v, but can not enter the subtree rooted at u_i; and $c(v) \geq 2$ after finishing at v. Similarly let $s_v(u_i)$, for $i \in [k]$, denote the maximum value of $F(T)$ if B starts at v, but can not enter the subtree rooted at u_i; and $c(v) \geq 1$ after visiting required vertices in T.

Now, we present a linear time algorithm to solve BUMBLEBEE VISITATION on trees. Consider a tree T' with vertex set V.

Algorithm 1: BUMBLEBEE VISITATION on Trees

1. Choose a vertex v_0 and build a Breadth First Search tree of T' rooted at v_0. Let us denote this BFS tree as T.
2. Compute tr_v and ts_v, for every $v \in V$ in a bottom up manner in T.
3. Compute r_{v_0}, s_{v_0} and $r_{v_0}(u_i)$, $s_{v_0}(u_i)$ for every child u_i of v_0 in T.
4. Compute r_v, s_v, for all v in T; and $r_v(u_i)$, $s_v(u_i)$ for every child u_i of v in T in the BFS order.
5. $F^*(T) = max(s_v)$, for all $v \in V$.

Now we will explain each step in detail.

Step 1

In the first step of Algorithm 1, the BFS tree T of T' can be found in $O(n)$ time.

Step 2

In the second step of Algorithm 1, we compute tr_v and ts_v for all v in T. First we compute tr_v and ts_v for leaf vertices of T. For a leaf vertex v, $tr_v = ts_v = c(v)$. Then we calculate tr_v and ts_v for other vertices in a bottom up manner using the following lemma.

Lemma 1. *Let T_v be a subtree rooted at v, and let $u_1, u_2, \ldots, u_k$ be the children of v. If we know $tr_{u_1}, \ldots, tr_{u_k}$ and $ts_{u_1}, \ldots, ts_{u_k}$, then we can compute tr_v and ts_v in $O(k)$ time.*

Proof. Computing tr_v: By definition of tr_v, B has to start from v, finish at v and B is restricted to T_v. Every time B enters and returns from subtree rooted at one of its children u_i, $c(v)$ is reduced by 2. Also, by definition of tr_v, $c(v) \geq 2$ in the end. So B can enter at most $l = \lfloor \frac{c(v)-2}{2} \rfloor$ of its children. When B enters and returns from subtree rooted at one of its children u_i, both $c(v)$ and $c(u_i)$ are reduced by 2. So B enters and returns only from its children u_i's whose $tr_{u_i} > 4$ (else if B visits a u_i with $tr_{u_i} \leq 4$, then the contribution of u_i i.e. $tr_{u_i} - 4$ is at most 0). By definition of tr_{u_i}, B can always enter and return u_i from v when $tr_{u_i} \geq 2$. We compute tr_v using Algorithm 2.

Algorithm 2: Compute tr_v

Find the lth order median for values tr_{u_i}. Let (one of) the corresponding
   vertex which attains this value be u_j;
Set count=0, i=1, $tr_v = 0$;
while *count* $< l$ *and* $i \leq k$ **do**
  | **if** $tr_{u_i} > tr_{u_j}$ *and* $tr_{u_i} > 4$ **then**
  | | $tr_v = tr_v + tr_{u_i} - 4$;
  | | $count = count + 1$;
  | **end**
  | $i = i + 1$;
end
if *count* $< l$ **then**
  | set $i = 0$;
  | **while** *count* $< l$ *and* $i \leq k$ **do**
  | | **if** $tr_{u_i} = tr_{u_j}$ *and* $tr_{u_i} > 4$ **then**
  | | | $tr_v = tr_v + tr_{u_i} - 4$;
  | | | $count = count + 1$;
  | | **end**
  | | $i = i + 1$;
  | **end**
end

In Algorithm 2, to achieve the maximum value of tr_v, we want to visit l children with largest tr_{u_i} values (that are greater than four). So we compute the lth order median, say tr_{u_j} (corresponding to vertex u_j). For every child u_i of v, with $tr_{u_i} > tr_{u_j}$ and $tr_{u_i} > 4$, B can visit that child and return contributing a value of $tr_{u_i} - 4$ to tr_v. In case the number of visited children is less than l, we need to visit those children u_i's whose $tr_{u_i} = tr_{u_j}$ and $tr_{u_i} > 4$. For every such child u_i of v, with $tr_{u_i} = tr_{u_j}$ and $tr_{u_i} > 4$, B can visit that child and return contributing a value of $tr_{u_i} - 4$ to tr_v.

In Algorithm 2, we can also keep track of the children that are visited by B; which we use later. We also compute the next value that would have been contributed if B could visit $l + 1$ subtrees and return to v. Let us call this value $next_v$ and we fix a vertex that contributes this value as u_{next}.

Computing ts_v: This is similar to computing tr_v. By definition, B has to start from v and is restricted to T_v; also $ts_v \geq tr_v$. If B finishes at v then $ts_v = tr_v$. Else it enters and returns from some of v's children and finally stays in a subtree rooted at one of v's children u_i. Every time B enters and returns from subtree rooted at one of its children u_i, $c(v)$ is reduced by 2. For entering and staying in one of the subtrees $c(v)$ is reduced by 1. Also, by definition of tr_v, $c(v) \geq 1$ in the end. So B can enter and return from at most $l = \lfloor \frac{c(v)-2}{2} \rfloor$ of its children. B can compute ts_v by following Algorithm 3.

Algorithm 3: Compute ts_v

Compute tr_v;
Set $ts_v = tr_v$ and $max = tr_v$;
for *every u_i* **do**
    if *if u_i was not contributing in tr_v* **then**
        if $ts_{u_i} > 2$ **then**
          | $max = ts_v + ts_{u_i} - 2$
        end
    else
        if $next_v > 4$ **then**
          | $max = ts_v - tr_{u_i} + next_v + ts_{u_i} - 2$
        end
        if $next_v \leq 4$ **then**
          | $max = ts_v - tr_{u_i} + 4 + ts_{u_i} - 2$
        end
    end
    if $max > ts_v$ **then**
        | $ts_v = max$
    end
end

Now we explain the steps in Algorithm 3. To achieve the maximum value of ts_v, we want B to enter and return l children of v and stay in another child of v. Algorithm 3 finds the maximum such value by considering l values from the largest $(l + 1)$ values of tr_{u_i} and one from all the remaining ts_{u_i} values whose corresponding tr_{u_i} values are not among the selected l values.

For each child u_i of v, we find the possible ts_v if B finally stays in the subtree rooted at u_i, and then maximize over all u_i's.

- If u_i was not contributing to tr_v, then we can enter and stay in u_i (if $ts_{u_i} > 2$) contributing a value of $ts_{u_i} - 2$ to ts_v.
- If u_i was contributing to tr_v, then B can enter and stay at u_i ($ts_{u_i} > 2$ as $ts_{u_i} \geq tr_{u_i} > 4$). u_i was contributing value $tr_{u_i} - 4$ to ts_v, but now it contributes $ts_{u_i} - 2$. Now it can also enter and return from the next largest tr_u value which is $next_v$.

- If $next_v \leq 4$, then it cannot enter and return from u_{next}. So ts_v is updated to $ts_v - (tr_{u_i} - 4) + (ts_{u_i} - 2)$.
- If $next_v > 4$, then it can enter and return from u_{next} which will contribute $next_v - 4$ to ts_v. So ts_v is updated to $ts_v - (tr_{u_i} - 4) + (ts_{u_i} - 2) + (next_v - 4) = ts_v - tr_{u_i} + next_v + ts_{u_i} - 2$.

If u_i, the vertex being used for staying was previously being used for returning (while computing tr_v), then we update $u_{next} = u_i$ and $next_v = tr_{u_i}$. This is done as u_i has the highest return value tr_{u_i} among vertices that are not used for entering and returning. Clearly we can compute tr_v and ts_v in $O(k)$ time, as Algorithms 2 and 3 takes $O(k)$ time.

This ends the proof of Lemma 1.   □

Step 3

In third step of Algorithm 1, we compute r_{v_0}, s_{v_0} and $r_{v_0}(u_i)$, $s_{v_0}(u_i)$ for every child u_i of v_0 in T.

Computing r_{v_0} and s_{v_0}: The value of r_{v_0} can be computed similar to tr_{v_0} as the only difference between the two is the following. As B starts from v_0, we need not ensure $c(v_0) \geq 2$ while calculating r_{v_0}. So we set $l = \lfloor \frac{c(v_0)}{2} \rfloor$ and follow the algorithm for calculating tr_{v_0}. Similarly for computing s_{v_0}, we need not ensure $c(v_0) \geq 1$; so we set $l = \lfloor \frac{c(v_0)-1}{2} \rfloor$ and follow the algorithm for calculating ts_{v_0}.

Computing $r_{v_0}(u_i)$: Now we will compute $r_{v_0}(u_i)$ for every child u_i of v_0 in T. Depending on whether u_i was contributing in tr_{v_0} or not we have the following two cases.

- If u_i was not contributing to tr_{v_0}, then $r_{v_0}(u_i) = tr_{v_0}$.
- Else, we have to remove the contribution of of c_i and add the contribution of $next_{v_0}$. So, $r_{v_0}(u_i) = tr_{v_0} - tr_{u_i} + next_{v_0}$ if $next_v > 4$; and $r_{v_0}(u_i) = tr_{v_0} - tr_{u_i} + 4$ otherwise.

Computing $s_{v_0}(u_i)$: Similarly we will compute $s_{v_0}(u_i)$ for every child u_i of v_0 in T. We can have the following cases depending on the contribution of u_i to ts_{v_0}.

1. If u_i was not contributing to ts_{v_0}, then $s_{v_0}(u_i) = ts_{u_i}$.
2. If u_i was contributing to ts_{v_0} such that B enters and stays in subtree rooted at u_i, then $s_{v_0}(u_i)$ will be equal to ts_v value without considering the subtree rooted at u_i. This can be computed in $O(k)$ time. This happens only once.
3. If u_i was contributing to ts_{v_0} such that B enters and returns from u_i, then one of the following sub cases occur.
   (a) If B was staying at u_{next}, then for every child u_i used for returning, $tr_{u_i} + ts_{u_{next}} \geq tr_{u_{next}} + ts_{u_i}$. So when one such vertex u_i is not considered (by definition of $s_{v_0}(u_i)$), B would still not stay at any of such u_i's. So among u_{next} and vertices not contributing to ts_{v_0}, B can enter and return from one vertex and enter and stay in another. We denote these vertices as u_x and u_y respectively, which we can compute in $O(k)$ time among these vertices. This has to be computed only once. So for these u_i's, we have these sub cases.

- If $tr_x > 4$ and $ts_y > 2$, then $s_{v_0}(u_i) = ts_{v_0} - tr_{u_i} - ts_{u_{next}} + tr_x + ts_y$.
- If $tr_x > 4$ and $ts_y \leq 2$, then $s_{v_0}(u_i) = ts_{v_0} - tr_{u_i} - ts_{u_{next}} + tr_x + 2$.
- If $tr_x \leq 4$ and $ts_y > 2$, then $s_{v_0}(u_i) = ts_{v_0} - tr_{u_i} - ts_{u_{next}} + 4 + ts_y$.
- If $tr_x \leq 4$ and $ts_y \leq 2$, then $s_{v_0}(u_i) = ts_{v_0} - tr_{u_i} - ts_{u_{next}} + 4 + 2$.

(b) Suppose B was not staying at u_{next}. Since u_i was used for enter and return, now B can enter and return from one more of children of v_0. Since u_{next} has the highest unused return value,

- if $next_v > 4$, then $s_{v_0}(u_i) = ts_{v_0} + next_v - tr_{u_i}$.
- if $next_v \leq 4$, then $s_{v_0}(u_i) = ts_{v_0} + 4 - tr_{u_i}$.

We can compute $s_{v_0}(u_i)$ in $O(k)$ time as we are taking $O(k)$ time twice (cases 2 and 3(a)) and $O(1)$ time in other cases. So total time required to compute $s_{v_0}(u_i)$'s for all u_i's is $O(k)$. Also in this step we have computed r_{v_0}, s_{v_0} and $r_{v_0}(u_i)$ in $O(k)$ time.

Step 4
In the fourth step of of Algorithm 1, we compute r_v, s_v and $r_v(u_i)$, $s_v(u_i)$ for every vertex v in T in a BFS order using the following lemma.

Lemma 2. *Let v be a vertex in T with parent p and children $u_1, u_2, \ldots, u_k$. If we know $tr_{u_1}, \ldots, tr_{u_k}$ and $ts_{u_1}, \ldots, ts_{u_k}$, and $r_p(v)$, $s_p(v)$ we can compute r_v and s_v, $r_v(u_i)$, $s_v(u_i)$ in $O(k)$ time, where u_i, for $i \in [k]$, are children of v.*

Proof. We root the tree at v: now p is also a child of v. Observe that tr_p in this tree is equal to $r_p(v)$ in T and ts_p in this tree is equal to $s_p(v)$ in T. Now v has $k + 1$ children. So we can compute r_v and s_v, $r_v(u_i)$, $s_v(u_i)$ in this tree by following the Step 3 of Algorithm 1 (this can be done as v is a root here). Following the time complexity analysis in Step 3 of Algorithm 1, we can do it in $O(k)$ time.   □

Step 5
In the final step of Algorithm 1, we can traverse the tree and find the maximum s_v among all vertices of T. This completes the description of Algorithm 1.

Time Complexity: In Algorithm 1, we do a constant number of traversals of the tree. In every traversal, we perform computations for each vertex only once and these computations take $O(k)$ time where the vertex has k children. So, our algorithm takes $O(n)$ time.

Correctness: The values of tr_v and ts_v (Step 2 of Algorithm 1) are calculated using Algorithms 2 and 3 respectively. Both of these use dynamic programming and hence their proof of correctness follows. These tr_v and ts_v are used to find the r_v and s_v values (Step 4 of Algorithm 1). This outlines the correctness of Algorithm 1.

4   Conclusion

We have presented a linear time algorithm for solving BUMBLEBEE VISITATION on trees. We have the following immediate questions in mind.

1. Solve BUMBLEBEE VISITATION on Outerplanar graphs.
2. Develop approximation algorithms for BUMBLEBEE VISITATION.

Acknowledgements. The authors would like to thank the referees for their valuable comments which lead to an improvement of the original manuscript. Authors would also like to thank Uma kant Sahoo for positive discussions.

References

1. Alspach, B.: Searching and sweeping graphs: a brief survey. Matematiche (Catania) **59**, 5–37 (2006)
2. Angelopoulos, S., Fraignaud, P., Fomin, F., Nisse, N., Thilikos, D.M.: Report on GRASTA 2017, 6th Workshop on GRAph Searching, Theory and Applications, 10–13 April 2017, Anogia, Crete, Greece (2017)
3. Fomin, F.V., Thilikos, D.M.: An annotated bibliography on guaranteed graph searching. Theor. Comput. Sci. **399**(3), 236–245 (2008)
4. Hefetz, D., Krivelevich, M., Stojaković, M., Szabó, T.: Positional Games. OWS. Springer, Basel (2014). https://doi.org/10.1007/978-3-0348-0825-5
5. Garey, M.R., Johnson, D.S.: Computers and Intractability: A Guide to the Theory of NP-Completeness. W. H. Freeman and Co., New York (1979)
6. West, D.B.: Introduction to Graph Theory, 2nd edn. Pearson Education, London (2001)

On Graphs with Minimal Eternal Vertex Cover Number

Jasine Babu[1]([⊠]), L. Sunil Chandran[2], Mathew Francis[3], Veena Prabhakaran[1], Deepak Rajendraprasad[1], and J. Nandini Warrier[4]

[1] Indian Institute of Technology Palakkad, Palakkad, India
111704003@smail.iitpkd.ac.in, {jasine,deepak}@iitpkd.ac.in
[2] Indian Institute of Science, Bangalore, India
sunil@csa.iisc.ernet.in
[3] Indian Statistical Institute, Chennai, India
mathew@isichennai.res.in
[4] National Institute of Technology Calicut, Calicut, India
nandini.wj@gmail.com

Abstract. The eternal vertex cover problem is a variant of the classical vertex cover problem where a set of guards on the vertices have to be dynamically reconfigured from one vertex cover to another in every round of an attacker-defender game. The minimum number of guards required to protect a graph from an infinite sequence of attacks is the eternal vertex cover number (evc) of the graph. It is known that, given a graph G and an integer k, checking whether $\text{evc}(G) \leq k$ is NP-Hard. However, for any graph G, $\text{mvc}(G) \leq \text{evc}(G) \leq 2\,\text{mvc}(G)$, where $\text{mvc}(G)$ is the minimum vertex cover number of G. Precise value of eternal vertex cover number is known only for certain very basic graph classes like trees, cycles and grids. Though a characterization is known for graphs for which $\text{evc}(G) = 2\,\text{mvc}(G)$, a characterization of graphs for which $\text{evc}(G) = \text{mvc}(G)$ remained open. Here, we achieve such a characterization for a class of graphs that includes chordal graphs and internally triangulated planar graphs. For some graph classes including biconnected chordal graphs, our characterization leads to a polynomial time algorithm to precisely determine $\text{evc}(G)$ and to determine a safe strategy of guard movement in each round of the game with $\text{evc}(G)$ guards.

Keywords: Eternal vertex cover · Chordal graphs · Connected vertex cover

1 Introduction

A vertex cover of a graph $G(V, E)$ is a subset $S \subseteq V$ such that for every edge in E, at least one of its endpoints is in S. A minimum vertex cover of G is a vertex cover of G of minimum cardinality and its cardinality is the minimum vertex cover number of G, denoted by $\text{mvc}(G)$. Equivalently, if we imagine that a guard placed on a vertex v can monitor all edges incident at v, then $\text{mvc}(G)$

© Springer Nature Switzerland AG 2019
S. P. Pal and A. Vijayakumar (Eds.): CALDAM 2019, LNCS 11394, pp. 263–273, 2019.
https://doi.org/10.1007/978-3-030-11509-8_22

is the minimum number of guards required to ensure that all edges of G are monitored.

The eternal vertex cover problem is an extension of the above formulation in the context of a multi-round game, where mobile guards placed on a subset of vertices of G are trying to protect the edges of G from an attacker. This problem was first introduced by Klostermeyer and Mynhardt [1]. In this formulation, guards are initially placed by the defender on some vertices with at most one guard per vertex. In each round of the game, the attacker gets the first turn when he attacks an edge of the graph and then the defender is allowed to let each guard remain in its current vertex or move it to a neighboring vertex, ensuring that at least one guard moves through the edge that was attacked in the current round. Then the game proceeds to the next round of attack-defense. The movement of guards in a round is assumed to happen in parallel, but no two guards are allowed to be on the same vertex at any time. Clearly, if the vertices occupied by the guards do not form a vertex cover at the beginning of each round, there is an attack which cannot be defended, namely an attack on an edge that has no guards on its end points.

If $\mathcal{C}$ is a family of vertex covers of G of the same cardinality, such that the defender can choose any vertex cover from $\mathcal{C}$ as the starting configuration and successfully keep on defending attacks forever by moving among configurations in $\mathcal{C}$ itself, then $\mathcal{C}$ is an *eternal vertex cover class* of G and each vertex cover in $\mathcal{C}$ is an *eternal vertex cover* of G. If S is an eternal vertex cover belonging to an eternal vertex cover class $\mathcal{C}$, we say that S is a configuration in $\mathcal{C}$. Eternal vertex cover number of G, denoted by $\mathrm{evc}(G)$, is the minimum cardinality of an eternal vertex cover of G.

Klostermeyer and Mynhardt [1] showed that, for C_n, a cycle on n vertices with $n \geq 3$, $\mathrm{evc}(C_n) = \mathrm{mvc}(C_n) = \lceil \frac{n}{2} \rceil$ and for any tree on n vertices with $n \geq 2$, eternal vertex cover number is one more than its number of internal vertices. In particular, for a path on an odd number of vertices, its eternal vertex cover number is twice its vertex cover number. They also showed that, for any graph G, $\mathrm{mvc}(G) \leq \mathrm{evc}(G) \leq 2\,\mathrm{mvc}(G)$. From the examples given above, it can be seen that both these bounds are tight even for bipartite graphs.

Fomin et al. [2] discusses the computational complexity and derives some algorithmic results for the eternal vertex cover problem. They use an eternal vertex cover problem model in which more than one guard can be placed on a single vertex. They showed that given a graph $G(V, E)$ and an integer k, it is NP-hard to decide $\mathrm{evc}(G) \leq k$. The paper also gave an exact algorithm with $2^{O(n)}$ time complexity and exponential space complexity. They also gave an FPT algorithm with eternal vertex cover number as the parameter, to solve the eternal vertex cover problem. They also describe a simple polynomial time 2-factor approximation algorithm for the eternal vertex cover problem, using maximum matchings. The above results also carry forward (with minor modifications in proofs) for the original model which allows at most one guard per vertex. It is not yet known if the problem is in NP. It is also unknown whether the eternal vertex cover problem for bipartite graphs is NP-hard. Some related graph parameters based on multi-round attacker-defender games and their relationship with eternal vertex cover number were investigated by Anderson et al. [3] and Klostermeyer et al. [4].

Klostermeyer and Mynhardt [1] gave a characterization for graphs G which have $evc(G) = 2\,mvc(G)$. The characterization follows a nontrivial constructive method starting from any tree T which requires $2\,mvc(T)$ guards to protect it. They also give a few examples of graphs G for which $evc(G) = mvc(G)$ such as complete graph on n vertices(K_n), Petersen graph, $K_m \,\square\, K_n$, $C_m \,\square\, C_n$ (where $\square$ represents the box product) and $n \times m$ grid, if n or m is even. However, they mention that an elegant characterization of graphs for which $evc(G) = mvc(G)$ seems to be difficult.

Here, we achieve such a characterization for a class of graphs that includes chordal graphs and internally triangulated planar graphs. For some graph classes including biconnected chordal graphs, our characterization leads to a polynomial time algorithm to precisely determine $evc(G)$ and to determine a safe strategy of guard movement in each round of the game with $evc(G)$ guards.

Klostermeyer and Mynhardt [1] proved that a graph G with two disjoint minimum vertex covers and each edge contained in some maximum matching has $evc(G) = mvc(G)$. They had posed a question whether it is necessary for every edge e of G to be present in some maximum matching, to satisfy $evc(G) = mvc(G)$. We also present an example which answers this question in negative.

2   Characterising Graphs with $evc(G) = mvc(G)$ for Some Graph Classes

Without loss of generality we may assume that the input graph $G(V, E)$ is connected and has at least two vertices. For any subset $U \subseteq V$, $G[U]$ denotes the induced subgraph of G on the vertex set U. Instead of trying to characterize graphs for which $evc(G) = mvc(G)$, we look at a slightly more general question.

For any subset $U \subseteq V$, let $mvc_U(G)$ be the minimum cardinality of a vertex cover of G that contains all vertices of U and let $evc_U(G)$ be the minimum number of guards required to provide eternal protection to G in such a way that in every configuration of the eternal vertex cover class, all vertices of U are occupied by guards. Note that when $U = \emptyset$, $mvc_U(G) = mvc(G)$ and $evc_U(G) = evc(G)$.

We first derive a necessary condition for $evc_U(G) = mvc_U(G)$ for any (connected) graph $G(V, E)$ (with $|V| \geq 2$) and $U \subseteq V$. Suppose $evc_U(G) = mvc_U(G)$. Let $\mathcal{C}$ be an eternal vertex cover class of G in which each configuration has size equal to $evc_U(G) = mvc_U(G)$ and contains U. Note that for every vertex $v \in V \setminus U$, there must be a configuration in $\mathcal{C}$ in which there are guards on all vertices of $U \cup \{v\}$. Otherwise, an attack on an edge incident to v cannot be defended, since after this attack, there must be a guard on v, implying that we must be in a configuration that contains $U \cup \{v\}$. Since every configuration in $\mathcal{C}$ has size equal to $mvc_U(G)$, we have the following observation.

Observation 1. *If* $evc_U(G) = mvc_U(G)$, *then for every vertex* $v \in V \setminus U$, $mvc_{U \cup \{v\}}(G) = mvc_U(G)$.

A vertex cover S of a graph G is called a connected vertex cover if $G[S]$ is connected. The *connected vertex cover number*, $\mathrm{cvc}(G)$, is the size of a minimum cardinality connected vertex cover of G. In the lemma below we show that if every vertex cover S of G, with $U \subseteq S$ and $|S| = \mathrm{mvc}_U(G)$, is connected, then the necessary condition mentioned in Observation 1 is also sufficient to get $\mathrm{evc}_U(G) = \mathrm{mvc}_U(G)$.

Lemma 1. *Let $G(V, E)$ be a connected graph with $|V| \geq 2$ and let $U \subseteq V$. Suppose that every vertex cover S of G of size $\mathrm{mvc}_U(G)$ that contains U is connected. Then $\mathrm{evc}_U(G) = \mathrm{mvc}_U(G)$ if and only if for every vertex $v \in V \setminus U$, $\mathrm{mvc}_{U \cup \{v\}}(G) = \mathrm{mvc}_U(G)$.*

Proof. Let $k = \mathrm{mvc}_U(G)$. Suppose every vertex cover S of G with $U \subseteq S$ and $|S| = k$ is connected.

If $\mathrm{evc}_U(G) = \mathrm{mvc}_U(G)$ then by Observation 1, the forward direction of the lemma holds.

To prove the converse, assume that for every vertex $v \in V \setminus U$, $\mathrm{mvc}_{U \cup \{v\}}(G) = k$. We will show the existence of an eternal vertex cover class $\mathcal{C}$ of G with exactly k guards such that in every configuration of $\mathcal{C}$, all vertices in U are occupied. We may take any vertex cover S of G with $U \subseteq S$ and $|S| = k$ as the starting configuration. It is enough to show that from any vertex cover S_i of G with $U \subseteq S_i$ and $|S_i| = k$, following an attack on an edge uv such that $v \notin S_i \ni u$, we can safely defend the attack by moving to a vertex cover S_j such that $(U \cup \{v\}) \subseteq S_j$ and $|S_j| = k$.

Consider an attack on the edge uv such that $v \notin S_i \ni u$. Let $\Gamma = \{S' : S'$ is a vertex cover of G with $|S'| = k$ and $(U \cup \{v\}) \subseteq S'\}$. We will show that it is possible to safely defend the attack on uv by moving from S_i to S_j, where $S_j \in \Gamma$ is an arbitrary minimum vertex cover such that the cardinality of its symmetric difference with S_i is minimized. Let $Z = S_i \cap S_j$, $S_i = Z \uplus X$ and $S_j = Z \uplus Y$. Since S_i is a vertex cover of G that is disjoint from Y, we can see that Y is an independent set. Similarly, X is also an independent set. Hence, $H = G[X \uplus Y]$ is a bipartite graph. Further, since $|S_i| = |S_j|$ we also have $|X| = |Y|$.

Claim 1. *H has a perfect matching.*

Proof. Note that $U \subseteq Z$. Consider any $Y' \subseteq Y$. Since $S_i = Z \uplus X$ is a vertex cover of G, we have $N_G(Y') \subseteq Z \uplus X$. If $|N_H(Y')| < |Y'|$, then $S' = Z \uplus (Y \setminus Y') \uplus N_H(Y')$ is a vertex cover of size smaller than k with $U \subseteq S'$, violating the fact that $\mathrm{mvc}_U(G) = k$. Therefore, $\forall Y' \subseteq Y$, $|N_H(Y')| \geq |Y'|$ and by Hall's theorem [5] H has a perfect matching.

Since $v \in S_j \setminus S_i$, we have $|X| = |Y| \geq 1$.

Claim 2. *$\forall x \in X$, the bipartite graph $H \setminus \{x, v\}$ has a perfect matching.*

Proof. If $H \setminus \{x, v\}$ is empty, then the claim holds trivially. Consider any non-empty subset $Y' \subseteq (Y \setminus \{v\})$. By Claim 1, $|N_H(Y')| \geq |Y'|$. If $|N_H(Y')| = |Y'|$, then $S' = Z \uplus (Y \setminus Y') \uplus N_H(Y')$ is a vertex cover of G with $|S'| = k$ and

$(U \cup \{v\}) \subseteq S'$. This contradicts the choice of S_j, since the symmetric difference of S' and S_i has lesser cardinality than that of S_i and S_j. Therefore, $|N_H(Y')| \geq |Y'| + 1$ and $|N_H(Y') \setminus \{x\}| \geq |Y'|$. Hence, for all subsets $Y' \subseteq (Y \setminus \{v\})$, $|N_H(Y') \setminus \{x\}| \geq |Y'|$ and by Hall's theorem, $H \setminus \{x, v\}$ has a perfect matching.

We will now describe how the attack on the edge uv can be defended by moving guards.

- Case 1. $u \in X$:
  By Claim 2, there exists a perfect matching M in $H \setminus \{u, v\}$. In order to defend the attack, move the guard on u to v and also all the guards on $X \setminus u$ to $Y \setminus v$ along the edges of the matching M.
- Case 2. $u \in Z$:
  Recall that $|X| = |Y| \geq 1$. By our assumption, the vertex cover $S_i = Z \uplus X$ is connected. Let P be a shortest path from X to u in $G[X \cup Z]$. By the minimality of P, it has exactly one vertex x from X and x will be an endpoint of P. Suppose $P = (x, z_1, z_2, \cdots, z_t = u)$ where $z_i \in Z$, for $1 \leq i \leq t$. By Claim 2, there exists a perfect matching M in $H \setminus \{x, v\}$. In order to defend the attack, move the guard on u to v, x to z_1 and z_i to z_{i+1}, $\forall i \in [t-1]$. In addition, move all the guards on $X \setminus \{x\}$ to $Y \setminus \{v\}$ along the edges of the matching M.

In both cases, the attack can be defended by moving the guards as mentioned and the new configuration is S_j.   □
We now show a necessary condition for a graph G to have $\mathrm{evc}(G) = \mathrm{mvc}(G)$.

Lemma 2. *Let $G(V, E)$ be any connected graph. Let $X \subseteq V$ be the set of cut vertices of G. If $\mathrm{evc}(G) = \mathrm{mvc}(G)$, then $\mathrm{evc}_X(G) = \mathrm{mvc}_X(G) = \mathrm{evc}(G) = \mathrm{mvc}(G)$.*

Proof. Suppose $\mathrm{evc}(G) = \mathrm{mvc}(G) = k$ and $\mathcal{C}$ be a minimum eternal vertex cover class of G. If $X = \emptyset$, the result holds trivially. If $X \neq \emptyset$, we will show that in any minimum eternal vertex cover class $\mathcal{C}$ of G, all cut vertices of G have to be occupied with guards in all configurations.

Let x be any cut vertex of G. Let H be a connected component of $G \setminus x$, $H_1 = G[V(H) \cup \{x\}]$ and $H_2 = G[V \setminus V(H)]$. Note that H_1 and H_2 are edge-disjoint subgraphs of G with x being their only common vertex. Let $k_1 = \mathrm{mvc}(H_1)$ and $k_2 = \mathrm{mvc}(H_2)$. It is easy to see that $k = \mathrm{mvc}(G) \in \{k_1 + k_2 - 1, k_1 + k_2\}$. Since $\mathrm{evc}(G) = \mathrm{mvc}(G)$, there must be a vertex cover configuration S in the eternal vertex cover class $\mathcal{C}$ such that $x \in S$. Either $|S \cap V(H_1)| = k_1$ or $|S \cap V(H_2)| = k_2$ or both. If both $|S \cap V(H_1)| = k_1$ and $|S \cap V(H_2)| = k_2$, then $k = k_1 + k_2 - 1$ and G has no minimum vertex covers without x. This would immediately imply that in every configuration of $\mathcal{C}$, x is occupied by a guard.

Therefore, without loss of generality, we need to consider the only case when H_1 has no minimum vertex cover containing x. If x is not occupied by a guard in a configuration $S' \in \mathcal{C}$, we must have $|S' \cap V(H_1)| = k_1$ and $|S' \cap V(H_2)| = k_2$. In this configuration, consider an attack on an edge ux in H_1. A guard must move to x from u and no guard from H_2 can move to H_1 at this point. This is impossible because H_1 has only k_1 guards on it and for occupying x, at least $k_1 + 1$ guards are required on $V(H_1)$. Hence, in this case also, x is occupied by a guard in every configuration of $\mathcal{C}$.

Since x was an arbitrary chosen cut vertex, this implies that all vertices of X must be occupied in all configurations of the eternal vertex cover class $\mathcal{C}$ and hence $\text{evc}_X(G) = \text{mvc}_X(G) = \text{evc}(G) = \text{mvc}(G)$.   □

The following is a corollary of Lemma 2 and Observation 1.

Corollary 1. *For any connected graph G with at least three vertices and minimum degree one, $\text{evc}(G) \neq \text{mvc}(G)$.*

The corollary holds because a degree one vertex and its neighbor (which is a cut vertex, if the graph itself is not just an edge) cannot be simultaneously present in a minimum vertex cover of G.

The following theorem, which follows from Lemmas 1 and 2, gives a necessary and sufficient condition for a graph G to satisfy $\text{evc}(G) = \text{mvc}(G)$, if every minimum vertex cover of G that contains all cut vertices is connected.

Theorem 1. *Let $G(V, E)$ be a connected graph with $|V| \geq 2$ and $X \subseteq V$ be the set of cut vertices of G. Suppose every minimum vertex cover S of G with $X \subseteq S$ is connected. Then $\text{evc}(G) = \text{mvc}(G)$ if and only if for every vertex $v \in V \setminus X$, there exists a minimum vertex cover S_v of G such that $(X \cup \{v\}) \subseteq S_v$.*

Proof. Suppose every minimum vertex cover S of G with $X \subseteq S$ is connected. Let $k = \text{mvc}(G)$.

If $\text{evc}(G) = \text{mvc}(G)$, by Lemma 2, $\text{evc}_X(G) = \text{mvc}_X(G) = k$. Hence, by Lemma 1, for every vertex $v \in V \setminus X$, there exists a minimum vertex cover S_v of G such that $(X \cup \{v\}) \subseteq S_v$.

For the converse, assume that for every vertex $v \in V \setminus X$, there exists a minimum vertex cover S_v of G such that $(X \cup \{v\}) \subseteq S_v$. This implies that $\text{mvc}_X(G) = k = \text{mvc}(G)$ and from our assumption, every vertex cover S of G with $X \subseteq S$ and $|S| = k$ is connected. Hence, by Lemma 1, $\text{evc}(G) = \text{evc}_X(G) = \text{mvc}_X(G) = k$.   □

Theorem 1 gives a method to determine $\text{evc}(G)$, if G is biconnected and all its minimum vertex covers are connected.

Theorem 2. *Let $G(V, E)$ be a biconnected graph for which every minimum vertex cover is connected. If for every vertex $v \in V$, there exists a minimum vertex cover S_v of G such that $v \in S_v$, then $\text{evc}(G) = \text{mvc}(G)$. Otherwise, $\text{evc}(G) = \text{mvc}(G) + 1$.*

Proof. Klostermayer et al. [1] showed that $\text{evc}(G)$ is at most one more than the size of a connected vertex cover of G. Hence, from our assumption that all minimum vertex covers of G are connected, we have $\text{evc}(G) \leq \text{mvc}(G) + 1$. Now, the theorem follows by Theorem 1, because G is biconnected and therefore has no cut vertices.   □

To illustrate the usefulness of Theorems 1 and 2, we will look at some well-known graph classes for which every vertex cover that contains all cut vertices is a connected vertex cover.

A graph G is *locally connected* if for every vertex v of G, its open neighborhood $N_G(v)$ induces a connected subgraph in G. A *block* in a connected graph G is either a maximal biconnected component or a bridge of G. The following is an easy observation.

Observation 2. *Let $G(V, E)$ be a connected graph and $X \subseteq V$ be the set of cut vertices of G. If every block of G is locally connected, then every vertex cover S of G with $X \subseteq S$ is connected.*

Proof. The restriction of a vertex cover S of G to a block will give a vertex cover of the block. Hence, to prove the observation, it is enough to show that all vertex covers of a biconnected locally connected graph G are connected.

For contradiction, suppose G is a biconnected locally connected graph and S is a vertex cover of G such that $G[S]$ is not connected. Then, there exists a vertex $v \in V \setminus S$ and two components C_1 and C_2 of $G[S]$ such that v is adjacent to vertices $v_1 \in V(C_1)$ and $v_2 \in V(C_2)$. Since S is a vertex cover that does not contain v, we have $N_G(v) \subseteq S$. Since G is locally connected, we know that $N_G(v)$ is connected and therefore, v_1 and v_2 must belong to the same component of $G[S]$, which is a contradiction. Hence, $G[S]$ is connected. $\square$

From Observation 2 and Theorem 1, we have the following result.

Corollary 2. *Let $G(V, E)$ be a connected graph with $|V| \geq 2$ and $X \subseteq V$ be the set of cut vertices of G. If every block of G is locally connected, then $\mathrm{evc}(G) = \mathrm{mvc}(G)$ if and only if for every vertex $v \in V \setminus X$, there exists a minimum vertex cover S_v of G such that $(X \cup \{v\}) \subseteq S_v$.*

A graph is *chordal* if it contains no induced cycle of length four or more. A graph is an *internally triangulated planar graph* if it has a planar embedding in which all internal faces are triangles. It can be easily seen that biconnected chordal graphs and biconnected internally triangulated planar graphs are locally connected. Hence, we have the following special case of Corollary 2.

Corollary 3. *For any graph G that is chordal or is an internally triangulated planar graph, $\mathrm{evc}(G) = \mathrm{mvc}(G)$ if and only if for every vertex v of G that is not a cut-vertex, there is a minimum vertex cover of G that contains v and all the cut-vertices.*

Chartrand and Pippert [6] proved that if G is a graph of order p such that for every pair of vertices u, v, $deg(u) + deg(v) > \frac{4}{3}(p - 1)$, then G is locally connected. Some other sufficient conditions for a graph to be locally connected were given by Vanderjagt [7]. Threshold phenomenon for local connectivity of a random graph was given by Erdös Palmer and Robinson [8]. Hence, it may be noted that Corollary 2 is applicable to more graph classes other than those mentioned in Corollary 3.

A trivial reduction from the vertex cover problem on connected graphs to the vertex cover problem of locally connected graphs is known. Given a connected graph $G(V, E)$ and an integer k, a locally connected graph G' can be constructed from G by inserting a new vertex and making it adjacent to all vertices in V. It is easy to see that G has a vertex cover of size k if and only if G' has a vertex cover of size $k + 1$. Hence, given a locally connected graph G and an integer k, deciding whether G has a vertex cover of size at most k is NP-Complete. Using the same construction, it can also be shown that given a locally connected graph G, a vertex v of G and an integer k, deciding whether G has a vertex cover of size at most k containing v is also NP-Complete. Hence, the characterization mentioned in Corollary 2 does not immediately give a polynomial time algorithm for deciding whether a locally connected graph G satisfies $\text{evc}(G) = \text{mvc}(G)$. In the remaining parts of this section, we study some graph classes where such a polynomial time algorithm can be derived.

A class of graphs $\mathcal{H}$ is called hereditary, if deletion of vertices from any graph G in $\mathcal{H}$ would always yield another graph in $\mathcal{H}$.

Theorem 3. *Let $\mathcal{H}$ be a hereditary graph class such that :*

- *for graphs in $\mathcal{H}$, their minimum vertex cover number computation can be done in polynomial time and*
- *for every biconnected graph H in $\mathcal{H}$, all vertex covers of H are connected.*

Then,

1. *for any connected graph G in $\mathcal{H}$, in polynomial time we can decide whether $\text{evc}(G) = \text{mvc}(G)$*
2. *for any connected graph G in $\mathcal{H}$ with $\text{evc}(G) = \text{mvc}(G)$, starting from any minimum vertex cover of G that contains all cut vertices of G as the first configuration, it is possible to determine a safe strategy of guard movements in each round of the game that uses exactly $\text{evc}(G)$ guards and requiring only polynomial time to compute the strategy in any round*
3. *for any biconnected graph G in $\mathcal{H}$, in polynomial time we can compute $\text{evc}(G)$. Further, it is possible to determine a safe strategy of guard movement in each round of the game that uses exactly $\text{evc}(G)$ guards and requiring only polynomial time to compute the strategy in any round.*

Proof. 1. For any connected graph G in $\mathcal{H}$, in polynomial time we can compute $\text{mvc}(G)$. Identifying the set of cut vertices X of G can also be done in polynomial time. By Theorem 1, to decide whether $\text{mvc}(G) = \text{evc}(G)$, it is enough to check for every vertex $v \in V \setminus X$ whether G has a minimum vertex cover $S_v \supseteq X \cup \{v\}$. Checking whether G has a minimum vertex cover containing $X \cup \{v\}$ is equivalent to checking whether $\text{mvc}(G) = \text{mvc}(G') + |X| + 1$, where $G' = G \setminus (X \cup \{v\})$. Since $G' \in \mathcal{H}$, we can compute $\text{mvc}(G')$ and perform this checking in polynomial time.

2. Consider a connected graph G in $\mathcal{H}$ with $\text{evc}(G) = \text{mvc}(G) = k$ and X be the set of cut vertices of G. By Lemma 2, $\text{evc}_X(G) = \text{mvc}_X(G) = k$. By our assumption, for any block of G, all its vertex covers are connected and hence,

every vertex cover S of G with $X \subseteq S$ is a connected vertex cover. Therefore, by Observation 1, for every vertex $v \in V \setminus X$, $\text{mvc}_{X \cup \{v\}}(G) = k$. We complete the proof by extending the basic ideas used in the proof of Lemma 1.

Take any minimum vertex cover S of G with $X \subseteq S$ as the starting configuration. It is enough to show that from any minimum vertex cover S_i of G with $X \subseteq S_i$, following an attack on an edge uv such that $v \notin S_i \ni u$, we can safely defend the attack by moving to a minimum vertex cover S_j such that $(X \cup \{v\}) \subseteq S_j$. Consider an attack on the edge uv such that $v \notin S_i \ni u$. To start with, choose an arbitrary minimum vertex cover S' of G with $(X \cup \{v\}) \subseteq S'$ as a candidate for being the next configuration.

i. Suppose $C = S_i \cap S'$, $S_i = C \uplus A$ and $S' = C \uplus B$. By similar arguments as in the proof of Lemma 1, $H = G[A \uplus B]$ is a non-empty bipartite graph with a perfect matching. If $\forall x \in A$, the bipartite graph $H \setminus \{x, v\}$ has a perfect matching, then we can choose S' to be the new configuration S_j and move guards as explained in the proof of Lemma 1. Otherwise, we describe a method to choose another configuration instead of S'.

ii. If the bipartite graph $H \setminus \{x, v\}$ does not have a perfect matching for some $x \in A$, in polynomial time we can identify a subset $B' \subseteq (B \setminus \{v\})$ for which $|B'| > |N_{H \setminus \{x,v\}}(B')|$, using a standard procedure described below. First find a max-matching M in $H \setminus \{x, v\}$ and identify an unmatched vertex $y \in B \setminus \{v\}$. Let B' be the set of vertices in $B \setminus \{v\}$ reachable via M-alternating paths from y in $H \setminus \{x, v\}$, together with vertex y. If $|B'| \leq |N_{H \setminus \{x,v\}}(B')|$, it would result in an M-augmenting path from y, contradicting the maximality of M. Thus, $|B'| > |N_{H \setminus \{x,v\}}(B')|$. (In fact, since H has a perfect matching, $|B'| \leq |N_H(B')|$ and this would mean $x \in N_H(B')$.) Now, let $S'' = C \cup \{x\} \cup (B \setminus B') \cup N_{H \setminus \{x,v\}}(B')$.

iii. It is easy to see that S'' is a minimum vertex cover of G with $(A \cup \{v\}) \subseteq S''$ and the symmetric difference of S'' and S_i is smaller than the symmetric difference of S' and S_i. Now we replace S' with S'' and iterate the steps above by redefining the sets C, B and A and the graph H.

We will repeat these steps until we reach a point when the (re-defined) bipartite graph $H \setminus \{x, v\}$ has a perfect matching, $\forall x \in A$. This process will terminate in less than n iterations, because in each iteration, the symmetric difference of the candidate configuration with S_i is decreasing.

The basic computational steps involved in this process are computing minimum vertex covers containing $X \cup \{v\}$, finding maximum matching in some bipartite graphs and computing some alternating paths. All these computations can be performed in polynomial time [9].

3. Let G be a biconnected graph in $\mathcal{H}$. By Theorem 2, $\text{evc}(G) \in \{\text{mvc}(G), \text{mvc}(G) + 1\}$. Therefore, by using part 1 of this theorem, $\text{evc}(G)$ can be decided exactly, in polynomial time. If $\text{evc}(G) = \text{mvc}(G)$, using part 2 of this theorem, we can complete the proof. If $\text{evc}(G) = \text{mvc}(G) + 1$, we will make use of the fact that every minimum vertex cover of G is connected. We will fix a minimum vertex cover S and initially place guards on vertices of S and one additional vertex. Using the method given by Klostermeyer et. al [1] to

show that $\text{evc}(G) \leq \text{cvc}(G) + 1$, we will be able to keep defending attacks while maintaining guards on all vertices of S after end of each round of the game.   □

It is well-known that chordal graphs form a hereditary graph class and computation of a minimum vertex cover of a chordal graph can be done in polynomial time [10]. Since biconnected chordal graphs are locally connected, by Observation 2, we have the following result.

Corollary 4.

(i) For any chordal graph G, we can decide in polynomial-time whether $\text{evc}(G) = \text{mvc}(G)$. Also, if $\text{mvc}(G) = \text{evc}(G)$, a polynomial-time strategy for guard movements as mentioned in Theorem 3 using $\text{evc}(G)$ guards, exists.

(ii) If G is a biconnected chordal graph, then we can determine $\text{evc}(G)$ in polynomial-time. Moreover, a polynomial-time strategy for guard movements as mentioned in Theorem 3 using $\text{evc}(G)$ guards exists.

3 A Graph G with an Edge Not Contained in Any Maximum Matching but $\text{mvc}(G) = \text{evc}(G)$

Klostermeyer et. al [1] proved that if a graph G has two disjoint minimum vertex covers and each edge is contained in a maximum matching then $\text{mvc}(G) = \text{evc}(G)$. They had asked if $\text{mvc}(G) = \text{evc}(G)$, is it necessary that for every edge e of G there is a maximum matching of G that contains e. Here, we give a biconnected chordal graph G for which the answer is negative. The graph G shown in Fig. 1 has $\text{mvc}(G) = 5$, a maximum matching of size 4 and the edge $(8, 4)$ not contained in any maximum matching. It can be shown that $\text{evc}(G) = 5$ because G has an evc class with two configurations, $S_1 = \{1, 8, 3, 5, 7\}$ and $S_2 = \{6, 8, 2, 4, 7\}$. Hence, even for a graph class $\mathcal{H}$ such that for all $G \in \mathcal{H}$, $\text{evc}(G) \leq \text{mvc}(G) + 1$, there could be a graph $G \in \mathcal{H}$ with $\text{mvc}(G) = \text{evc}(G)$ and an edge not present in any maximum matching of G.

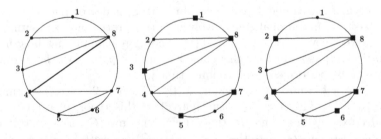

Fig. 1. $\text{mvc}(G) = \text{evc}(G) = 5$ and size of maximum matching is 4. Edge $(8, 4)$ is not contained in any maximum matching.

References

1. Klostermeyer, W., Mynhardt, C.: Edge protection in graphs. Australas. J. Comb. **45**, 235–250 (2009)
2. Fomin, F.V., Gaspers, S., Golovach, P.A., Kratsch, D., Saurabh, S.: Parameterized algorithm for eternal vertex cover. Inf. Process. Lett. **110**(16), 702–706 (2010)
3. Anderson, M., Carrington, J.R., Brigham, R.C., Dutton, D.R., Vitray, R.P.: Graphs simultaneously achieving three vertex cover numbers. J. Comb. Math. Comb. Comput. **91**, 275–290 (2014)
4. Klostermeyer, W.F., Mynhardt, C.M.: Graphs with equal eternal vertex cover and eternal domination numbers. Discret. Math. **311**, 1371–1379 (2011)
5. Diestel, R.: Graph Theory. GTM, vol. 173. Springer, Heidelberg (2017). https://doi.org/10.1007/978-3-662-53622-3
6. Chartrand, G., Pippert, R.E.: Locally connected graphs. Časopis pro pěstování matematiky **99**(2), 158–163 (1974)
7. Vanderjagt, D.W.: Sufficient conditions for locally connected graphs. Časopis pro pěstování matematiky **99**(4), 400–404 (1974)
8. Erdös, P., Palmer, E.M., Robinson, R.W.: Local connectivity of a random graph. J. Graph Theor. **7**(4), 411–417 (1983)
9. Hopcroft, J.E., Karp, R.M.: An $n^{5/2}$ algorithm for maximum matchings in bipartite graphs. SIAM J. Comput. **2**(4), 225–231 (1973)
10. Gavril, F.: Algorithms for minimum coloring, maximum clique, minimum covering by cliques, and maximum independent set of a chordal graph. SIAM J. Comput. **1**, 180–187 (1972)

On Chordal and Perfect Plane Triangulations

Sameera M. Salam, Daphna Chacko$^{(\boxtimes)}$, Nandini J. Warrier,
K. Murali Krishnan, and K. S. Sudeep

Department of Computer Science, National Institute of Technology Calicut,
Kozhikode 673601, Kerala, India
{sameera_p120091cs,daphna_p120038cs,nandini_p120037cs,kmurali,
sudeep}@nitc.ac.in

Abstract. We investigate a method of decomposing a plane near-triangulation G into a collection of induced component subgraphs which we call the W components of the graph. Each W component is essentially a plane near-triangulation with the property that the neighbourhood of every internal vertex induces a wheel. The problem of checking whether a plane near-triangulation G is chordal (or perfect) is shown to be transformable to the problem of checking whether its W components are chordal (or perfect). Using this decomposition method, we show that a plane near-triangulated graph is chordal if and only if it does not contain an internal vertex whose closed neighbourhood induces a wheel of at least five vertices. Though a simple local characterization for plane perfect near-triangulations is unlikely, we show that there exists a local characterization for perfect W components that does not contain any induced wheel of five vertices.

Keywords: Plane triangulated graphs ·
Plane near-triangulated graphs · Chordal graphs ·
Perfect graphs

1 Introduction

A plane embedding of a (planar) graph G is called a triangulation if the boundary of every face is a cycle of length three. A plane embedding of a simple maximal planar graph is a plane triangulation. If the above definition is relaxed to permit the boundary of the outer face to be a cycle of length exceeding three, the embedding is called a plane near-triangulation. A graph G is said to contain a chordless cycle if it has an induced cycle of four or more vertices. A graph that does not contain any chordless cycle is called a chordal graph [1]. A graph G is perfect if for each induced subgraph G' of G, the number of colours required for a proper colouring of G' is equal to the number of vertices in a maximum clique in G' [1]. It is known that a graph G is perfect if and only if neither G nor its complement contain any induced odd cycle of length 5 or more [7].

© Springer Nature Switzerland AG 2019
S. P. Pal and A. Vijayakumar (Eds.): CALDAM 2019, LNCS 11394, pp. 274–285, 2019.
https://doi.org/10.1007/978-3-030-11509-8_23

Investigation of the structural properties of maximal planar graphs and some of their subfamilies like Apollonian networks have been elaborately undertaken in the literature [2–5] owing to their rich and interesting geometric structure. Here we investigate structural characterizations for chordal and perfect plane near-triangulations. The results hold true for plane triangulations as well.

In Sect. 3, we describe a simple decomposition method for plane near-triangulations which will be used to prove our results. We show that a plane near-triangulation G can be decomposed into a unique set of induced component subgraphs which we call the W components of the graph. Each W component is essentially a plane near-triangulation with the property that the neighbourhood of every internal vertex induces a wheel. The problem of checking whether a plane near-triangulation G is chordal (or perfect) is shown to be transformable to the problem of checking whether the collection of its W components are chordal (or perfect).

Using the W component decomposition described above, in Sect. 4, we derive a simple characterization for chordal plane near-triangulations that depends solely on the structure of the local neighbourhood of individual vertices.

Though structural characterizations for perfect plane triangulations were investigated in the literature [6], a local characterization for perfect plane triangulations (or plane near-triangulations) appears unlikely. We show in Sect. 6 that W components that do not contain any induced W_5 (See Definition 1) admits a simple local characterization. A proof for the characterization is given in Sect. 6.

The following section establishes some notation and definitions.

2   Preliminaries

Let $G = (V, E)$ be a simple undirected graph. A *drawing* of a graph maps each vertex $u \in V$ to a point $\varepsilon(u)$ in $\mathbb{R}^2$ and each edge $uv \in E$ to a path with endpoints $\varepsilon(u)$ and $\varepsilon(v)$. The drawing is a plane embedding if the points are distinct, the paths are simple and do not cross each other and the incidences are limited to the endpoints. The well known Jordan Curve Theorem states that if J is a simple closed curve in $\mathbb{R}^2$, then $\mathbb{R}^2 - J$ has two components $(Int(J)$ and $Ext(J))$, with J as the boundary of each. Given a plane triangulation G, $Int(G)$ $(Ext(G))$ denotes the set of vertices in the interior (and respectively, exterior) of the closed curve defined by the boundary of the triangulation. The notation C_n will be used for a cycle of n vertices.

Definition 1 (Wheel). *A wheel on n ($n \geq 4$) vertices, W_n, is the graph obtained by adding a new vertex v to a cycle C_{n-1} and making it adjacent to all vertices in C_{n-1}. The cycle C_{n-1} is called the rim of the wheel, the vertex v is called the center of the wheel and the added edges joining v and vertices in C_{n-1} are called spokes of the wheel.*

Definition 2 (Hole). *Any induced cycle of length at least four in a graph is called a hole. A hole with odd number of vertices is known as odd hole.*

A separator in a connected graph is a set of vertices, the removal of which disconnects the graph. A clique in a graph is a set of pairwise adjacent vertices.

A clique separator is a separator which is a clique. Next we define the notion of separating triangle in a plane near-triangulation.

Definition 3 (Separating triangle). *A separating triangle in a plane near-triangulation is a clique separator with three vertices whose interior contains at least one vertex.*

Definition 4 (W near-triangulation). *A plane near-triangulation G is called a W near-triangulation if either G is a K_4 or G contains neither:*

1. *a separating triangle, nor*
2. *a chord connecting any two vertices lying on the boundary of the external face.*

In particular, if G is a plane triangulation, then G is called a W triangulation.

Observation 1. *Wheel graphs are W near-triangulations.*

Observation 2. *A W near-triangulation G does not contain an induced K_4 unless $G = K_4$.*

Recall that in a plane triangulation G, $Int(G)$ ($Ext(G)$) denotes the set of vertices in the interior (and respectively, exterior) of the closed curve defined by the boundary of the triangulation.

Definition 5 (Even W near-triangulation). *A W near-triangulation G is called an even W near-triangulation if the degree of every vertex in $Int(G)$ is even.*

Let G be any graph and $v \in V(G)$. Define the open neighbourhood of the vertex v as $N(v) = \{v' \in V : (v', v) \in E\}$ and the closed neighbourhood of v as $N[v] = \{v\} \cup N(v)$.

Definition 6 ($C-$ components). *Let $C = uv$ be a chord connecting any two vertices on the boundary of the external face of a plane near-triangulated graph. Let $P_1 = uu_1u_2\ldots v$ and $P_2 = vv_1v_2\ldots u$ be the two paths connecting u and v on the boundary of the external face of G. Then the chord $C = uv$ separates G into two connected induced near-triangulations say G_1 and G_2 with $P_1 \cup C$ and $P_2 \cup C$ as the boundary of the external face of G_1 and G_2 respectively. We call the component graphs G_1 and G_2 as the $C-$ components of G.*

Definition 7 ($T-$ components). *Let $T = (u, v, w)$ be a separating triangle in a plane near-triangulated graph G. Then T divides G into two components say, $G_1 = Int(T) \cup T$ and $G_2 = Ext(T) \cup T$ where G_1 is a plane triangulation and G_2 is a plane near-triangulation. We call the component graphs G_1 and G_2 as the $T-$ components of G.*

Observation 3. *Let the edge $C = uv$ be a chord connecting two vertices u and v on the boundary of the external face of a plane near-triangulated graph G. Then, G is chordal (or, perfect) if and only if both the $C-$components of G, G_1 and G_2 (see Definition 6) are chordal (or, perfect).*

Observation 4. *Let the vertices u, v and w form a separating triangle T in a plane near-triangulated graph G. Then G is chordal (respectively, perfect) if and only if both the $T-$components of G, G_1 and G_2 (see Definition 7) are chordal (respectively, perfect).*

The following section explains a method of decomposition of plane near-triangulations based on Observations 4 and 3. This decomposition will be used later to characterize plane chordal near-triangulations and plane perfect near-triangulations.

3 W Decomposition

Let G be a plane near-triangulated graph. If G contains a chord C connecting any two vertices on the boundary of the external face, then by Observation 3, the chord divides G into two $C-$components. If any one of the component graphs contain a chord connecting two vertices on the boundary of the external face, then we can further divide the component graph into two components. We can repeat this process until we obtain a decomposition of G into induced subgraphs, none of which contains any chord connecting two vertices on the boundary of external face. Let $G_1, G_2, \ldots, G_k$ be the induced subgraphs of G obtained through such decomposition process. If any G_i for $1 \leq i \leq k$ is not a W near-triangulation, then G_i contains a separating triangle T and by Observation 4, T divides G_i into two $T-$components - say G_i' and G_i''. Note that these $T-$ components do not have any chord connecting two vertices on the boundary of the external face. We can continue to find T components of G_i' and G_i'' to eventually decompose G_i into a collection of W near-triangulations - say $G_{i,1}, G_{i,2}, \ldots G_{i,n_i}$ for some $n_i > 0$. Note that $\bigcup_i \bigcup_j G_{i,j} = G$. We call this process of decomposition of a plane near-triangulated graph into a collection of induced component subgraphs $\{G_{i,j}\}$ such that each $G_{i,j}$ is a W near-triangulation as a W **decomposition** of G. The subgraphs $G_{i,j}$ will be called the W components of G. It is not hard to see that the W components of a given plane near-triangulated graph G determined as above are unique and does not depend on the order in which decompositions are performed.

Lemma 1. *Every plane near-triangulation can be decomposed into W components in finite number of steps. Moreover, G is chordal (respectively, perfect) if and only if all its W components are chordal (respectively, perfect).*

The following section characterises chordal plane near-triangulations.

4   Chordal Plane Near-Triangulations

We first prove the following Lemma:

Lemma 2. *If G is a W near-triangulation with at least five vertices then for all $u \in Int(G)$, $N[u]$ induces a wheel W_k for some $k \geq 5$.*

Proof. Let G be a W near-triangulation, $u \in Int(G)$. As G is a plane near-triangulation, it is easy to see that $|N(u)| \geq 3$. Moreover as G is a W near-triangulation and $G \neq K_4$, by Observation 2, G does not contain any induced K_4. Further, $|N(u)| \neq 3$ for otherwise vertices in $N(u)$ will form a separating triangle. Hence, we may assume without loss of generality that $N(u) = \{u_0, u_1, u_2, \ldots u_{k-1}\}$ for some $k \geq 4$, such that $uu_0, uu_1, \ldots, uu_{k-1}$ is the clockwise ordering of the edges incident with u. We claim that $u_i u_{i+1} \in E(G)$, where index $i \in 0, 1, \ldots, k-1$ is taken modulo k. Indeed, if $u_i u_{i+1}$ is not an edge, then uu_i and uu_{i+1} will be on the boundary of a face of length greater than three, contradicting that G is a W near-triangulation. Now suppose that there exists an edge $u_i u_j$ with $j \notin \{i+1, i-1\}$. Then $\{u, u_i, u_j\}$ will be a separating triangle. Therefore $N[u]$ is a wheel.   □

The following observation is directly verifiable and covers the base cases of the inductive argument that follows.

Observation 5. *Every near-triangulations with five or fewer vertices except W_5 is chordal. Every near-triangulation having no internal vertex is chordal.*

Lemma 3. *A W near-triangulation is not chordal iff it contains at least one internal vertex.*

Proof. Let G be a W near-triangulation. If $|V(G)| \leq 5$ then by Observation 5, G is not chordal iff G is a W_5, which has an internal vertex. If $|V(G)| > 5$ and there is no internal vertex in G then by Observation 5 G is chordal. If $|V(G)| > 5$ and G contains at least one internal vertex say u, then by the Lemma 2, $N[u]$ will induce a wheel say W_k for $k \geq 5$. As the rim of W_k is a chordless cycle of length $(k-1) > 3$, G is not chordal.   □

The following theorem shows that chordal plane near-triangulations are characterized by the closed neighbourhoods of internal vertices.

Theorem 1. *A plane near-triangulated graph is not chordal iff it contains an induced wheel W_k for some $k \geq 5$.*

Proof. Let $G(V, E)$ be a near-triangulated graph. If G contains an induced W_k for some $k \geq 5$ then the rim of W_k is a chordless cycle and G is not chordal.

Conversely, if G is not chordal, by Lemma 1, we can decompose G into a unique set of W components - say $G_1, G_2, \ldots, G_t$ for some $t > 0$ such that G is not chordal if and only if at least one G_i, $1 \leq i \leq t$ is not chordal. Let G_k be a non-chordal W component of G. Since G_k is a near-triangulation which is not chordal, by Lemma 3, G_k contains at least one internal vertex. By Lemma 2, neighbours of each internal vertex must induce a wheel W_k for some $k \geq 5$. The result follows from the fact that the internal vertex of a component is also an internal vertex of G.   □

Corollary 1. *A plane triangulated graph is not chordal iff it contains an induced wheel W_k for some $k \geq 5$.*

5   Perfect Near-Triangulations

Our next objective is to investigate the problem of providing a local characterization for near-triangulated perfect graphs similar in spirit to Theorem 1. It is
easy to see that the complement of cycle C_n for $n \geq 7$ is not planar. Moreover,
the complement of C_5 is isomorphic to C_5. Thus, it follows from the strong perfect graph theorem [7] that to prove a plane triangulated graph G is perfect, it
is enough to prove that G does not contain an induced odd hole.

Let G be a plane near-triangulated graph. If G contains an induced even
wheel then clearly G is not perfect. However the absence of an induced even
wheel is not sufficient to guarantee the perfectness of a plane near-triangulation.
For example, the graph shown in Fig. 1 does not contain any induced even wheel.
But the vertices on the boundary of external face induce an odd hole.

A simple local characterization for plane perfect triangulations (or plane perfect near-triangulations) appears elusive. We believe that a promising approach
to the problem could be to try to locally characterize W triangulations. By
Lemma 1, we know that a plane near-triangulation G is perfect if and only if all
of its component W near-triangulations are perfect. Consequently, the problem
of characterizing perfect near-triangulations reduces to the problem of characterizing perfect W near-triangulations. Clearly, a simple local characterization
for perfect W near-triangulations is unlikely as this would solve the problem of
characterizing plane perfect near-triangulations. In this work, we characterize a
subclass of W near-triangulations that indeed admits a simple local characterization. In the next section, we derive a simple local structural characterization
for W near-triangulations that does not contain any induced W_5.

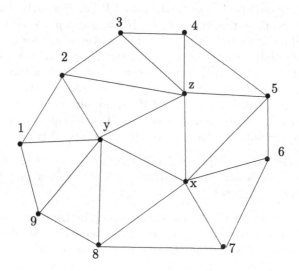

Fig. 1. An even wheel-free non-perfect plane near-triangulation

6 W_5 Free W Near-Triangulations

In this section, we prove that any non-perfect W_5 free W near-triangulation G contains either an even wheel or contains three vertices forming an internal face such that the open neighbourhood of these vertices induce an odd hole. Throughout this section, we use the notation $W(u)$ to denote a wheel with vertex u at the centre. We first establish some properties of W near-triangulations that will be useful for deriving the characterization.

Lemma 4. *Let a W near-triangulation G contain e edges, f internal faces and t edges on the boundary of external face, then $f \equiv t$ mod 2. That is, f is odd if and only if t is odd.*

Proof. Each internal face is bounded with exactly three edges and each edge except those in the boundary of the external face is shared by two faces. This implies $3f = 2e - t$. Hence t is odd if and only if f is odd.   □

Definition 8 (Face intersecting wheels). *Let $W(u)$ and $W(v)$ $(u \neq v)$ be any two wheels in a W near-triangulation. $W(u)$ and $W(v)$ are said to be face intersecting if they share at least one face.*

Lemma 5. *Let G be a W near-triangulation and $W(u)$ and $W(v)$ $(u \neq v)$ be any two face-intersecting wheels in G, then $W(u)$ and $W(v)$ share exactly two faces with the edge uv as one of its boundaries.*

Proof. Since $W(u)$ and $W(v)$ are face-intersecting and $u \neq v$, u should be on the rim of $W(v)$. Similarly v should lie on the rim of $W(u)$. Hence the edge uv should be a spoke in both the wheels. As u is on the rim of $W(v)$, u will have exactly two neighbours (say x, y) on the rim of $W(v)$. Similarly v also have two neighbours (say p, q) on the rim of $W(u)$. If $p \neq x$ and $p \neq y$ then the edge pu will be a chord on the wheel $W(v)$ and the vertices u, p, v forms a separating triangle in G, which is a contradiction to the definition of W near-triangulation. Hence $p = x$ or $p = y$ Similarly $q = y$ or $q = x$. This implies that either $p = x$ and $q = y$ or $p = y$ and $q = x$. So x and y are the only vertices in $N(u) \cap N(v)$ and the edges ux and uy on the rim of $W(v)$ are also spokes of $W(u)$ and vx and vy on the rim of $W(u)$ are also spokes of $W(v)$. That is, $\{ux, xv, vu\}$ and $\{uy, yv, vu\}$ are the only two faces shared by $W(u)$ and $W(v)$.   □

Corollary 2. *Let $W(x)$, $W(y)$ and $W(z)$ (with $x \neq y \neq z$) be three face-intersecting odd wheels in a W near-triangulation G. Then they share exactly the common face $\{xy, yz, xz\}$.*

Proof. Since $W(x)$ and $W(y)$ are face-intersecting, by Lemma 5, they share two faces (faces which has the edge xy as one of its boundary). Similarly $W(y)$ and $W(z)$ share two faces (faces which has the edge yz as one of its boundary) and $W(x)$ and $W(z)$ share two faces (faces which has the edge xz as one of its boundary). This implies that xy, yz and xz forms either a separating triangle or a face which is shared by $W(x)$, $W(y)$ and $W(z)$. But as G is a W near-triangulation, the edges xy, yz and xz can not form a separating triangle.   □

Definition 9 (W_Δ). *Let G be a W triangulation and $W(x), W(y)$ and $W(z)$ be three face intersecting even wheels in G. If $N[x] \cup N[y] \cup N[z] \setminus \{x, y, z\}$ induces an odd hole in G, then the subgraph induced by $N[x] \cup N[y] \cup N[z]$ is called a W_Δ. The graph shown in Fig. 1 is an example of W_Δ.*

If a W near-triangulation G with at least five vertices contains an internal vertex u of odd degree exceeding 3, then the rim of $W(u)$ induces an odd hole in G and thus G cannot be perfect. Since an internal vertex of degree 3 would induce a separating triangle, a W near-triangulation with at least 5 vertices cannot contain an internal vertex whose degree is 3. Consequently, the non-trivial case to handle is to characterize perfect W triangulations whose internal vertices are all of even degree.

Lemma 6. *Let G be a W_5 free even W near-triangulation and $W(x), W(y)$ and $W(z)$ are three face intersecting wheels in G. Then $N[x] \cup N[y] \cup N[z]$ induces a W_Δ.*

Proof. Let x, y and z be three vertices of G such that $W(x), W(y)$ and $W(z)$ are face intersecting wheels in G. Let G_1 be the subgraph of G induced by the vertices x, y, z and their neighbours. That is, G_1 is a subgraph of G induced by $N[x] \cup N[y] \cup N[z]$. Let G_2 be the subgraph of G_1 induced by $V(G_1) \setminus \{x, y, z\}$. If G_1 does not induce a W_Δ then there exists at least one chord in G_2. Without loss of generality we may assume that there exists two non consecutive vertices p and q on the rim of wheels $W(x)$ and $W(y)$ respectively such that pq is a chord in G_2. We may further assume without loss of generality that there is no chord between the vertices of the clockwise boundary of G_1 from p to q (see Fig. 2).

Let $P = pq_1q_2 \ldots q_rq$ (where $r \geq 1$) be the path joining p and q in G_1 (see Fig. 2). As $W(x)$ and $W(y)$ are face intersecting, there must be at least one vertex, say q_i, $1 \leq i \leq r$) in P that lies on the rim of both the wheels $W(x)$ and $W(y)$ (see Fig. 2).

Let s be the neighbour of p on the rim of $W(x)$ in the anti clockwise direction and t be the neighbour of q on the rim of W_y in the clockwise direction (see Fig. 2). Let G_3 be the subgraph of G_1 induced by the vertices $p, q_1, \ldots, q_i, \ldots, q_r, q$ and their neighbours except x, y, s and t. That is, G_3 is the subgraph of G_1 induced by the vertices $(N[p] \cup N[q_1] \cup ..N[q_i] \cup N[q]) \setminus \{x, y, s, t\}$ (see Fig. 3).

Let e, f, n_e and n_i be the number of edges, internal faces, external vertices and internal vertices in G_3 respectively. Let $n = n_e + n_i$ be the total number of vertices in G_3. Since G_3 is internally triangulated, by Lemma 4 we have,

$$3f = 2e - n_e \tag{1}$$

Using Euler's formula [1] we get:

$$3n_e + 3n_i + 3f = 3e + 3 \tag{2}$$

from (1) and (2) we get:

$$e = 2n_e + 3n_i - 3 \tag{3}$$

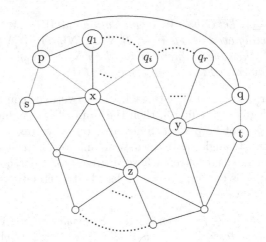

Fig. 2. G_1

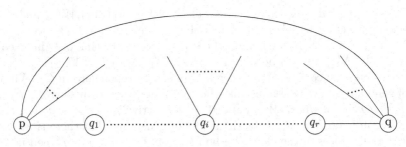

Fig. 3. G_3

Let V_i and V_e be the set of internal and external vertices in G_3 respectively. As G_3 is an induced subgraph of G_1 and pq a chord in G_1, all vertices except p and q in V_e are internal vertices of G_1 (see Figs. 2 and 3). Also every vertex in V_e except q_i must either be on the rim of W_x or on the rim of W_y, but not on both. That is, for all external vertices in G_3 except p, q and q_i, all but one of their neighbours in G_1 must be in the graph G_3. It follows that the degree of all vertices on the external face of G_3 except p, q and q_i will be at least five. This is true because we have assumed that G_1 is a W_5 free even W near-triangulation and hence has no internal vertex of degree below six.

The vertices p, q and q_i have two neighbours on the cycle $p, q_1, \ldots, q_i, \ldots, q_r, q$ and as G_3 is plane near-triangulated, they must have at least one neighbour in V_i. So the degree of p, q and q_i will be at least three in G_3. Since every neighbour (in G_1) of vertices in V_i is also present in G_3, degree of all vertices in V_i must be at least six in G_3. Counting the degree of vertices, we get:

$$2e \geq 6n_i + 5(n_e - 3) + 9 \tag{4}$$

Substituting (3) we get,

$$4n_e + 6n_i - 6 \geq 6n_i + 5n_e - 6 \implies 0 \geq 2n_e \implies 0 \geq n_e \qquad (5)$$

which is a contradiction. □

The following lemma shows that Lemma 6 characterizes all non-perfect W_5 free even W near-triangulations.

Lemma 7. *Every W_5 free even W near-triangulation G that contains an induced odd hole must contain an induced W_Δ.*

Proof. Let C be an induced odd hole in G. As G is a plane near-triangulation, there must exist at least one vertex in $Int(C)$. Let $G' = (V', E')$ be the subgraph induced by the vertices in $Int(C)$. If $|V'| = 1$ then $V' \cup V(C)$ will have to induce an odd wheel which is impossible as G is an even W near-triangulation. The case $|V'| = 2$ is also not possible as two face intersecting odd wheels will not induce an odd hole (See Lemma 4). Thus we may assume that $|V'| \geq 3$. Let G'' be the subgraph of G induced by the vertices $V' \cup V(C)$. We have to consider the following cases.

1. G' contains an induced triangle $\Delta = (xyz)$: In this case, $W(x)$, $W(y)$ and $W(z)$ are face intersecting odd wheels and by Lemma 6, $N(x) \cup N(Y) \cup N(z)$ induces an odd hole, proving the lemma.
2. G' does not contain any induced triangles: In this case, as G is a plane near triangulation, the only possibility is that V' induces a tree T of at least 3 vertices. (See Fig. 4. T could possibly a path as in Fig. 4.) Let $v_0, v_1, \ldots, v_r$ for some $r \geq 2$ be the vertices in T ordered in such a way that v_0 is the root of the tree and each node v_j for $j > 0$ is a child of some unique v_i, $i < j$ in T. Note that every neighbour of a vertex v_i in T except its children and its parent in the tree T must be a vertex in the odd hole C. Let $W(v_i)$ be the wheel induced by $N[v_i]$. Since G is an even W_5 free near-triangulation, v_i must have even degree (greater than 4) for each $i \in \{0, 1, \ldots, r\}$. Further, each edge incident on v_i ($0 \leq i \leq r$) is shared by exactly two internal faces in G'' (See Fig. 4). Hence, for each $i, j \in \{0, 1, \ldots, r\}$, if v_i is the parent of v_j in the tree T, the wheels $W(v_i)$ and $W(v_j)$ must be face intersecting odd wheels sharing exactly two faces. Using this observation, we count the the the total number of internal faces in G'' to be $f = \sum_{i=0}^{r} deg(v_i) - 2(r - 1)$. As the degree of every internal vertex in G'' is even, f must be even. Then by Lemma 4, the number of external vertices of G'' should be even. That is, $|V(C)|$ must be even. However, this contradicts the assumption that C is an odd hole. □

Lemmas 6 and 7 enable us to conclude the following:

Theorem 2. *A W_5 free plane triangulated W near-triangulation G is perfect if and only if the following conditions hold*

- *G does not contain an even wheel*
- *G does not contain an induced W_Δ.*

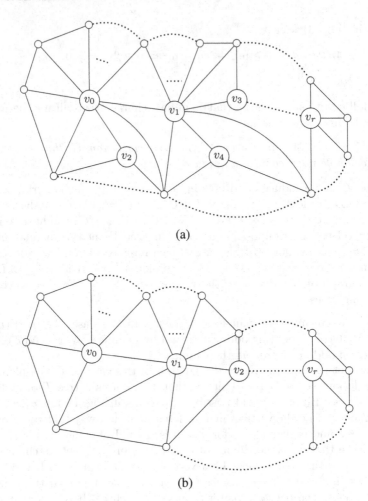

Fig. 4. (a) V' induces a tree (b) V' induces a path

7   Conclusion

Decomposition of a plane triangulation into its W components appears to be a promising way for characterizing perfect near-triangulations as the question of determining whether a triangulation is chordal (or perfect) can be transformed to the corresponding problem on a collection of W triangulations. The simplicity of W triangulations makes structural investigations on them easier than that in general triangulations. The present work gives a local characterization for W triangulations in which induced W_5 is forbidden. Further investigation in this direction may be aimed at locally characterizing larger classes of W near-triangulations, with the eventual hope of deriving a simple structural characterization for all or almost all perfect near-triangulations.

Acknowledgment. We would like to thank Dr. Ajit A Diwan, IIT Bombay and Dr. Jasine Babu, IIT Palakkad for their comments and suggestions.

References

1. West, D.B.: Introduction to Graph Theory, pp. xvi+512. Prentice Hall Inc., Upper Saddle River (1996)
2. Laskar, R.C., Mulder, H.M., Novick, B.: Maximal outerplanar graphs as chordal graphs, path-neighborhood graphs, and triangle graphs. Australas. J. Combin. **52**, 185–195 (2012)
3. Biedl, T., Demaine, E.D., Duncan, C.A., Fleischer, R., Kobourov, S.G.: Tight bounds on maximal and maximum matchings. Discret. Math. **285**(1), 7–15 (2004)
4. Kumar, P.S., Veni Madhavan, C.E.: A new class of separators and planarity of chordal graphs. In: Veni Madhavan, C.E. (ed.) FSTTCS 1989. LNCS, vol. 405, pp. 30–43. Springer, Heidelberg (1989). https://doi.org/10.1007/3-540-52048-1_30
5. Cahit, I., Ozel, M.: The characterization of all maximal planar graphs, Manuscript. http://www.emu.edu.tr/~ahit/prprnt.html
6. Benchetrit, Y., Bruhn, H.: h-perfect plane triangulations, arXiv preprint arXiv:1511.07990
7. Chudnovsky, M., Robertson, N., Seymour, P., Thomas, R.: The strong perfect graph theorem. Ann. Math., JSTOR, 51–229 (2006)

Author Index

Printed in the United States
By Bookmasters